New World of Textiles
Shaping for the Future

New World of Textiles
Shaping for the Future

Editors

K.N. Guruprasad
Ranjit Kumar Datta
Tapan Kumar Sinha
Swapnil Jain

Shri Vaishnav Vidyapeeth Vishwavidyalaya
Shri Vaishnav Institute of Textile Technology
Indore

ALLIED PUBLISHERS PVT. LTD.

New Delhi • Mumbai • Kolkata • Chennai • Bangalore • Hyderabad

ALLIED PUBLISHERS PRIVATE LIMITED

1/13-14 Asaf Ali Road, **New Delhi**–110002
Ph.: 011-23239001 • E-mail: delhi.books@alliedpublishers.com

17 Chittaranjan Avenue, **Kolkata**–700072
Ph.: 033-22129618 • E-mail: cal.books@alliedpublishers.com

15 J.N. Heredia Marg, Ballard Estate, **Mumbai**–400001
Ph.: 022-42126969 • E-mail: mumbai.books@alliedpublishers.com

No. 25/10, Commander-in-Chief Road, Ethiraj Lane (Next to Post Office)
Egmore, **Chennai**–600008
Ph.: 044-28223938 • E-mail: chennai.books@alliedpublishers.com

P.B. No. 9932, No. 15, 3rd Floor (Next to Vijaya Bank), 5th Cross, Gandhinagar, Karnataka, **Bangalore**–560009, Ph.: 080-41530285 / 22386239
• E-mail: bngl.journals@alliedpublishers.com / apsabng@airtelmail.in

Sri Jayalakshmi Nilayam, No. 3-4-510, 3rd Floor (Above More Super Market)
Barkatpura, **Hyderabad**–500027
Ph.: 040-27551811, 040-27551812 • E-mail: hyd.books@alliedpublishers.com

Website: www.alliedpublishers.com

ISBN: 978-81-942903-9-1

Foreword

Combating the Impact of COVID-19

The global economic fallout due to COVID-19 pandemic will continue to remain a challenge and India has been looking at it not only as a constraint but as an opportunity also to reform its economy as well as rebalance international trade. The breakdown of the global supply chains has, perhaps, for the first time created an opportunity for India to become a market leader globally. The Indian textile industry is considered as mother of all other industries of this country. It provides direct as well as indirect employment to 45 and 60 million people respectively. The Indian textile industry is one of the world's largest producers of textiles and apparel. The sector accounts for 13% of India's industrial production, 2.3% of GDP, 12% of export earnings and has approximately 5% share of global export.

During COVID-19 in the mid of 2020, the development of an indigenous supply-chain for manufacturing Personal Protective Equipment (PPE) kits in six weeks is an example of creative resilience of the Indian textile industry with the support of the Ministry of Textiles. The various proactive measures have been taken by Central Government to combat the impact of COVID-19 to reboot Indian textiles and apparel sectors.

- *Announcement in Union Budget 2021–22*: The Mega Investment Textile Parks, under which seven textile parks will be established over a period of three years, to enable the industry to become globally competitive, attract large investments and boost employment generation. This will increase the scale of production and help Indian players to produce faster and at a lower cost with maximum efficiency in operation.
- *Under Aatmanirbhar Bharat Abhiyan*: Implementation of Production-Linked-Initiatives (PLI) scheme to 10 key manufacturing sectors, including textiles for enhancing India's manufacturing capabilities and exports is a challenge. The PLI scheme will make Indian textile manufacturing globally competitive with cutting edge technology, ensuring efficiency and quality of global standard, create economies of scale, attract large foreign investments to boost domestic manufacturing and enhance export especially in man-made fibre and technical textiles.
- On September 2, 2020, the Union Cabinet approved signing the Memorandum of understanding between textile committee, India and M/s Nissenken Quality Evaluation Centre (NQEC), Japan for improving quality and testing Indian textiles and clothing for the Japanese market. This India-Japan pact on cooperation in textiles shall facilitate Indian exporters to meet the requirements of Japanese importers as per the latter's technical regulations.

- *Under Union Budget 2020–21*, a National Technical Textiles Mission is proposed for the period from 2020–21 to 2023–24 at an estimated outlay of ₹ 1,480 crore (US$ 211.76 million).
- This year new Textiles Policy is expected to be released by the Ministry of Textiles.
- The Directorate General of Foreign Trade (DGFT) has revised the rates for incentives under Merchandise Exports from India Scheme (MEIS) for two subsectors of Textile Industry—readymade garments and made-ups—from 2% to 4%.
- The Government announced a special package of US$ 31 billion to boost export, create one crore job opportunities and attract investment worth ₹ 80,000 crore (US$ 11.93 billion) during 2018–2020.
- *The Government of India* has taken several measures including Amended Technology Upgradation Fund Scheme (A-TUFS), estimated to create employment for 35 lakh people and enable investment worth ₹ 95,000 crore (US$ 14.17 billion) by 2022.
- *Integrated Wool Development Programme* (IWDP) was approved by Government of India to provide support to the wool sector, starting from wool rearer to end consumer, with an aim to enhance quality and increase production during 2017–18 and 2019–20.
- *Foreign Direct Investment (FDI)*: The Indian government has come up with number of export promotion policies for the textile sector, such as 100% FDI under the automatic route to increase the ease of doing business for foreign investors, and the single-window clearances.
- *Increasing Export*: Initiatives to sign trade agreements with developed countries to enhance export opportunities, such as Free Trade Agreements (FTAs) to gain duty-free access to large textile markets e.g., the European Union, Australia and the United Kingdom.

The modern economy has been built on the foundation of the industrial revolution with innovative initiatives for maximizing national interests. Emerging technologies are exciting and bring innovation and new opportunities across the globe. Indian textile industry needs to adopt latest technology and build smart factories that are digitally enabled with value chain. Industry is going through a paradigm shift in technology by adopting "Industry 4.0". This transformation that makes it possible to gather and analyze data across machines, while enabling faster, more flexible and efficient process to produce goods at reduced cost. These are Big-data, Artificial Intelligence, Advanced Robotics, Cloud data, 3D printing and Internet of Things. The textile and apparel industry shall have to adapt to Industry 4.0 and ensuing Industry 5.0 to survive and open up wide range of possibilities in redefining the textile and garment industry.

There is need for building a more sustainable, and circular value chain. Transforming the industry to usher in a new textile economy requires system-level change with an unprecedented degree of commitment, collaboration, and innovation. Such an approach would rally key industry players and other stakeholders behind the objective of a new textile economy, set ambitious joint commitments, kick-start cross value chain projects, and orchestrate and reinforce complementary initiatives.

These technological transformations are not only reshaping the textile manufacturing processes but are also helping the plants and factories become smarter, connected and more efficient. To understand the nuances of changing face of textile industry, this conference organized on the theme "*New World of Textiles: Shaping for the Future*" had received number of papers for presentation during *Texcon-2021. The papers* have been thoroughly edited and presented in this book for the benefit of delegates. The readers may find this book useful.

Date: 10.02.2021

Dr. Upinder Dhar
Vice Chancellor
Shri Vaishnav Vidyapeeth Vishwavidyalaya,
Indore

Prologue

"New World of Textiles: Shaping for the Future"

Globally the COVID-19 pandemic has impacted nearly every aspect of life whether it is economy or health. It has, however, also created significant new growth opportunities. Across the world, each and every country and firms are reevaluating their manufacturing and trade strategies. In this opportunity, India can serve as the economic engine of the world, considering its inherent competitive advantages that it possesses.

Even though textile industry is one of the oldest industries of the Indian economy, popularly known as "Sunrise Sector", but the fact is that the textiles industry finds it difficult to compete with big giants like China, recently even with much smaller players such as Bangladesh and Vietnam. This is mainly because of policy disparities and less focus on making these sectors an economic driver. In textile sector there are some deep-rooted problems and challenges which need to be addressed on priority for long-term sustainable revival as global leader. The Indian textile industry has strong and diverse raw material base, whether it is natural or synthetics fibres, strong presence in the entire value chain to finished garments. India is the largest producer of cotton, accounting for 25% of the global output, the second-largest producer of man-made fibres (MMF), lowest cost producer of spun yarns and the world's second-largest producer of textiles and garments after China. More importantly, India has a strong demographic advantage of population, increasing urbanisation, growing disposable income and higher market penetration of organized retail *(www.wazir.in)*.

In spite of these advantages, the Indian textile industry currently is uncompetitive. The challenges are Indian textile sector is highly fragmented, small size of operation, outdated technology, inflexible labour laws, lower productivity and quality issues from unskilled workforce, tax structure disparity predominantly focus on cotton based while world is with MMF, the GST (Goods and Service Tax) makes the garments expensive and uncompetitive in domestic as well as international markets, stagnant exports - the export from the sector has been remained blow at the $40-billion level for the last six years , lack of global scale, lack of foreign investment, high cost of working capital and power, logistic cost as well as longer turnaround time at port. Among all the issues, the weaving sector remains one of the weakest links of the Indian textile industry along with fibre policy/tax disparity and global scale of operation.

It is a fact that globally fiber consumption is predominated toward man-made fiber (MMF) having approx. 71% (79.3 million MT) share in total fiber consumption of 115.1 million MT while natural fibers constitute only 29% (32.2 million MT). Contrary to this global trend, in India fiber consumption is skewed towards cotton 65%~70%. In order to achieve the desired growth target of 300 billion USD market, it becomes crucial for India to focus on MMF textiles.

Sensing all these challenges, Government of India has taken a series of steps in quick succession over the past few months like, *Aatmanirbhar Abhiyan* under which a special economic package of Rs. 20 lakh crore (equivalent to 10% of India's GDP) has announced with the aim of making the country independent against the tough competition in the global supply chain and to help in empowering the poor, labourers, migrants who have been adversely affected by COVID-19 pandemic. Further, in order to reduce India's dependence on China, the Prime Minister Mr. Narendra Modi has given an approval to introduce the *Production-Link- Initiatives (PLI)* scheme to 10 key manufacturing sectors, including textiles and apparel for enhancing India's manufacturing capabilities and exports. The total approved financial package of *PLI* is Rs. 1,45,980 Crores while for Textile sector is Rs. 10,683 Crores. The *PLI* scheme will make Indian textile manufacturing globally completive with cutting edge technology, ensure efficiency and global standard quality, create economies of scale, attract large foreign investment to boost domestic manufacturing and enhance export especially in MMF segments and technical textiles.

This year *Union Budget 2021-22,* has also announced to setup *Seven Mega Textile Parks* over a period of three years, to enable the textile industry to attract large investments, boost employment generation, and globally competitive. This scheme will be in addition to existing *Production-Linked Initiatives (PLI*) schemes in order to create world class infrastructure, with plug-and-play facilities to enable and create global champions in export. The Budget has also proposed to reduce the basic customs duty on caprolactam, nylon chips, nylon fiber and yarn to 5%, which will encourage the growth of the Man Made Fibre (MMF) sector especially the MSMEs. The imposition of a 10% basic customs duty (BCD) raw cotton which will make imports of extra-long staple cotton costlier, especially *Giza Cotton* from Egypt and *Supima* Cotton from the US.

In order to boost textile sectors, it is recommended to remove anti-dumping duty and also slotting the entire MMF value chain under 5% GST rate at par with cotton value chain apart from addressing inverted duty structure at processing and finished goods. Last year, removal of anti-dumping duty on PTA (raw material for Polyester) and on acrylic fiber were positive steps but the proposed withdrawal of anti-dumping duty on MEG and PSF are still pending. These altogether will thus enable global competitiveness for Indian textile sector. This year Ministry of Textiles is expected to announce its new textile policy with an ambitious target of achieving 20 per cent share of the global textile trade and helping the domestic industry attain a size of $650 billion by 2024–25 by focusing on investments, skill development and labour law reforms along with promoting innovation and R&D activities.

Technical textiles are engineered products with a definite functionality. These products find diverse end-use application across multiple non-conventional textile sectors such as healthcare, construction, automobile, aerospace, defence, agriculture geotextiles, sports & fitness and many more. The technical textiles are an emerging, high-growth area and have a huge potential to grow at unprecedented rates in upcoming time. Indian technical textiles market is currently worth 19 billion USD and is expected to reach a size of 42 billion

USD in 2025 registering growth rate of 12% (*KPMG & FICCI - Technical textiles: Growth engine of Indian textile sector*, August 2019).

During this COVID-19 crisis, Ministry of Textiles, under the "Vocal for Local" initiative of the Government of India, with support from the Ministry of Health and Family Welfare, was able to create an end-to-end indigenous supply chain set-up to manufacture Personal Protective Equipment (PPE) kits as per WHO standards in six week time. From manufacturing 0 PPE kits in March, India now stands to produce around 4.5 lakh PPEs and more than 1.5 crores masks per day. The Indian PPE market, the second largest in the world now, is expected to reach beyond Rs. 7.8 billion by the end of 2020, with a compounded annual growth rate of 19% (*www.Invest India.Gov.in*, Technical Textiles: The Future of Textiles, August 2020). The manufacturing of PPE kits during the COVID-19 pandemic is an example of creative resilience of the Indian textile industry.

With a growing world population and rapid growth in textile consumption in developing countries, a whole range of new application areas for textiles are constantly rising to meet the user requirements in terms of functionality, variety, precision, performance, reliability and environmental friendliness of textile products. This transformation process will focus on the exploitation of research, innovation and knowledge-orientation across all business functions and sub-sector activities driven through multi-disciplinary approach that are shaping the future of textiles in novel application fields.

Keeping this in the view, to bring academia, research minds and industry together Shri Vaishnav Institute of Textile Technology (SVITT) under SVVV, Indore has initiated a National level Conference *TEXCON-2021*. The conference theme ***"New World of Textiles: Shaping for the Future"*** is focused on latest technological advancements and innovations in the entire value chain for sustainable growth driven through multi-disciplinary approach that are shaping the future of textiles.

This book has contribution from a number of people with wide experience in academic institutions, research associations, industry and consulting firms. We would specially like to express our gratitude to all the authors. We would also like to thank the conveners, editorial team and members of organizing committee of Texcon-2021 for their un-conditional support. Last but not the least, we acknowledge the help and cooperation received from all the faculty and staff members of Shri Vaishnav Institute of Textile Technology, Shri Vaishnav Vidyapeeth Vishwavidyalaya, Indore.

Date: 02.02.2021

Dr. Ranjit Kumar Datta
Director
Shri Vaishnav Institute of Textile Technology,
Indore

About the Editors

Dr. K.N. Guruprasad is currently working as Director, Shri Vaishnav Institute of Science, Shri Vaishnav Vidyapeeth Vishwavidyalaya, Indore. He works in the area of Plant Science and specializes in Photobiology and Magneto-biology. He has an experience of teaching and research in Life Science of over thirty five years. He was Professor and Head of Life Science Department of Devi Ahilya Vishwavidyalaya, Indore. He has supervised the work of twenty five research students who have worked under him for obtaining their Ph. D degree. He has been the Principal Investigator of various research projects funded by UGC, CSIR, DST, ICAR, BRNS, MPCST and BASF India. He has worked as a Visiting Scientist in Kobe University, Japan and Moscow University, Russia. He has over eighty five research papers published in research journals of National and International repute.

Dr. R.K. Datta is Director of Shri Vaishnav Institute of Textile Technology under SVVV having 35 years of working experiences, 25 years in manufacturing of various corporate companies at top management level and 10 years in R&D activities in Research Institutes. He had also worked as Vice President (Technical) in world Class Integrated Polyester Plant Indo Rama Synthetics Ltd., Nagpur. He has a wide experience in Project implementation and Setting up of CP plants and working with world best technology suppliers such as Dupont/Chemtex USA, Toyobo Japan, Oerlikon Barmag Germany, Tejin Seiki Japan, Zimmer AG Germany, Autefa Germany, LVD Belgium, Foxbro Invensys Singapore for process optimization. Since interception of Indo Rama Plant operation, he played a leadership role in implementation of Quality System both ISO – 9000 & 14000 as a MR and Products Specifications. Before that, he worked at Ahmadabad Textiles Industries Research Association (ATIRA) as a Principal Scientist for R & D and Technical Services and Man Made Textiles Research Association (MANTRA), Surat as a Sr. Scientist for R & D and Technical Services. He completed MOT project on Micro-Fibes and worked on UNDP project–Kawa-bata system to assessed fabric feel and handle, Completed of two MOT (Ministry of Textile) Project on Melt Spinning, Air Jet Texturizing and TFO. He had published 15 Research papers in National & International Journal and presented 25 Research & Technical Papers. He is a well travelled professional who has visited USA, UK, Switzerland, Japan, Indonesia, Syria, Iran, Turkey, and Nepal to understand latest technologies, production sites, international seminars/ conferences and customer visits for products promotion. He is also a lead assessor for achieving CII-EXIM Bank award for Business Excellence. After serving 23 years in Indo-Rama Synthetics India Ltd., since June'19, he has joined Shri Vaishnav Institute of Textile Technology (SVITT) under SVVV, Indore as Director.

T.K. Sinha is Professor of Shri Vaishnav Institute of Textile Technology under the umbrella of Shri Vaishnav Vidyapeeth Vishwavidyalaya, Indore. He has wide 15 years of industrial experience as Manager (R&D) and thereafter 21 years of Teaching Experience in the field of Textile Technology. He obtained M. Tech in Textile Engineering from IIT Delhi and presently pursuing Ph.D. from SVVV, Indore. His specialization is in textile engineering in the field of yarn manufacturing and machineries. Besides taking various industrial consulting assignments, he has the wide experience in technical assignments to the various textile Industry in the country as well as in the abroad for problem solving, technical seminar and exhibitions. He has authored/ co-authored more than 31 papers and a number of presentations all over the country. He has supervised more than 80 under-graduate and 02 M. Tech dissertations. He is the Chairman of Board of Studies of Textile Engineering of Rajiv Gandhi Proudyogiki Vishwavidyalaya, Bhopal and also in the Shri Vaishnav Vidyapeeth Vishwavidyalaya, Indore. Prof. T.K. Sinha was awarded by Institute of Engineers in 2014 for his lifetime achievement in developing and spreading the Textile Engineering education in the state of Madhya Pradesh and also by Textile Association of India, M.P. unit in the year of 2015. He is also a member as subject expert in the textile goods purchasing committee of DAVV University. His area of interest is yarn manu-facturing, textile testing, fibre science and Technical textile. He is also a life member of Textile association of India.

Dr. Swapnil Jain is working as Assistant Professor in Department of Electrical and Electronics Engineering at Shri Vaishnav Vidyapeeth Vishwavidyalaya, Indore, India. He has completed B.E. in Electronics Engineering from Nagpur University and M. Tech in Digital Communication from Rajiv Gandhi Proudyogiki Vishwavidyalaya, Bhopal. He has received his Ph.D. from MGCGV, Chitrakoot in year 2016. His research interests are data and network security, Quantum cryptography, advance communication, Vehicular ad-hoc network. He has published and presented more than 20 papers in various international conferences and peer reviewed journals. He has teaching experience of more than 18 years. He is a life member of ISTE.

Contents

SECTION III: Garment and Fashion

SECTION IV: Technical Textiles and Composite Materials

SECTION V: Environment and Sustainability

Section-I

Textile Fibre

Agro-Jute: A Solution for Better and Green Cultivation

Surajit Sengupta
ICAR-National Institute of Research on Jute and
Allied Fibre Technology, Kolkata, India
E-mail: drssengupta42@gmail.com

ABSTRACT: *Soil covering is an age-old practice in agriculture, where the field is covered for weed control and moisture retention of soil during cultivation. Presently, the thin synthetic sheet is widely used for this purpose. In this study, a uniform, porous and bulky mechanically entangled fibrous sheet (called needle punched nonwoven) from eco-friendly and natural, low-grade/process waste jute fibre has been designed for strawberry cultivation. The agricultural performance of this engineered eco-friendly textile material has been compared with the commonly used plastic sheet and traditionally used sawdust, straw, and banana leaf in the same strawberry field under same conditions using random complete block design with three replicas. It was observed that processed waste jute nonwoven shows better performance than other materials under experimentation. The 500 g/m^2 nonwoven shows an improvement of 13% in plant dry weight, 9% in plant height, 14% in fruit weight, 28% in soil temperature, 11% in yield and 68% weed control over the synthetic sheet. Results with other natural materials are not consistent because of too much irregularity on laying and influenced in the atmospheric condition like rain, strong wind etc. Among them, straw gives good results. Hence, the application of jute nonwoven promotes the bio-waste utilization for the sustainable and green agriculture and is a promising alternative as a soil cover.*

Keywords: Nonwoven, Plastic Sheet, Soil Cover, Strawberry, Waste Jute.

1. INTRODUCTION

Use of eco-friendly materials in farming is coming up in western countries in a great way. Big farmers and organized farming sectors are also encouraging environment-friendly farming as they are earning more profit out of this agricultural produce though the initial cost is higher. Traditionally, agricultural practices were eco-friendly using natural products. With the advancement of technology, farmers used to apply synthetic materials for higher production. In agriculture and gardening, mulching is a protective cover placed over the soil to retain moisture, reduce erosion, suppress weed growth and seed germination. From ancient age, farmers were using various naturally available waste materials like straw, wood chips/dust, banana leaf etc., whenever it was required for cultivation. Presently, the synthetic sheet is being used for this purpose in place of natural materials due to insufficient availability (Xie *et al.*, 2005). This is very much useful in dry land areas where yearly rainfall is low or other sources of water for cultivation is insufficient.

Worldwide, different types of technical textiles are coming up in a big way (Horrocks & Anand, 2000). Among these, the agro-textile is a promising area by which we can control the productivity and quality of the crop. Being agriculture based country; India may exploit the different avenues of agriculture using tailor-made/engineered textiles with natural fibres as it is abundantly available in India.

A nonwoven is a textile structure produced by the bonding or interlocking of fibres, or both, accomplished by mechanical, chemical, thermal, or solvent means and combinations thereof. In needle punched nonwoven preparation, a fibre web i.e., a thin semi-transparent layer of fibres is prepared in which fibres are attached by surface cohesion only and then the layer is laid many-fold (several webs on top of each other) to form a butt, which goes directly to bonding by mechanical means using barbed needles. Needle punched nonwoven is bulky, uniform and possesses excellent hydraulic and thermal insulation property with lower but sufficient strength (Lunenschloss & Albrecht, 1985). This is called mechanically-bonded fibre sheet or more technically needle punched nonwoven.

Jute is annually renewable, eco-friendly, and totally biodegradable. In the jute yarn making system, about 5% material is not utilized and called soft waste. Out of that, 2% are mixed with good fibre to make low-grade yarn. Moreover, twisted waste is generated in fabric making process. Recycled fibre can be made from this twisted waste and unusable products (called hard waste). For diversified use and value addition of process waste jute and recycled fibres (about 10%), an attempt has been made to use this bio-waste as the soil cover in strawberry cultivation. The engineered textile material has been compared with the commonly used plastic sheet and traditionally used sawdust, straw, and banana leaf in the same field under same conditions using random design to explore this application in agriculture. Success in this attempt will encourage its use in other high-valued crops.

2. MATERIALS AND METHOD

In the area of covering, black polyethylene is conventionally used. The agricultural performance of needle punched nonwoven from processed waste jute fibre with scrim cloth (woven jute cloth with very high thread to thread spacing having 30–40 g/m^2 area density) reinforcement is compared with black polyethylene (7 microns), straw, sawdust and banana leaf and black polyethylene of 7 microns has been procured from local market. The straw, sawdust and banana leaf have been collected from the local sources. The nonwoven has been prepared in the laboratory from the 80% machine waste (procured from industry) and 20% TD_6 grade jute. The construction and laying parameters of different materials are shown in Table 1. Figure 1 shows the use of these materials in the field.

Table 1: Covering Materials

Sample Code	*Type*	*Punch Density Punches/cm^2*	*Needle Penetration (mm)*	*Area Density g/m^2*	*Thickness mm*
S1	Straw	–	–	479	10.29
S2	Sawdust	–	–	494	3.98
S3	Banana leaf	–	–	482	6.37
S4	Synthetic sheet	–	–	80	0.022
S5	Jute waste nonwoven	220	14	489	4.88

(a) Straw (b) Sawdust

(c) Banana Leaves (d) Jute Soil Cover in Strawberry

Fig. 1

2.1 Nonwoven Fabric Preparation

The middle portion of TD_6 jute reed (after removal of root and tip portion) was subjected to softening treatment. Water was sprayed to maintain nearly 35% moisture on the weight of fibre. The reeds were then kept in a bin for piling or conditioning for 24 hours as commonly practiced (Pal *et al.*, 1997) in jute mills. Then it was processed through jute softener. Before processing in breaker card 20% of this low-grade jute is mixed well with 80% of process waste jute. The breaker card sliver was fed to Dilo nonwoven plant comprising a card, a camel back cross-lapper and a needle loom (model number OD II/6) to prepare the pre-needled web of 50 g/m^2 with 25 punches/cm^2 and 12 mm depth of needle penetration. The fabric mass per unit area was achieved by needling the required number of layers of pre-needled fabric considering mass per unit area loss due to needling (Sengupta *et al.*, 1999). A jute scrim cloth (very open and low area density jute cloth) is given in the back for reinforcement. This sandwiched nonwoven has been made using 21 gauge needles with 220 punches/cm^2 and 14 mm depth of needle penetration as shown in Figures 2&3 shows the schematic diagram of needle punching system and Figure 4 shows the flowchart to prepare the nonwoven fabric.

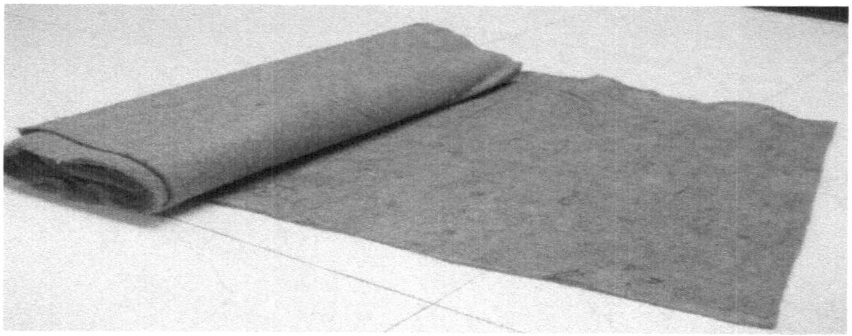

Fig. 2: Nonwoven Sheet

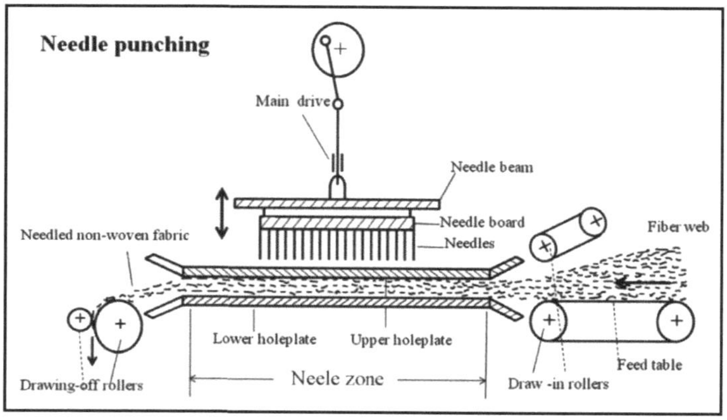

Fig. 3: Schematic Diagram of Nonwoven Preparation

Nonwoven Process
⇩
Softening of fibres
⇩
Opening in breaker card
⇩
Web formation in roller & clearer card
⇩
Cross laying of the web in cross lapper
⇩
Bonding or interlocking needle punching machine
⇩
Nonwoven fabric

Fig. 4: Nonwoven Preparation Flowchart

2.2 Measurement of Thickness, Area Density, and Bulk Density

The thickness of fabrics was measured by the Prolific Thickness Tester taking the average of ten tests from different places of sample fabric. The reading after the 30 second in the application of 1.55 kPa pressure by the top anvil (diameter 2.54 cm i.e., area 5.067 cm^2)

was considered (Debnath & Madhusoothanan, 2007). The actual area density was measured following the ASTM standard (ASTM D6242-98, 2003). Bulk density was measured multiplying area density by thickness.

2.3 Measurement of Air Permeability

A Shirley Air Permeability Tester was used for measuring the air permeability of the fabric samples. Air permeability was calculated by dividing the flow meter reading in cc/s at 10 mm pressure drop by the test area (5.07 cm^2 or 1 $inch^2$). The results were expressed as the units of volume of air in cubic meter passed per min through 1 m^2 of fabric at a pressure difference of 10 mm head of water following ASTM Standard (ASTM D734-04, 2012). Fifteen pieces were cut from the different parts of each of the fabric being tested for air permeability. The mean air-permeability of each sample has been reported. Sectional air permeability is measured multiplying air permeability by thickness.

2.4 Measurement of Thermal Insulation

The thermal resistance of samples was measured using guarded two-plate thermal resistance instrument (BS 4745, 1974). The area of a test specimen used is 706.85 cm^3 (diameter 30 cm). Average of readings of 5 samples from different places randomly was considered. The measurement was done under pressure of 0.3352 kPa. Specific thermal resistance was determined dividing thermal resistance value in terms of Km^2/W by mean thickness in meter (Debnath & Madhusoothanan, 2011).

2.5 Measurement of Tensile Behavior

The tensile properties of the fabrics, in the machine direction, were determined on an Instron Tensile Testing Machine, model number 5567. The test conditions (Sengupta & Bhattacharya, 2008) were: test length, 10 cm; cross-head speed, 5 cm/min; and strip width, 2.5 cm. The fabric tenacity and elongation-at-break were determined as follows:

$$\text{Tenacity (cN/tex)} = \text{Breaking load (cN)}/[\text{Specimen width (mm)} \times \text{Fabric mass per unit area (g/m}^2)]$$

$$\text{Extension-at-break, \%} = \text{Elongation-at-break (cm)} \times 100/\text{Gauge length (cm)}$$

Average of twenty tests for each sample has been reported.

2.6 Measurement of Water Absorbency

Single or stack of fabrics containing plies of known dimensions is weighed on a balance to three significant figures. The stack is then placed gently, but in one continuous smooth motion, on the surface of a pool of water in a tray (0.4 m × 0.3 m × 0.2 m is convenient and suitable) and a timer is started. At the last disappearance of dryness from the upper surface of the topmost ply, the timer is stopped. The stack of nonwovens is taken out of the water, allowed to drain for the 60s and then reweighed. The extrinsic absorptive capacity (ESC), i.e., the volume of liquid absorbed per unit area and extrinsic rate of sorption (ERS), i.e., the volume of water absorbed by unit area in unit time were calculated using the following relationships, considering the density of water as 0.997 g/ml at 25°C (Mattina & Oathout, 1995).

$$\text{ESC (mL/m}^2\text{)} = [(m_w - m_d)/\ 0.997]/[n \times l \times w] \text{ and}$$
$$\text{ERS (mL/m}^2\text{/s)} = \text{ESC}/[t_n/n] = [(m_w - m_d)/\ 0.997]/[l \times w \times t_n]$$

Where, m_d is the mass of dry nonwovens (g); m_w, the mass of wet nonwovens (g); n, the number of plies; l, the length of sample (m); w, the width of sample (m); and t_n, the time to complete sorption (s).

2.7 Measurement of Degradation of Fabric

Ten pieces of wet fabric strip with 20 cm × 5 cm dimensions were kept above the moist soil (in the field) for 1 month. An equal amount of water is spread every day for complete wetting. After 1 month these fabric strips were taken out from the soil and tested for tensile behaviour in the Instron Tensile Tester following the same procedure as normal tensile tests, stated above. The percent loss of strength shows the degradation. Similarly, degradation after two months was also measured. Different properties of waste jute nonwoven fabric and a synthetic sheet is shown in Table 2.

Table 2: Properties of Jute Nonwoven Fabric and Synthetic Sheet

Sample	*Waste Jute Nonwoven Fabric*	*Synthetic Sheet*
Bulk density (g/cm^3)	0.105	0.92
Tenacity	1.83 g/tex	23.1 MPa
Elongation (%)	15.32	392
Sectional Air permeability (cm^3 /s/cm)	56.41	0
Extrinsic rate of sorption (mL/m^2/S)	8.14	0
Extrinsic absorptive capacity (mL/m^2)	2715	0
Specific Thermal resistance (K m/W)	19.20	3.03
Degradation after 1 month above moist soil in wet condition	15.18%	2.84%
Degradation after 2 months above moist soil in wet condition	38.54%	7.45%
Drape coefficient	78.53%	%

2.8 Agricultural Practice

2.8.1 *Field Experiment*

Field experiments were conducted in the semi-arid zone (in the state of Punjab of India) with strawberry for consecutive three years (2011–14) and the average data has been presented in this paper. Experiments were set up in a randomized complete block design, with three replications. The condition is: type of soil, sandy & clayey loam & alkaline; average temperature, 40°C in summer and 10°C in winter; annual rainfall, 95 cm.

Photograph of different covering material and strawberry cultivation is shown in Figure 1.

2.8.2 *Yield and Fruit Characteristics*

The yield was taken from 44 plants in the middle row, while whole plants were randomly sampled from the outer rows. The average fruit was determined by compiling the average of 25 marketable fruit from each harvest date per season. All other standard cultural programs were followed according to recommended practices (Poling, 1993).

2.8.3 *Growth Parameters*

Whole plants were harvested every 4–5 weeks. At each whole plant harvest date, four whole plants from each treatment were harvested. Roots were washed over fine mesh sieve to separate soil from root tissue. Plants were then divided into roots, crowns, leaves, flowers, and fruit. Leaf area of fresh leaves was determined using a leaf area meter. All plant parts were bagged separately, then placed in a drying oven at 158°F (70°C) for 10 d. Individual plant yield was calculated on each harvest date by dividing total yield (grams) per plot by the number of plants in each plot (Al-Ramamneh *et al.*, 2013).

3. RESULTS AND DISCUSSION

3.1 Covering Materials and its Properties

Properties of needle punched nonwoven fabric made from the waste of jute has been shown in Table 2. Its density, permeability, water absorption and water retention, thermal insulation, degradation behavior the some important properties in favour of soil cover in addition to its biodegradability, eco-friendliness, and nutrient improvement. If it is compared with existing and widely used synthetic sheet, Table 2 shows nonwoven fabric is having low density due to voids in the structure which are responsible for high air permeability, absorptive capacity, and thermal insulation. Its degradation rate is much higher than synthetic. The main disadvantages of synthetic sheets are nonporous, impermeable, and hydrophobic. Tenacity and elongation of nonwoven are sufficient to withstand the laying forces. After laying and watering, it is basically of no use. The drape is important to touch the soil surface by the jute fabric in higher extent. The higher surface friction of jute nonwoven compared to synthetic sheet helps in better anchoring with the soil. Higher drape and surface friction are responsible for reduction of soil erosion and nutrient loss.

Other traditional materials like straw, sawdust and banana leaf are not a compact and continuous material hence, difficult to maintain density and thickness. However, an approximate value of density and thickness has been reported in Table 1. These materials are having a good permeability (though in wet condition sawdust reduces permeability significantly), moderate water absorption (sawdust shows the highest value in this regard) and high thermal insulation (straw shows the highest value), biodegradable (banana leaf degrades quicker than others) and eco-friendly. Except for synthetic sheets, all the materials are bio-waste and hence low cost. The only nonwoven fabric includes processing cost. Plant, fruit, soil, and other characteristics of cultivation have been tabulated in Table 3. The percent increase of biomass and other characteristics are shown in Figures 5&6.

Table 3: Plant, Fruit, Soil and Other Characteristics of Cultivation

	Control	*Synthetic Sheet*	*Straw*	*Sawdust*	*Banana Leaf*	*Nonwoven*
Biomass and Growth						
Plant dry weight (g)	19.12	22.47	25.26	21.58	21.08	25.54
Shoot dry weight (g)	8.26	10.37	11.50	9.72	9.94	11.30
Root dry weight (g)	9.98	10.55	13.17	10.07	10.21	13.61
Plant height (cm)	21.04	24.34	26.45	23.19	22.53	26.50
Root depth (cm)	11.97	13.84	15.13	12.85	12.26	15.06
No. of leaves. $plant^{-1}$	6.98	8.08	8.37	7.75	7.96	9.13
Leaf area (cm^2)	117.5	136.0	145.8	125.3	123.6	148.8
Leaves dry weight. $plant^{-1}$ (g)	5.26	6.08	6.59	5.68	5.85	6.51
Fruit Characteristics						
Number of fruits. $plant^{-1}$	2.72	3.11	4.10	2.97	2.74	3.90
Average fruit fresh weight (g)	6.23	7.22	7.88	6.89	6.82	8.23
Fruit length (cm)	3.08	3.56	3.89	3.35	3.29	3.95
Soil Moisture and Temperature Condition						
3 days after irrigation (%)	31.4	44.53	45.15	41.52	40.37	46.31
7 days after irrigation (%)	24.7	38.61	38.91	34.62	35.71	40.55
14 days after irrigation (%)	19.3	32.26	36.35	30.84	31.50	35.37
Soil temperature, °F	72	80	52	61	63	57
Yield and Weed Suppression						
Yields, t/ha	19.73	22.84	24.50	21.38	20.52	25.36
Weed suppression, %	–	72.92	57.09	51.75	48.35	68.25

3.2 Biomass and Growth Characteristics

Table 3 shows the data of plant dry weight, shoot dry weight, root dry weight, plant height, root length and leaf characteristics of strawberry cultivation on control (without covering), synthetic sheet, jute nonwoven, sawdust, straw, and banana leaf soil covering. The straw shows better growth followed by jute nonwoven fabric in comparison with control and the synthetic sheet covering. Straw and waste jute nonwoven show better performance than synthetic sheet probably due to low bulk density, higher permeability, higher water absorbency and higher thermal resistance. Figure 5(a)&(b) shows the percent growth improvement in waste jute nonwoven fabric over control and the synthetic sheet covering. It is already established that soil cover improves the plant growth hence, significant improvement of jute nonwoven cover over control is quite obvious. It also shows the improvement of 14% in plant dry weight, 9% in shoot dry weight, 9% in plant height, 7% in leaf dry weight in comparison to synthetic sheet. The straw shows improvement over jute nonwoven of 3% in root dry weight and 9% in number of leaves with the marginal increase in other parameters.

3.3 Fruit Characteristics

Table 3 shows the data of the number, fresh weight, and length of fruit of control (without covering), synthetic sheet, jute nonwoven, straw, sawdust, and banana leaf covering. In comparison to control, synthetic sheet and other covering, jute nonwoven/straw show better performance probably due to low bulk density, higher permeability, higher water absorbency and higher thermal resistance. Figure 5(c) shows the percent fruit characteristics improvement of jute nonwoven fabric over control, synthetic sheet, and straw covering. The improvement of fruit by jute covering over synthetic is 11% in length, 25% in number 14% in fresh weight per plant, whereas it is inferior to straw covering by 5% in number and 4% in weight per plant.

3.4 Soil Moisture and Temperature

The soil moisture condition after 3, 7 and 14 days of irrigation and average soil temperature in control (without covering), synthetic sheet, jute nonwoven and other traditional covering have been reported in Table 3. In comparison to control and other systems of covering, jute nonwoven and straw again show better water retention and temperature control of soil probably due to low bulk density, high water retention, and thermal insulation. Jute nonwoven shows improved moisture condition by 5% after 7 days of irrigation and 9% after 14 days of irrigation over the synthetic sheet, whereas inferior moisture condition by 8% and 12% after 7 and 14 days after irrigation respectively over straw covering (Figure 5(d)).

The average soil temperature in jute covering is lower by about 21% over control, 28% over synthetic sheet covering and higher by 9% over straw (Figure 5(e)). Here, the main point to note is that synthetic sheet covering increases the soil temperature which is detrimental to the plant and that may be the cause of lower growth. But as jute fabric or straw has better insulation property, it is useful in extreme temperature of summer as well as winter.

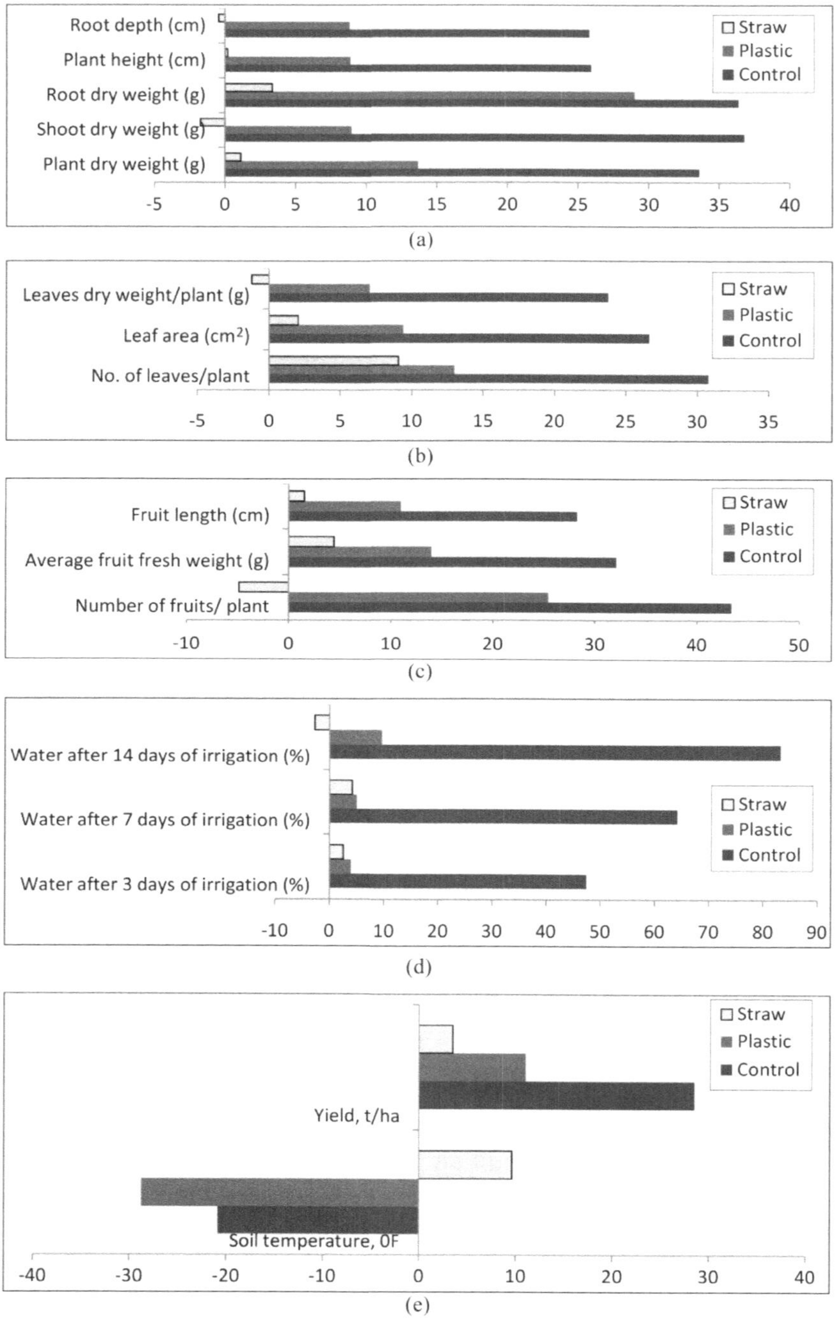

Fig. 5: Percent increase in waste jute nonwoven over control, plastic and straw; (a) Plant dry weight, shoot dry weight, Root dry weight, Plant height & Root depth; (b) No. of leaves/plant, Leaf area & Leaves dry weight/plant; (c) Number of fruits/plant, Average fruit fresh weight & Fruit length; (d) Water required after 3, 7 & 14 days of irrigation; (e) Soil temperature & Yield.

3.5 Yield

The yield of strawberry under control and different types of covering material has been shown in Table 3. It shows that straw produces the highest yield followed by jute nonwoven due to better permeability, moisture retention and thermal insulation property of those covering material. Yield of strawberry increases by 11% over the synthetic sheet and 28% over control (Figure 5(e)). Straw covering shows 3% better yield over jute covering.

3.6 Weed Suppression

Weed reduction of the strawberry field in synthetic sheet, jute nonwoven and other covering systems over control has been shown in Table 3. Figure 6 showed that synthetic sheet produces maximum control of weed which is around 73% followed by jute nonwoven (68%) due to compact structure (high bulk density) and a better cover of fabric which restricts penetration of sunlight and obstructs the growth of weed. Weed suppression by straw (57%) is much lower than the synthetic sheet or nonwoven.

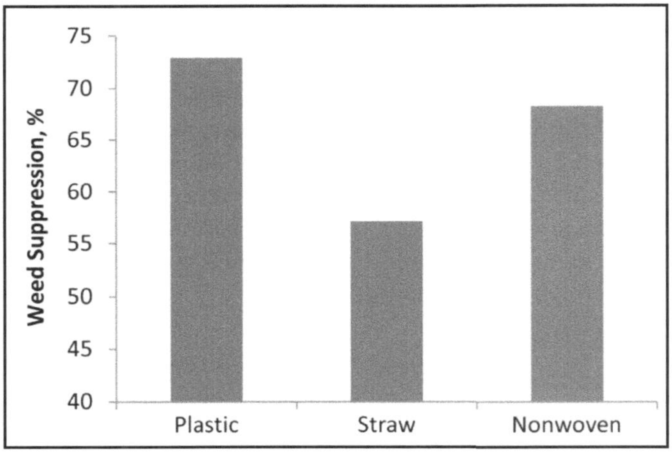

Fig. 6: Percent Weed Suppression in Plastic, Straw and Nonwoven Soil Cover

3.7 General Observations with Respect to Jute Nonwoven

Nonwoven fabric from waste jute looses strength (38%) after two months and sustained about six months but makes a solid coating over the surface which performs well up to 8–12 months depending on area density, structure of fabric, and soil characteristics. Jute fabric holds the loose soil particles in its place and resists the erosion and nutrient loss considerably during rain/thunderstorm/blowing of strong air/watering due to its rough and sticky surface. Water is not accumulated on jute fabric cover. Water logging over synthetic sheet is detrimental for strawberry. Jute absorbs and transmits/drains excess water keeping the surface free from water logging. It also spreads evenly throughout the fabric in all directions. Jute nonwoven fabric laying for covering is much easier than synthetic sheet due to higher area density and rough surface.

Strawberry is winter crop and at the beginning of summer, the plants are dried up. It was observed that with waste jute nonwoven covering the fruit is available in the field more than a month than other alternatives due to cooling effect to the soil, resulting in high production and rate during the additional period (off season) which fetches more financial gain to the farmers.

Waste jute nonwoven is costlier proposition compared to other covering material. Hence, the initial cost is higher. But at the end, farmers earn more profit using waste jute nonwoven soil cover which is 70% higher than without cover and 38% than plastic cover due to less requirement of pesticide, fertilizer, de-weeding cost etc; better yield and higher revenue from fruit. Therefore, it is economically viable for high-valued crops.

4. CONCLUSION

Waste jute needle-punched engineered nonwoven of 500 g/m^2 is a promising alternative to existing polyethylene sheet or other occasionally used traditional materials in strawberry cultivation. It was observed that waste jute nonwoven showed the remarkable improvement over the synthetic sheet with 13% in plant dry weight, 9% in plant height, 14% in fruit weight, 28% in soil temperature and 11% in yield. The straw performance is also comparable to jute nonwoven as covering materials under this trial in terms of biomass/ growth characteristics, fruit characteristics, soil moisture retention, soil temperature and yield. But the availability of huge amount of straw and its laying is difficult. Jute nonwoven reduces weed by 68% which is better than other biodegradable covers. As jute nonwoven fabric is costlier proposition than other covers, this is economically viable for high-valued crops. Jute fabric holds the loose soil particles in its place and resists the erosion and nutrient losses considerably during rain/thunderstorm/blowing of strong air/watering due to its rough and sticky surface.

REFERENCES

[1] Al-Ramamneh, E., Al-Rawashdeh, Z., Karajeh, M. and Abu-Romman, S. (2013). Plant Response of Strawberry to Intra-row Spacing and Growing Conditions in South of Jordan, *Asian Journal Plant Sc.*, (12), 201–207.

[2] ASTM, D 6242-98 (2003). Test Method for Mass Unit Area of Nonwoven Fabrics, *Annual Book of ASTM Standards.*

[3] ASTM, D 737-04 (2012). Test Method for Air Permeability of Textile Fabrics, *Annual Book of ASTM Standards.*

[4] BS 4745 (1974). Methods for determining of Thermal Resistance of Textiles, *British Standards Institute, Berkshire.*

[5] Debnath, S. and Madhusoothanan, M. (2007). Compressional Behaviour of Jute Polypropylene Blended Needle Punched Nonwoven Fabrics, *Indian J. Fibre Text. Res,* 32(4), 427–432.

[6] Debnath, S. and Madhusoothanan, M. (2011). Thermal Resistance Behaviour of Polyester Needle-Punched Nonwoven, *J. Inst. Eng. (I)*-TX, (91), 27–31.

[7] Fisher, P.D. (1995). An Alternative Plastic Covering System for Improved Water Management in Dry Land Maize Production, *Agricultural Water Management*, 27(2), 155–166.

[8] Horrocks, A.R. and Anand, S.C. (2000). Handbook of Technical Textiles, *Wood Head Publishing Limited, England,* 1–23.

[9] Laura, M.B., Gina, E.F. and Frank, J.L. (2002). Strawberry Plant Growth Parameters and Yield among Transplants of Different Types and from Different Geographic Sources, Grown in a Plasticulture System, *Hort. Technology,* 12(1), 100–103.

[10] Lunenschloss, J. and Albrecht, W. (1985). Nonwoven Bonded Fabrics, John Wiley and Sons, New York.

[11] Mattina, C.F. and Oathout, J.M. (1995). A New Method for Determining the Rate of Sorption of Wiping Material, *International Nonwoven Journal*, 7(1), 48.

[12] Pal, S.N., Burman-Roy, U., Bhattacharya, S. and Samanta, D. (1997). Norms for the Jute Industry, Part-I, Quality Parameters up to Spinning, *Indian Jute Industries Research Association*, Kolkata.

[13] Poling, E.B. (1993). Strawberry Plasticulture in North Carolina: II. Pre-plant, Planting and Post-plant Considerations for Growing Chandler Strawberry on Black Plastic Cover, *Horticulture Technology,* (3), 383–393.

[14] Purdy, A.T. (1983). Developments in Nonwoven Fabrics, *Textile Progress*, 12(4), 31.

[15] Sengupta, S. and Bhattacharya, G.K. (2008). Needle Punched Nonwoven Fabric—A New Dimension to Jute Diversification, *ICAR-National Institute of Research on Jute and Allied Fibre Technology*, Kolkata, 40.

[16] Sengupta, S., Samajpati, S. and Ganguly, P.K. (1999). Air-permeability of Jute-based Needle-Punched Nonwoven Fabrics, *Indian Journal Fibre Text. Res.*, 24(2), 103–110.

[17] Xie, Z., Wang, Y. and Li, F. (2005). Effect of Plastic Covering on Soil Water Use and Spring Wheat Yield in Arid Region of Northwest China, *Agricultural Water Management*, (75), 71–83.

Enzymatic Degumming *vis-á-vis* Study of Synergetic Effects on Decorticated Hemp Fibre

S.R. Shah[1] and K.J. Umrigar[2]

Department of Textile Chemistry, Faculty of Technology and Engineering,
The M.S. University, Baroda, Gujarat, India

E-mail: [1]shailesh.r.shah-tchem@msubaroda.ac.in; [2]k.j.umrigar-tchem@msubaroda.ac.in

ABSTRACT: *Cannabis sativa, Popular as Hemp fibre, has been little known in textile and allied industries. The main reason for the lack of exposure of the hemp fibre is due to the perfect chemical processing technology. The said fibre has got certain unique characteristics namely; high wet strength, anti-bacterial property, biodegradability, excellent breathability, superior UV blocking attributes and so on. In order to produce quality textile products from hemp fibre, the pre-treatment is considered as the central processing of hemp fibre. The efficient pre-treatment is possible to prepare high valued textile and allied products of hemp fibre. Therefore, in the present study, enzymatic pre-treatment (synergetic effect) of decorticated hemp fibre has been scientifically examined. The optimal conditions developed for the process has been justified through characterization by various techniques namely, chemical composition, moisture regain, whiteness index, yellowness index, dyeability and SEM. The results indicate that the said eco-friendly process can be explored at pilot scale.*

Keywords: Degumming, Enzyme, Hemp, SEM, Sustainability.

1. INTRODUCTION

Hemp–Cannabis sativa is a natural cellulosic bast fibre. This fibre has great potential to become alternative of cotton fibres. Hemp fibre is environment friendly which requires less pesticide, lesser time for growth and water for cultivations compared with cotton (Idler *et al.*, 2011, Singh, *et al.*, 2018). The hemp fibre has certain unique characteristics such as high wet strength, anti-bacterial property, biodegradability, excellent breathability, superior UV blocking attributes and so on. The hemp fibre is used for domestic as well as industrial application. The hemp fibre is used for paper making, rope or in composite with other fibres (Denialewicz & Surma-slusarska, 2017). Hemp is also used for textiles because of some superior characteristics such as higher tensile strength, low deformation under influence of axial, flexural, tensional forces, low dry and wet elongation and higher absorbency of perspiration, anti-electrostatic and bacterio-static properties, hygroscopicity, dye absorption and UV protection (Hebbal *et al.*, 2006). The fibre used for apparel application should be finer quality yarn which can be spun on cotton spinning system. The cotton spinning systems require some important properties such as uniform staple length, softness, fineness, low stiffness, and strength. The hemp fibre has sufficient strength but lack in other properties which make it difficult to be spun on cotton spinning system. The traditional processing of hemp fibre is too old and new cottonized method has been developed (Nabel, 1995). The hemp fibre comprises a high amount of non-cellulosic components. The processing of hemp fibre is centric on pre-treatment processes. Enzymatic pre-treatment of hemp fibre followed by mechanical treatment removes the non-cellulosic gummy substance and change liner density and length of the fibre (Sedelnik, 2004). The retting process of hemp fibre also plays an important role in processing of hemp fibre. The traditional methods of retting take 24 to 30 days to degrade the non-cellulosic gummy substances. The combination of enzymatic and chemical retting process is most efficient (Poonsawat *et al.*, 2016). The bio-chemical combined pre-treatment was also efficient for removal of non-cellulosic impurities which make the finer quality of hemp fibre suitable for cotton spinning system (Fand *et al.*, 2017). The hemp fibre can be converted to cotton like fibre by different spinning process in combination with cottonization process. This produces lower liner density fibre (Cierpucha *et al.*, 2004). The finishing treatment with softening agents also improves the spin ability of the hemp fibre. The finishing treatments with softening agent increase the uniformity of length, flexibility, surface friction and moisture absorption (Jin-qiu & Jian-chun, 2010).

2. MATERIALS AND EXPERIMENTAL METHOD

2.1 Materials

The hemp fibre in decorticated form purchased from local supplier was used in the present study.

All the chemicals used were of laboratory reagent or analytical reagent grade. Three different enzymes used in present study with their details are reported in Table 1.

Table 1: Specification of Enzyme

Enzyme Code	*Name of the Enzyme*	*Activity on*	*Supplier*
E–1	Hemi cellulose PDR	Hemi cellulose	Rossari Biotech Limited, Ahmedabad
E–2	Green boost Liquid	Pectin and Wax	Rossari Biotech Limited, Ahmedabad
E–3	Genecel JNI Liquid	Cellulose	Rossari Biotech Limited, Ahmedabad

2.2 Experimental Method

2.2.1 *Enzymatic Pre-treatment of Fibre*

The decorticated hemp fibre was pre-treated with different enzymes at various process parameters. The Process condition for different enzymatic pre-treatment is reported in Table 2. A synergetic effect of enzymes was also studied. The Process condition for synergetic effect of enzymes is reported in Table 3. The enzyme was deactivated after the stipulated time of application; samples were thoroughly washed with water and finally dried.

Table 2: Process Condition of Enzymatic Pre-treatment

Enzyme (% OWF)	5, 10, 15 and 20
Buffer	To maintain pH 4.5–5.0
Temperature (°C)	50
Time (h)	0.5, 1, 2, 4, 6, 12, 18, 24, 36 & 48

Table 3: Process Condition of Synergetic Effect of Enzyme Pre-treatment

E–1 (% OWF)	20
E–2 (% OWF)	20
E–3 (% OWF)	10
Buffer	To maintain pH 4.5–5.0
Temperature (°C)	50
Time (h)	0.5, 1, 2, 4, 6, 12, 18, 24, 36 & 48

2.2.2 *Dyeing of Hemp Fibre*

The dyeing of hemp fibres, before and after treatment was carried out with reactive dye as per the prescribed standard method (Chakraborty, 2010). After the samples were washed with water and soaped at 60°C for 10 minute using 1 gm/L non-ionic detergent they were finally washed to neutral pH and dried.

2.3 Analytical Method

The Chemical constituents of the treated and untreated hemp fibre were analyzed as per the method suggested by Turner and Doree (Garner, 1967).

The percentage weight loss of the samples after enzymatic treatment was determined using standard methods (Bhattacharya & Shah, 2002).

Wettability of various samples were measured by putting a droplet of water on surface of the fibre and measure the time taken by water droplet to be absorbed into the fibre using stopwatch. Four readings from different parts of sample were taken and average is reported.

The color strength in terms of K/S, whiteness and yellowness indices using formula ASTM313 of various samples were measured on Spectra Scan-5100 spectrophotometer (Model: Premier color scan, India). The structural changes of samples were examined at magnification of 1000x using SEM technique. (Model: JSM 5610LV 1.0, Jeol, Japan)

3. RESULTS AND DISCUSSION

3.1 Effect of the Individual Enzymes on Decorticated Hemp Fibre

Raw hemp in decorticated form has been eco-friendly pre-treated with individual enzymes. The effect of various process parameters has been examined in terms of weight loss percentage. The weight loss percentage of enzyme E–1 and E–2 is shown in Figures 1&2, respectively.

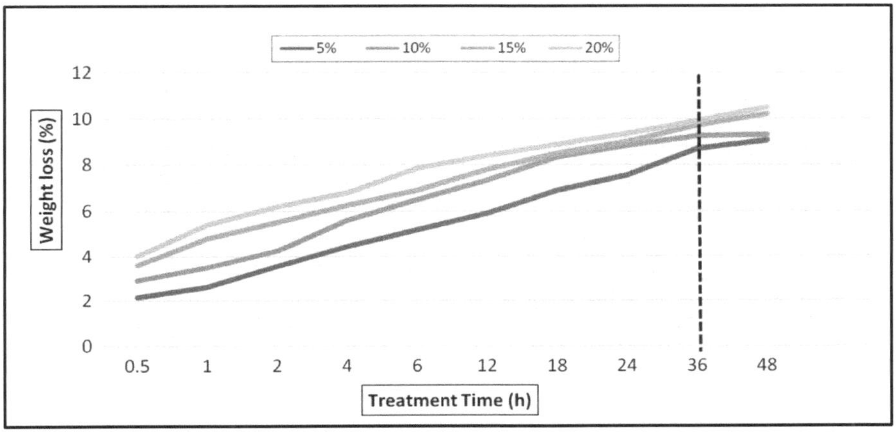

Fig. 1: Weight Loss (%) of Hemp Fibre Pre-treated with E–1 Enzyme

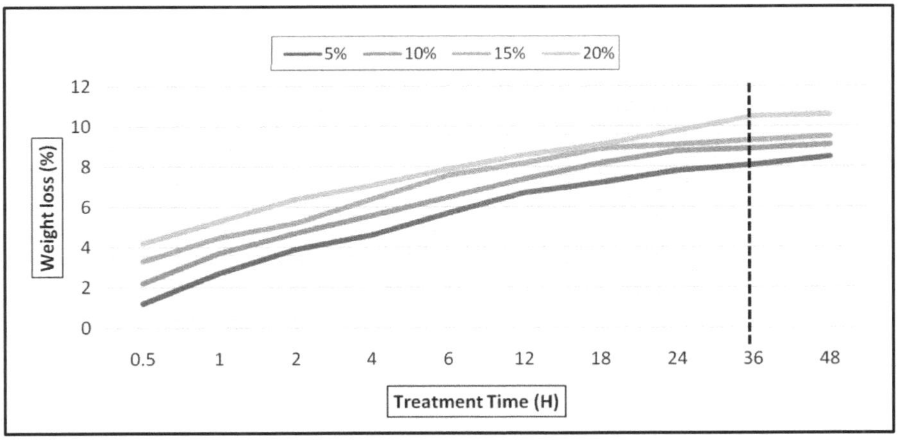

Fig. 2: Weight Loss (%) of Hemp Fibre Pre-treated with E–2 Enzyme

The enzyme E–1 is active on hemi cellulose and enzyme E–2 is active on pectin and wax component. The result shows that as the concentration and duration of treatment increase, removal of the amount of non-cellulosic impurities increases.

In case of enzyme E–1, which is active on hemi cellulose component, the rate of hydrolysis was high at initial stage, increase steadily and become stagnant at about

15%–20% concentration with 48 hours duration. Similar trend can be seen with enzyme E–2, which is active on pectin and wax component. The hydrolysis rate becomes almost constant at 15%–20% concentration with duration of 36 hours.

This can be explained as the major parts of non-cellulosic impurities are present on the outer surface of fibre, which are removed faster in milder or earlier condition. The non-cellulosic components present in intrinsic of fibre structure are bound with each other strongly can be removed partially. The optimized condition of enzymes E–1 and E–2 is reported in Table 4 and Table 5, respectively.

Table 4: Optimized Condition of E–1 Enzyme

Sample Code: E–1	
Concentration of Hemicellulase Enzyme (%)	20
Treatment time (hr)	48
Temperature (°C)	50

Table 5: Optimized Condition of E–2 Enzyme

Sample Code: E–2	
Concentration of Hemicellulase Enzyme (%)	20
Treatment time (hr)	36
Temperature (°C)	50

3.2 Effect of Mix Enzymes on Decorticated Hemp Fibre

Both the enzymes E–1 and E–2 are active in same condition, therir effects on decorticated hemp fibre has been examined synergestically along with the enzyme E–3 which is active on cellulose component. The process condition for synergetic effect is mentioned in Table 3. The process efficiency was examined in terms of percentage weight loss and represnted in Figure 3.

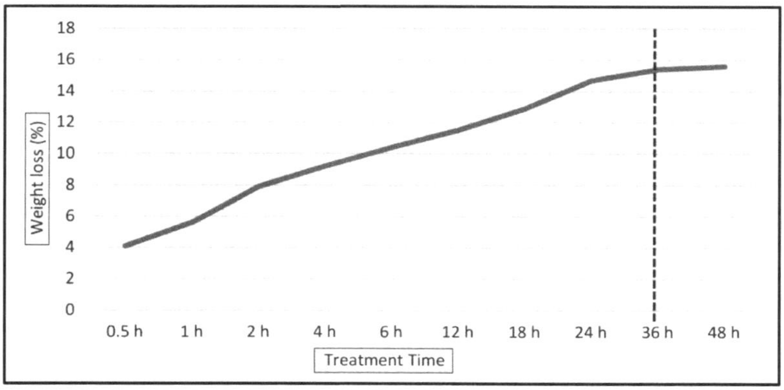

Fig. 3: Weight Loss (%) of Hemp Fibre Pre-treated with Mix Enzyme

The reults indicate that the rate of hydrolysis of non-cellulosic components increases with that of time from 30 minute to 24 hours and becomes alomost constant. This is because the mix enzyme can effectively remove synergestically all the components up to certain level. The faster removal of non-cellulosic component may not be possible with the enzymatic treatment. The optimized condition for mix enzymes is reported in Table 6.

Table 6: Optimized Condition of Mix Enzyme

Enzyme Code: M–1	
Concentration of Hemicellulase Enzyme (%)	20
Concentration of Green boost liquid Enzyme (%)	20
Concentration of Genecel JNI liquid Enzyme (%)	10
Treatment time (hr)	36
Temperature (°C)	50

3.3 Characteristics of Enzymatic Pre-treated Hemp Fibre at Optimized Condition

Chemical composition and all other important characteristics of optimal pre-treated hemp fibres have been evaluated and compared with that of raw hemp fibre which is reported in Table 7.

Table 7: Chemical Composition of Hemp Fibre

Chemical Composition (%)	*Raw Hemp*	*E–1*	*E–2*	*M–1*
Water soluble component	3.03	1.80	1.50	0.90
Fat and Waxes	0.75	0.61	0.50	0.40
Pectin	5.62	5.22	2.63	2.25
Hemi cellulose	24.55	17.18	21.91	15.25
Lignin	6.53	5.53	5.27	4.86
Cellulose	59.52	70.89	68.57	74.03
Absorbency (s)	> 180	10	17	9
Whiteness Index (HUNTER)	42.03	49.17	47.12	51.89
Yellowness Index (ASTM313)	39.92	35.61	37.55	32.47
Dye strength (K/S)	2.93	5.22	3.86	6.89

Chemical constituents of various hemp fibres are reported in table 07. In case of E–1, activated only on hemicelluloses component, there is significant reduction in hemi-celluloses content. Similarly, E–2, activated only on pectin and fats, reduction in particular component in greater extent. M–1 shows major removal of gummy components with respect to E–1 and E–2. The cellulose content is significantly improved in enzyme treated fibre.

Whiteness and yellowness indices are changed to some extend by the enzymatic action on hemp fibre. It can be seen from the Table 7 that the whiteness of the pre-treated hemp fibres is improved and yellowness is decreased in comparison with decorticated hemp fibre. The improvement in whiteness is due to the removal of the lignin component along with the other non-cellulosic components. The maximum whiteness is observed with M–1 which correlated with lower lignin content among various enzyme pre-treated fibres. The yellowness is inversely proportional to the whiteness. This trend was also seen from the Table 7.

Wettability of various hemp fibre at optimal condition and decorticated fibre is reported in table 07. Wettability of decorticated hemp fibre is more than 180 second. Decorticated hemp fibre consisting of layer of wax, makes fibre hydrophobic. Also, the fibre is joined together by gum like cementing materials consisting non-cellulosic components. Enzymatic pre-treatment on hemp fibre removed this hydrophobic layer of non-cellulosic impurities, resulting in improving the wet ability. The maximum wet ability in M–1 may be due to the removal of more gummy components from the fibre in comparison with E–1 and E–2.

Dye strength of various hemp fibre at optimal condition and decorticated fibre is reported in Table 7. The lowest dye strength is observed in case of raw hemp fibre, indicating hydrophobic nature of fibre due to the presence of non-cellulosic impurities. The dye strength is improved in case of enzyme pre-treated fibre and observed maximum in case of M–1. This result shows that fibre becomes more absorptive. This result is also correlated with wet ability of the fibres.

3.4 Characterization of Hemp Fibre at Optimized Condition through SEM Analysis

The SEM photograph of untreated and optimal pre-treated hemp fibres were examined at magnification of 1000x using scanning electron microscope. It is clearly seen from the SEM photograph that the fibre surface morphology is changed due to the application of enzymes. It can be seen from the SEM photograph of raw hemp in Figure 4(a) that fibres

Fig. 4(a): SEM of Raw Hemp

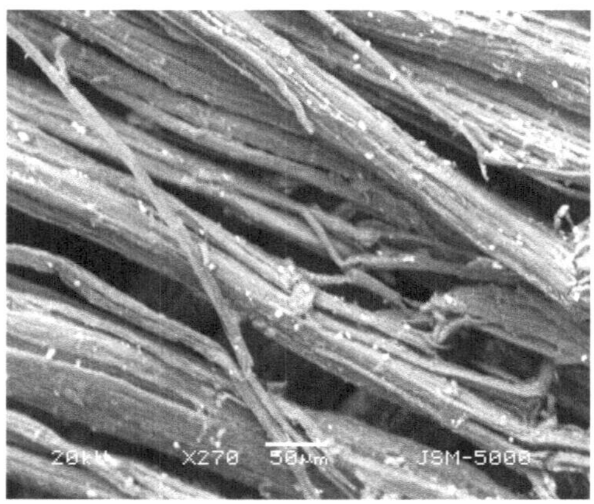

Fig. 4(b): SEM of E–1

are having smooth surface due to coating of gummy substance. The fibres are joined together indicating the presence of cementing materials. The SEM photographs of enzyme treated samples Figure 4(b), 4(c) and 4(d) show separation of fibre to some extent from the gummy substance. This indicates that the non-cellulosic substances are removed partially. The separation of fibres is more in case of M–1 indicating that greater amount of gummy substances are removed in comparison of E–1 and E–2.

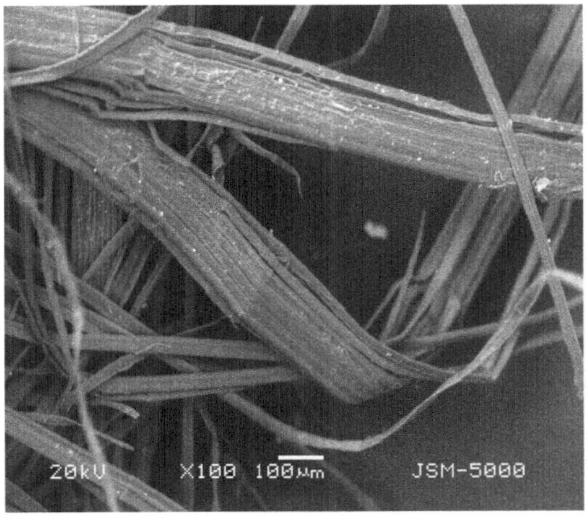

Fig. 4(c): SEM of E–1

Fig. 4(d): SEM of M–1

4. CONCLUSION

Decorticated hemp fibre was pre-treated eco-friendly with individual and mixture of enzymes. The individual enzymes act only on the specific component of non-cellulosic impurities keeping others intact. However, mix enzymes removed all the impurities in considerable amount. The removal of non-cellulosic components was supported by improved wettability, whiteness and dyeability. The SEM photograph also supports the better separation of fibre after mixing enzymatic treatment. Further work will be carried out to get comparative results with that of conventional alkaline pre-treatment process.

REFERENCES

[1] Bhattacharya, S.D. and Shah, S.R. (2002). Degumming of ramie—Role of Individual Constituents of Gummy Materials, *Colouration Technology,* 118, 295–299.

[2] Cierpucha, W., Kozłowski, R., Mańkowski, J., Józef Waśko, J. and Tomasz Mańkowski, T. (2004). Applicability of Flax and Hemp as Raw Materials for Production of Cotton-like Fibres and Blended Yarns in Poland, *Fibre and Textile in Eastern Europe,* 12, 3(47), 13–18.

[3] Chakraborty, J.N. (2010). Fundamentals and Practices in Coloration of Textiles, *India: Wood Head Publishing India Pvt. Ltd.*

[4] Danielewicz, D. and Surma-slusarska, B. (2017). Properties and Fibre Characterization of Bleached Hemp, *Birch and Pine Pulps: A Comparison, Cellulose,* 24, 5173–5186.

[5] Fang, G., Chen, H., Chen, A., Mao, K. and Wang, Q. (2017). An Efficient Method of Bio-Chemical Combined Treatment for Obtaining High Quality Hemp Fibre, *Bio Resources*, 12(1), 1566–1578.

[6] Garner, W. (1967). *Textile Laboratory Manual* (3rd edition) New York: Elsevier.

[7] Hebbal, O.D., Reddy, K.V. and Rajagopal, K. (2006). Performance Characteristics of a Diesel Engine with Deccan Hemp Oil. *Fuel,* 85(14–15), 2187–2194.

[8] Idler, C., Pecenka, R., Fürll, C. and Gusovius, H. (2011). Wet Processing of Hemp: An Overview, *Journal of Natural Fibres,* 8(2), 59–80.

[9] Jin-qiu, Z. and Jian-chun, Z. (2010). Effect of Finishing Treatment with Softening Agent on Spin Ability of Hemp Fibre, *Indian Journal of Fibre and Textile Research,* 135, 115–120.

[10] Jin-qiu, Z. and Jian-chun, Z. (2010). Effect of Refined Processing on The Physical and Chemical Properties of Hemp Bast Fibres, *Textile Research Journal,* 80(8), 744–753.

[11] Nabel, K.M. (1995). New Processing Strategy of Hemp Fibre, Journal of the International Hemp Association, 2(1), 6–9.

[12] Poonsawat, T., Sae-Bae, P., Bunphami, M., Wetchaiyoc, T. and Jatamaneerat, P. (2016). Effect of Water and Chemical Retting on Properties of Hemp Fibre and Hybrid Hemp/Cotton Spun Yarn, *Journal of Engineering and Applied Sciences,* 11(9), 1991–1995.

[13] Sedelnik, N. (2004). Properties of Hemp Fibre Cottonized by Biological Modification of Hemp Hackling Noils, *Fibre and Textile in Eastern Europe,* 12, 1(45), 58–60.

[14] Singh, M., Mamania, D. and Sinde, V. (2018). The Scope of Hemp (Cannabis Sativa L.) Use in Historical Conservation in India, *Indian Journal of Traditional Knowledge*, 17(2), 314–321.

Effect of Retting Conditions upon Optical Property of Fibres Extracted from 120 Day Old Okra Plants

Pavan Kumar Gupta and Shamayita Patra*

Shri Vaishnav Institute of Textile Technology, Shri Vaishnav Vidyapeeth Vishwavidyalaya, Indore, Madhya Pradesh, India

E-mail: *shamayitapatra@rediffmail.com

ABSTRACT: *Okra plant is one of the underutilized and sustainable natural resource for bast fibre production. In this study okra fibre was extracted from the 120±10 day old okra plant harvest waste and subjected to different retting conditions. Okra fibre was extracted at different time intervals (7–35 day) in two different water conditions i.e., stagnant and running water. Optical property is one of the crucial factors for deciding the cost of raw fibre, final product, and dyeing behaviors of their blends, etc. Optical property such as whiteness index of the obtained fibres was measured by spectro-photometer. The conducted study revealed the fact that the whiteness index is not only dependent upon the retting time but also upon the retting condition.*

Keywords: Okra Fibre, Okra Plant, Pectin, Water Retting, Whiteness Index.

1. INTRODUCTION

Okra (*Abelmoschus esculentus*) is a multi-purpose vegetable crop which is mainly grown in tropical and sub-tropical regions (Varmudy, 2011). Due to the abundant availability of harvested okra plant, it has attracted the attention of researchers as a potential fibre source (Kocak *et al.*, 2018). Okra fibre properties are compatible with jute fibre (Nabaneeta *et al.*, 2017). Therefore, okra fibre can be utilized for various applications same as the application of jute (Ammayappan *et al.*, 2017). Similar to jute fibre, okra fibre is also extracted through the retting process (Nabaneeta *et al.*, 2017). Water retting is one of the easiest and common method. Retting process parameters influence the quality of fibre as it influences the residual content of the gummy material in the extracted fibres. One of them is optical property (Ali *et al.*, 2015). No study has been reported about the effect of the retting condition (stagnant and running water) on optical properties of okra fibres. Therefore, the main objective of this study is to identify the effect of retting condition upon the optical appearance of the yielded fibres.

2. MATERIALS AND METHOD

2.1 Materials

2.1.1 *Raw Material*

Material: Okra plant harvest wastes after 120 days.
Source: Village-Hingonia, Indore, Madhya Pradesh, India.

2.1.2 *Retting Medium*

River water is used for running water retting condition. Ghambhir River located near Village Hingonia was used for running water retting.

While tap water was used for stagnant water during retting condition, water tank [3 m (L) × 3 m (W) × 2.5 m (H)] made of concrete was used as stagnant water retting pot.

2.1.3 *Instrument for Optical Characterization*

Spectrophotometer (Model-SS5100A) was used to assess the optical properties.

2.2 Method

2.2.1 *Retting Process*

Green okra plants leftover in bundle form were immersed into a concrete tank as well as river for 7 to 35 days at different time intervals. After the completion of the pre-decided time period fibre reeds were separated by hand picking. The separated fibre reeds were then washed with sufficient quantity of water and dried in ambient condition for 6–7 days in open air without direct Sunlight. The retted fibres are shown in Figure 1.

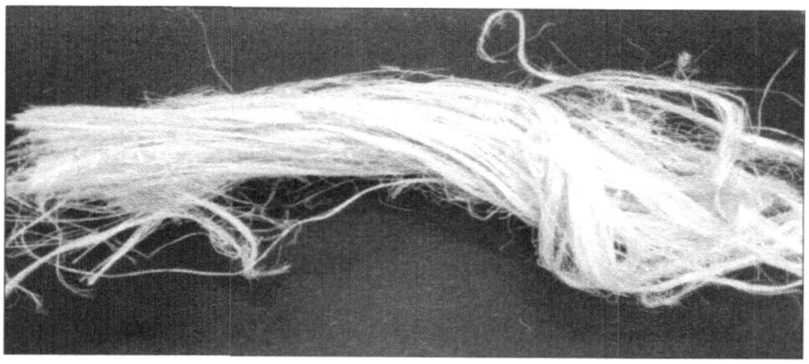

Fig. 1: Okra Fibre

2.2.2 *Optical Property Assessment Procedure*

(a) ***Sample Preparation:*** Samples for the spectrophotometric analysis were prepared by wrapping samples of 2 g around a white paper template backed with a hard board. The white template was standardized (Table 1) first against standard white tile for further evaluation.

Table 1: Measured Values of Standard White Tile and White Paper Template

Particulars	*Whiteness Index (HUNTER)*	
	Average	*Standard Deviation*
Standard White Tile	93.85	0.05
White Paper Template	91.22	0.17

(b) ***Assessment Process:*** To assess the optical properties of okra fibre, the spectrophotometer (Premier, Model-SS5100A) was calibrated with standard white tile. A total of five readings were taken for analysis purpose for each sample. The retted reed has shown very clear heterogeneous optical appearance throughout its length (Figure 2). Therefore, the upper portion and the bottom portion of the reed have been investigated separately.

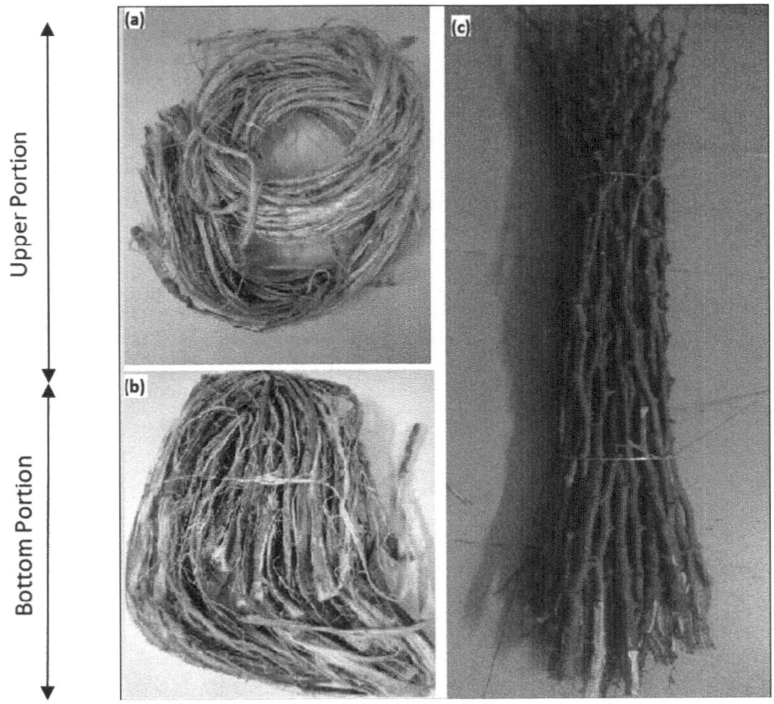

Fig. 2: The Optical Appearance of Retted Reed: (a) Upper Portion, (b) Bottom Portion and (c) Raw Okra Plant Stem

3. RESULTS AND DISCUSSION

Soluble pectin of plant is removed in retting process by the action of bacterial and fungal pectinases. The insoluble pectin remains with the fibre and determines the compactness of the fibre. These extracted fibre strands are mainly composed of cellulose, hemicellulose, and lignin with smaller amount of other components such as fats and waxes, pectin, coloring compounds and inorganic matters. The Retting process parameters influence the extracted fibre quality in terms of tensile strength, fineness, and optical properties, etc. (Ammayappan *et al.*, 2017). Only the effect retting condition on optical properties of okra fibres has been discussed in this paper.

3.1 Assessment of Whiteness Index (WI)

Whiteness is an attribute by which an object is judged to approach the preferred white. Whiteness index is used to indicate the degree of whiteness as the distance in color space from the sample to that of standard (Zarubica *et al.*, 2005). Different formulas for the whiteness assessment are currently used, such as Berger, Hunter, and Stensby formula. The formula for whiteness index $_{(\text{Hunter})}$ is expressed as,

$$W_{(\text{Hunter})} = L - 3b \qquad \ldots \text{(i)}$$

Where L is the vertical coordinate of a three-dimensional system of colors, which has values from 0 (black) to 100 (for white); and b is the horizontal coordinate the values of which range from –80 (blue) to +80 (yellow) (Zarubica *et al.,* 2005). Extracted fibre bundles obtained from different retting conditions and retting time were investigated to measure their $WI_{(\text{Hunter})}$. The obtained results from running water retting and stagnant water retting are represented in Figures 3 and 4 respectively. After 35 days of retting the fibre strength reduced significantly, therefore, further retting time was not taken into consideration in this research.

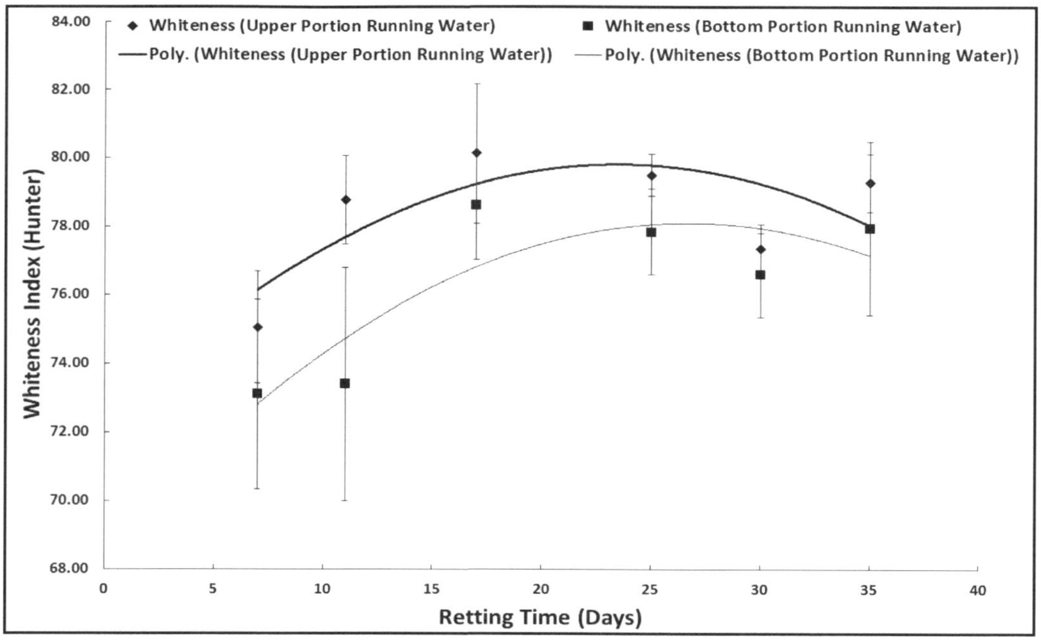

Fig. 3: Effect of Retting Time in Running Water Retting Condition on Okra Fibre Whiteness Index

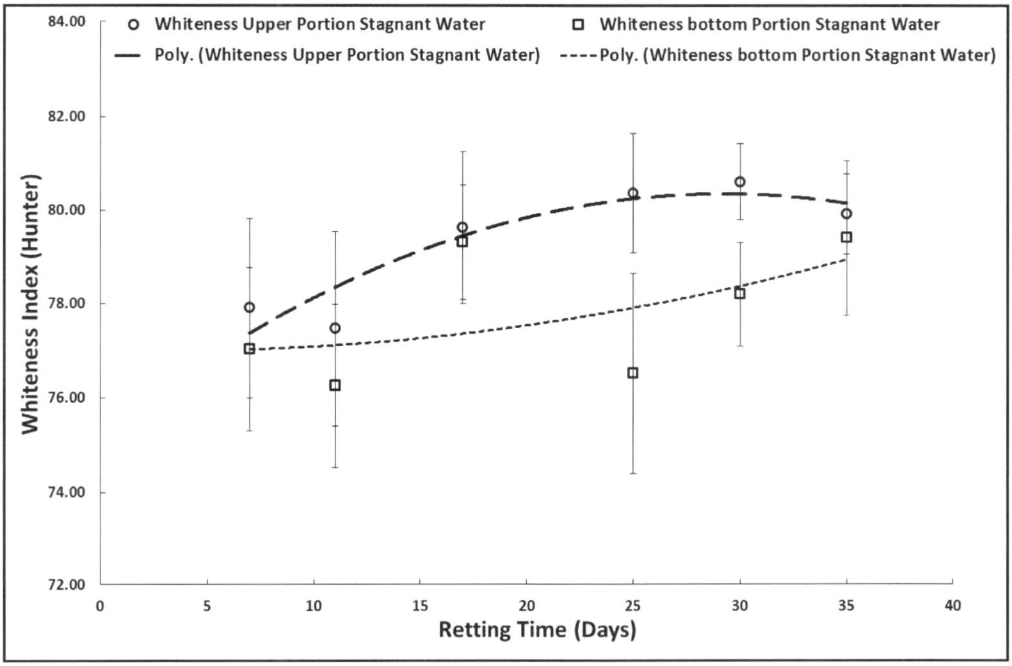

Fig. 4: Effect of Retting Time in Stagnant Water Retting Condition on Okra Fibre Whiteness Index

From Figures 3 and 4, for both the retting conditions, the whiteness index has reached its peak after a certain retting time and afterwards it has declined. Now, this increase in whiteness can be explained by the increasing detachment of the fibres from the bark due to elimination of soluble pectin. However, after reaching a peak it has declined, it may be due to the detrimental effect of oxidization of the components which has created some unsaturation (Chen, 2014). That unsaturation has imparted some hue to the substances. Throughout the retting process both of these reactions i.e., dissolution and oxidation, goes hand in hand. But, initially elimination of soluble pectin was more prominent therefore whiteness has increased but afterwards oxidation becomes more prominent and whiteness has declined. For stagnant water also the whiteness has declined, as it is stagnant water, there is a limitation of the major mobile phase (water). Due to the reduction in the mobile phase the same effect in term of whiteness is achieved for the stagnant water in a delayed time.

The upper portion of the fibre reed showed better whiteness than the bottom portion at every retting time and retting conditions. It may be due to the fact that the thickness of bark is higher at bottom portion, so it takes longer retting time to attain a similar result. In this research study, for both retting conditions, 17 days retted fibre showed minimum gap/variation between upper and bottom portion of the fibre reed as well as highest or very close to highest value of whiteness. Therefore, 17 days of retting period can be taken as ideal time for both the retting conditions.

4. CONCLUSION

From this study, it can be concluded that the optical appearance in terms of whiteness index not only depends upon the number of retting days but also on the retting condition, especially, it depends upon the amount of retting medium and its flow-ability. Running water retted fibre showed better homogeneity for both upper and bottom portion rather than the stagnant water retted fibre. The upper portion of okra fibre reed showed better whiteness than the bottom portion at every retting time and retting conditions. Retting time of 17 days can be considered as ideal time for both the retting conditions.

REFERENCES

[1] Agrawal, A.K., Behera, B.K. and Abbas, J. (2004). Development of Reinforced Paper Laminates Using Bhindi Fibres, *Indian Journal of Fibre and Textile Research*, 29(3), 49–56.

[2] Ali, M.R., Kozan, O., Rahman, A., Islam, K.T. and Hossain, M.I. (2015). Jute Retting Process: Present Practice and Problems in Bangladesh, *Agricultural Eng. International,* 17(2), 243–247.

[3] Ammayappan, L., Samanta, K.K. and Ghosh, R.K. (2017). Training Manual on Recent Advances in Processing Technologies for Value Addition of Jute and Allied Fibres, *ICAR-National Institute of Research on Jute and Allied Fibre Technology*, Kolkata.

[4] Camciuc, M., Deplagne, M., Vilarem, G. and Gaset, A. (1998). Okra-*Abelmoschus esculentus* L (Moench.) A Crop with Economic Potential for Set Aside Acreage in France, *Industrial Crops and Products*, 7(2–3), 257–264.

[5] Chen, H. (2014). Chemical Composition and Structure of Natural Ligno Cellulose, In *Biotechnology of Ligno Cellulose*, 25–71, Springer, Dordrecht.

[6] Department of Biotechnology (2011). Ministry of Science and Technology, Govt. of India, *Biology of Abelmoschus Esculentus L. (Okra),* Retrieved from*http://envfor.nic.in/divisions/csurv/geac/Biology_of_Okra(1.pdf.*

[7] De-Rosa, I.M., Kenny, J.M., Maniruzzaman, M., Moniruzzaman, M., Monti, M., Puglia, D., Santulli, C. and Sarasini, F. (2011). Effect of Chemical Treatments on the Mechanical and Thermal Behaviour of Okra (*Abelmoschus Esculentus*) Fibres, *Compo. Sci. and Tech.*, 71(2), 246–254.

[8] Fathima, M. and Balasubramanian, A. (2006). Effect of Plant Growth Regulators on the Quality of Bast Fibres in *Abelmoschus Esculentus* (Linn.) Moench, *Acta Botanica Croatica*, 65(1), 101–112.

[9] Fortunati, E., Puglia, D., Monti, M., Santulli, C., Maniruzzaman, M., Foresti, M. L., Vazquez, A. and Kenny, J.M. (2013). Okra (*Abelmoschus Esculentus*) Fibre-based PLA Composites: Mechanical Behaviour and Biodegradation, *Journal of Polymers and the Environment,* 21(3), 726–737.

[10] Isa, M.T., Usman, S., Ameh, A.O., Ajayi, O.A., Omorogbe, O. and Ameuru, S.U. (2014). The Effect of Fibre Treatment on the Mechanical and Water Absorption Properties of Short Okra/Glass Fibres Hybridized Epoxy Composites, *Inter. Jou. of Materials Eng.,* 4(5),180–184.

[11] Jahan, M.S., Rahman, M. and Rahman, M.M. (2012). Characterization and Evaluation of Okra Fibre (*Abelmoschus Esculentus*) as a Pulping Raw Material, *Journal of Science & Technology for Forest Products and Processes*, 2(5), 12–17.

[12] Kalia, S., Kaith, B.S. and Kaur, I. (2011). Cellulose Fibres: Bio- and Nano-Polymer Composites: Green Chemistry and Technology, *Springer Science and Business Media.*

[13] Khan, G.A., Abedin, S.M.A., Choudhury, M.J., Gafur, M.A. and Alam, M.S. (2014). Renewable Okra Bast Fibre-Reinforced Phenol-Formaldehyde Resin Composites: Mechanical and Thermal Studies, *Research and Reviews: Journal of Material Sciences*, 2(1), 32–36.

[14] Kocak, E.D., Merdan, N., Mistik, I. and Sahinbaskan, B.Y. (2018). Fibre Extraction from Okra Plant Agricultural Wastes, Their Characterizations and Surface Modifications by Environmental Methods, *In Sustainable Innovations in Textile Fibres* (53–80), Springer, Singapore.

[15] Kozłowski, R.M. (ed. 2012). Handbook of Natural Fibres: Types, Properties and Factors Affecting Breeding and Cultivation, *Elsevier.*

[16] Makhadmeh, I.M., Ereifej, K.I. (2004). Geometric Characteristics and Chemical Composition of Okra (*Hibiscus Esculentus* L.) Grown Under Semi-Arid Conditions, *International Journal of Food Properties*, 7(1), 83–90.

[17] Martin, F.W. (1982). Okra, Potential Multiple-Purpose Crop for The Temperate Zones and Tropics, *Economic Botany*, 36(3), 340–345.

[18] Nabaneeta, G., Minti, G. and Swapna, C. (2017). Utilization of Agro Waste-Okra and Its Potentiality, *Asian Journal of Home Science*, 12(1), 250–256.

[19] Onyedum, O., Aduloju, S.C., Sheidu, S.O., Metu, C.S., Owolabi, O.B. (2015). Comparative Mechanical Analysis of Okra Fibre and Banana Fibre Composite used in Manufacturing Automotive Car Bumpers, *American Jour. of Eng., Technology and Society*, *2*(6), 193–199.

[20] Paridah, M.T., Basher, A.B., Saiful, A.S. and Ahmed, Z. (2011). Retting Process of Some Bast Plant Fibres and its Effect on Fibre Quality: A Review, *Bio. Resources*, *6*(4), 5260–5281.

[21] Rai, S., Hossain, M. and Hossain, F. (2012). Evaluation of Okra (*Abelmoschus Esculentus* (Moench) L.) as Bast Fibre Crop, *Journal of Crop and Weed*, 8(1), 101–104.

[22] Sen, K.K., Saha, S.C., Mojumder, P., Ray, P.K. (1988). Some Physical Properties of Alkali□Treated Bhindi Fibre (*Hibiscus Esculentus* Linn.), *Journal of Applied Polymer Science*, 36(7), 1601–1605.

[23] Sen, K.K., Saha, S.C. and Ray, P.K. (1987). Physical and Structural Properties of Raw and Alkali-Treated Bhindi (*Hibiscus Esculentus L.*) Fibre, *Indian Journal of Textile Research*, 12(9), 152–153.

[24] Srinivasababu, N., Rao, K.M.M. and Kumar, J.S. (2009). Tensile Properties Characterization of Okra Woven Fibre Reinforced Polyester Composites, *International Jour. of Eng.* 3(4), 403–412.

[25] Srinivasababu, N., Rao, K. and Kumar, J.S. (2009). Experimental Determination of Tensile Properties of Okra, Sisal and Banana Fibre Reinforced Polyester Composites, *Indian Journal of Science and Technology*, 2(7), 35–38.

[26] Sule, U., Isa, M.T., Ameh, A.O., Ajayi, O.A. and Omorogbe, O. (2014). Studies on the Properties of Short Okra/Glass Fibres Reinforced Epoxy Hybrid Composites, *The International Journal of Science and Technoledge*, 2(7), 260–265.

[27] T 1216 sp-98, Indices for Whiteness, Yellowness, Brightness and Luminous Reflectance Factor, Approved by the Optical Properties Committee of the Process and Product Quality Division TAPPI, Retrieved from *http://www.balibago.org/Files/Tappi/DOCS/T1216.PDF.*

[28] Vanishree, S. and Mahale, G. (2016). Influence of Wet Processing on Physico-Chemical Properties of Okra Bast Fibre, *Journal of Farm Sciences*, 29(3), 412–414.

[29] Varmudy, V. (2011). Need to Boost Okra Exports, *Factsfor You*, 31(5), 21–23.

[30] Zarubica, A.R., Miljković, M.N., Purenović, M.M. and Tomić, V.B. (2005). Colour Parameters, Whiteness Indices and Physical Features of Marking Paints for Horizontal Signalization, *Facta Universitatis-Series: Physics, Chemistry and Technology*, 3(2), 205–216.

Preparation of Nonwoven using Fibres Extracted from *Saccharum munja* Grass and its Application in Composites

Girendra Pal Singh*, Pallavi Vishwas Madiwale and R.V. Adivarekar

ICT, Mumbai, Maharashtra, India

*E-mail: singhgirendra2000@gmail.com

ABSTRACT: *Fibres from Saccharum munja grass were extracted and converted into nonwoven mat for reinforcement with unsaturated polyester resin. The composites were prepared using compression moulding technique with various fibre loading. The effects of fibre loading on mechanical properties such as tensile, flexural and impact and water absorption were investigated. The composite with 50% of fibre loading gives excellent mechanical properties in terms of tensile strength, impact strength and flexural strength.*

Keywords: Composite Material, Dilo Nonwoven, Methyl Ethyl Ketone Peroxide (MEKP), Polyethylene Terephthalate (PET), *Saccharum munja* Grass.

1. INTRODUCTION

Application of natural fibres in composites are increasing due to their excellent properties such as renewability, lightweight, low density, high specific strength, less abrasivity, low cost and biodegradability compared to conventional reinforcing fibres like glass fibres. Besides conventional ligno-cellulosic fibres like Jute, hemp, kenaf, flax, sisal, coir, cotton, wood and bamboo some non-conventional ligno-cellulosic natural fibres were explored to their utilization as reinforcement in composites like curaua, henequen, olive husk, rice husk, pineapple, bagasse, banana, wheat straw, oil palm, abaca, aloevera, pine needle. The limitation of cellulosic fibres in use as reinforcement in composites is its incompatibility with resin because of swelling of these due to their hydrophilicity. This limitation can be overcome by various chemical treatments to the fibres to reduce hydrophilicity of these fibres. Many factors like length of the fibre, orientation of the fibres as reinforcement, surface characteristics of the fibres and percentage loading of the fibres as reinforcement affect the mechanical properties of composites (Fu *et al.*, 2008). Reinforced fibres can be used in different forms like short fibres, woven fabrics, knitted fabrics, braided structures and nonwoven fabrics. Nonwoven fabrics are more suitable for reinforcement in composites to get desired fibre loading and properties.

In the current work random oriented nonwoven fabrics were prepared for using the fibres extracted from *Saccharum munja* grass on conventional jute processing machinery. These fabrics were used as reinforcement in composites along with unsaturated polyester resin. The percentage fibre loading was varied to optimize tensile properties and to achieve maximum fibre loading.

2. EXPERIMENTAL

2.1 Extraction of Fibres from *Saccharum munja* Grass

The fibres were extracted as explained in our previous work (Singh *et al.*, 2014) briefly grass was cut and dried and this was thrashed by wooden hammer and chemical retting using 2% NaOH for 72 hours were used to extract fibres from the grass.

2.2 Nonwoven Preparation from Extracted *Saccharum munja* Fibre

Extracted fibres were cut to 12 inches of length. These fibres were treated with 30% (o.w.f) oil in water emulsion using spray techniques and kept for 48 hours to impart softness for mechanical processing. These fibres were carded using jute carding machine for opening and cleaning. After carding, fibres were processed in Dilo Nonwoven card for individualization and web was prepared. This web was then fed to needle punching machine (fibre locker machine) to prepare Nonwoven fabrics with different GSM. Various Nonwoven samples were prepared using 100% *Saccharum munja* fibres. These Nonwoven sheets were used as reinforcing material with unsaturated polyester resin to prepare composites.

Table 1: Recipe for Emulsion

Castor oil	2% o.w.f
Emulsifier or plasticiser	0.5% o.w.f
Water	27.5% o.w.f

Fig. 1: (a) Extracted fibres, (b) Fibres after opening, (c) Manufacturing of Nonwoven, (d) Prepared Nonwoven fabric

2.3 Composite Preparation from *Saccharum munja* Fibre

A stainless steel mould was used to fabricate composites. Size of mould was 190 mm × 190 mm × 3 mm. A polyethylene terephthalate (PET) sheet was placed in the bottom of the mould and a thin layer of mould releasing agent was also applied on it. Pre-weighed dried Nonwoven fabric of *Saccharum munja* fibre was laid on PET sheet in the mould. Proper amount of unsaturated polyester resin was taken and 3% cobalt accelerator and 1% methyl ethyl ketone peroxide (MEKP) hardener were mixed properly in a cup manually. Resin was poured uniformly on fibres manually and a plastic squeeze was used to spread resin uniformly in the mould. A PET sheet was placed on the mould and a silicon roller was rolled to remove entrapped air from the fibres. The mould was placed between upper and lower plate of compression moulding machine left for 2 hours at 140°C temperature with 65 kg/cm^3 pressure for curing of composite. After 2 hours the mould was taken out, cooled and fibre reinforced composite sheet were separated from mould. Specimen for tensile, flexural, impact strength and water absorption testing were cut on vertical cutting machine according to required dimensions. Fibre loading was varied from 10% to 50% in various composites and properties of these composites were further investigated.

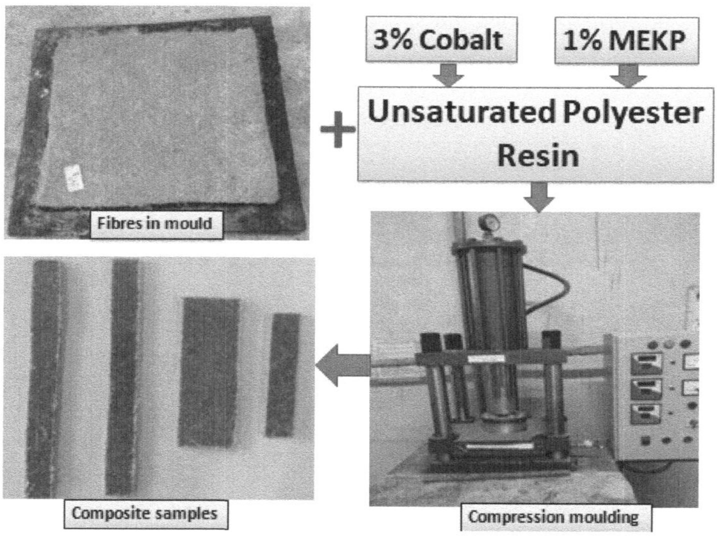

Fig. 2: Composite Preparation from *Saccharum munja* Fibre

3. CHARACTERIZATION

3.1 Tensile Strength

For comparison of tensile properties of different composites made by different compositions, they tested on Lloyd universal testing machine according to ASTM D 638. The tensile tests gave conducted at a crosshead speed of 5 mm/min at room temperature (27°C). To obtain a statistically significant result for each composite sample, five samples were tested to evaluate the mechanical properties. The test samples were cut in strip shape according to ASTM standard dimensions.

3.2 Flexural Strength Test

Three-point bending (flexural) tests were carried out according to ASTM D 790 on Lloyd LR tensile testing machine as shown in Figure 3.4. Standard dimensions of the samples 120 × 13 × 3 mm was used. The specimen was placed onto two supports having a 60 mm span length between the supports. The speed of the crosshead was set to 2 mm/min. Five samples of each group were tested to evaluate the flexural property.

3.3 Izod Impact Strength Test

Izod impact tests were conducted on Tinius Olsen impact tester model impact 104 for all specimens at room temperature to evaluate the resistance of the material to fracture. The dimensions of the samples used were 65 × 13 × 3 mm according to ASTM D 256.

3.3 Water Absorbency

Testing was done according to ASTM D570 standards. Samples were cut 3″ × 1″ according to standards and dried at 50°C temperature in an oven for 24 hours followed by cooling in the desiccator for 10 min and immediately weighed. It is then immersed in water for 24 hrs and wiped with a cloth and weighed. The difference between initial weight and final weight were noted and water absorption percentage was calculated.

$$\text{Water absorbency in \%} = \frac{W_i - W_f}{W_i} \times 100$$

W_i = Initial weight of the sample
W_f = Final weight of the sample

4. RESULTS AND DISCUSSIONS

4.1 Tensile Strength

Results of tensile strength of composite are shown in Figure 3 for both *Saccharum munja*/UPR composites. As it can be seen from the figure, the tensile strength of the fibre reinforced sample is more than the standard (polyester resin) which shows reinforcement increases the tensile strength of the composite samples. The tensile strength of 10% fibre loading composite is 48.2 MPa which increased 43.9% as compared to UPR sample without reinforcement. Further with increase in the fibre loading in composites tensile strength value also increases. Tensile strength of 20%, 30%, 40% and 50% fibre loading composites were observed 56.5 MPa, 68.2 MPa, 75.3 MPa and 80.6 MPa respectively. The percentage increase in tensile strength for 20%, 30%, 40% and 50% fibre loading was 17.2%, 16.6%, 10.4% and 7% respectively with respect to increase in every 10% of fibre loading. Up to 30% fibre loading increase in tensile strength was significant because fibres were capable to transferring load to one another. For higher fibre loading percentage increase in tensile strength was less because, the Fibres are not sufficiently wetted by matrix, and the increased population of Fibres leads to agglomeration and stress transfer gets reduced (Öztürk, 2010). Fibre loading more than 50% could not be achieved because amount of UPR resin for more than 50% loading was not enough to wet the fibres completely.

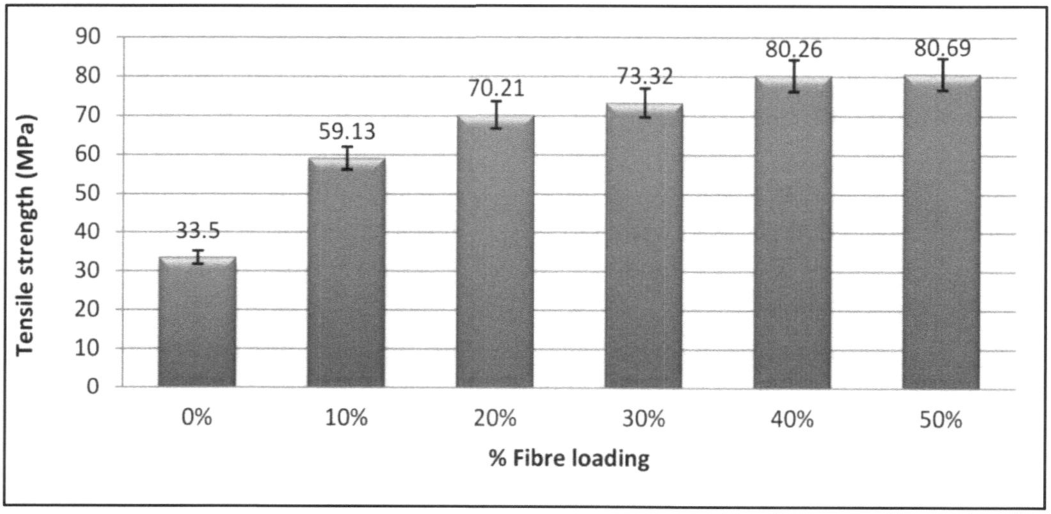

Fig. 3: Effect of Percentage Fibre Loading on Tensile Strength of SMF/UPR Composites

Results of tensile modulus are demonstrated in Figure 4. Modulus of UPR sample was found to be 1652 MPa. After 10% of fibre loading tensile modulus of *Saccharum munja* composite was 2415 MPa. The increase in tensile modulus after reinforcement of *Saccharum munja* fibres shows the reinforcing capability of fibres. Improvement in tensile modulus after 10% fibre loading was found 46%. Further increment in the fibre loading improves the tensile modulus. Tensile modulus for 20%, 30%, 40% and 50% fibre loading was found to be 2815 MPa, 3310 MPa, 4032 MPa and 4315 MPa, respectively. Maximum improvement in percentage (21%) tensile modulus was found at 40% fibre loading with respect to every 10% addition of fibres. Further improvement in tensile modulus for 50% fibre loading was found to be 7% because more addition of fibres causes improper wetting of reinforcing fibre which reduces stress transfer capability of the reinforcing fibres.

Tensile strength and modulus of composites increased after addition of *Saccharum munja* fibres as reinforcement. Upto 40% of fibre loading increase in tensile strength and tensile modulus was significant. Further increase in tensile strength and modulus for 50% loading was less. So we can say 40% fibre loading is optimum for *Saccharum munja*/UPR composites. More than 50% fibre loading cannot be achieved because of larger volume of fibres due to this proper wetting of fibres was not possible in pre-weighted resin.

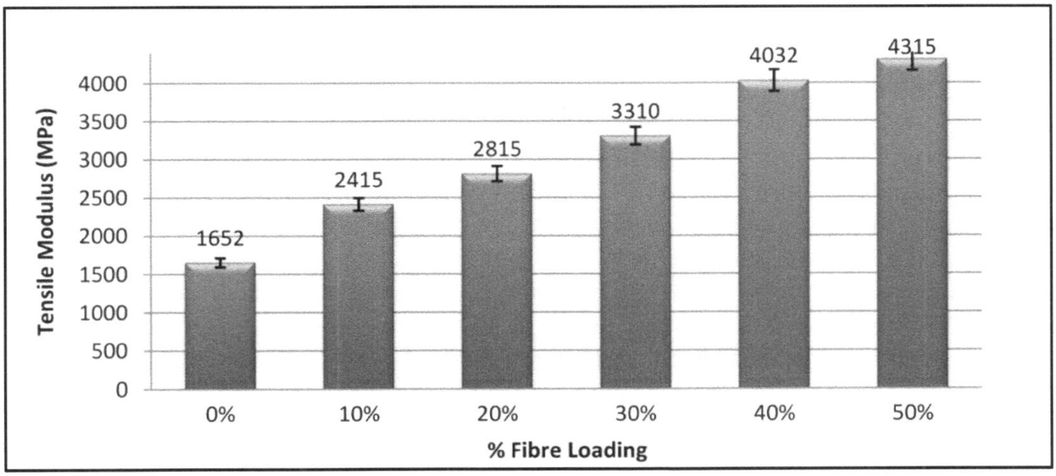

Fig. 4: Effect of Fibre Loading on Tensile Modulus of SMF/UPR Composites

4.2 Flexural Strength Test

Figure 5 shows the effect of fibre loading on the flexural modulus of SMF/UPR composites. It is seen that the improvement of the flexural modulus emerges when the fibre loading increases. The modulus of UPR sample without reinforcement was 2016 MPa while at 10% fibre loading tensile modulus increased (76%) upto 3516 MPa which can be attributed to the effect of *Saccharum munja* fibres as reinforcement in composites. Further increase in the fibre loading from 10% to 20% there was increment of 23% in tensile modulus (4315 MPa). For 30%, 40% and 50% of the fibres tensile modulus observed was 5425 MPa, 6789 MPa and 7502 MPa as shown in Figure 5 which

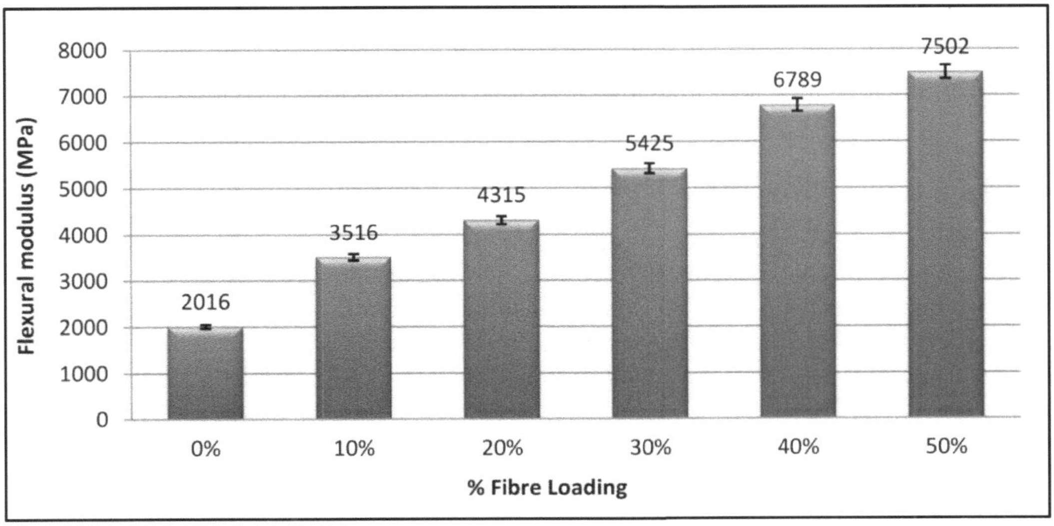

Fig. 5: Effect of % Fibre Loading on Flexural Modulus of SMF/UPR Composites

demonstrates increment in tensile modulus 20%, 25% and 10.5% respectively. From 40% to 50% fibre loading increment in tensile modulus was lesser than other samples because amount of reinforcing fibres was largest which causes agglomeration of fibres attributed to improper distribution of load during testing.

The flexural strengths of SMF are shown in Figure 6. From the table we can observe that flexural strength of UPR is 40.24 N/mm^2 without reinforcement. After reinforcement of 10% fibres flexural strength was observed 60.21 N/mm^2 with increment of almost 50% as compared to without reinforcement. With further addition of reinforcing fibres flexural strength increases for 20%, 30%, 40%, and 50% fibre loading flexural strengths were 84.23 N/mm^2, 99.6 N/mm^2, 120.4 N/mm^2 and 130.14 N/mm^2 respectively. This increase in flexural strength can be attributed to stiffness of the fibres. This stiffness increases as the % fibre loading increases in composites.

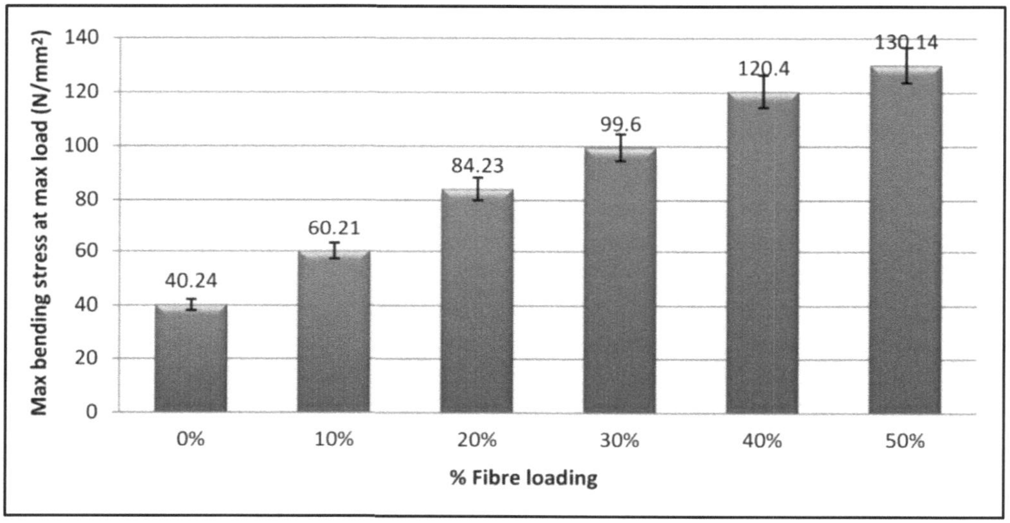

Fig. 6: Effect of Percentage Fibre Loading on Flexural Strength of SMF/UPR Composites

4.3 Izod Impact Strength

Effect of fibre loading on izod impact strength of composites is shown in Figure 7. Impact strength of composite increases as the reinforcing fibre loading increases from 0% to 50%. Impact strength of UPR sample without reinforcing fibres was 151 J/m^2 after addition of 10% fibre loading impact strength increases upto 235 J/m^2. For 20%, 30%, 40% and 50% fibre loading impact strength was 260.5 J/m^2, 283.5 J/m^2, 297.2 J/m^2 and 302.1 J/m^2 respectively. Fibres play a vital role in impact strength of the composites as they act as stress transferring medium and resist crack formation by bearing the load during impact and distribute it among the reinforcing fibre (Goud & Rao, 2012). The increase in impact strength of the composites with respect to increase in fibre loading is due to the increase in number of fibres to carry the load.

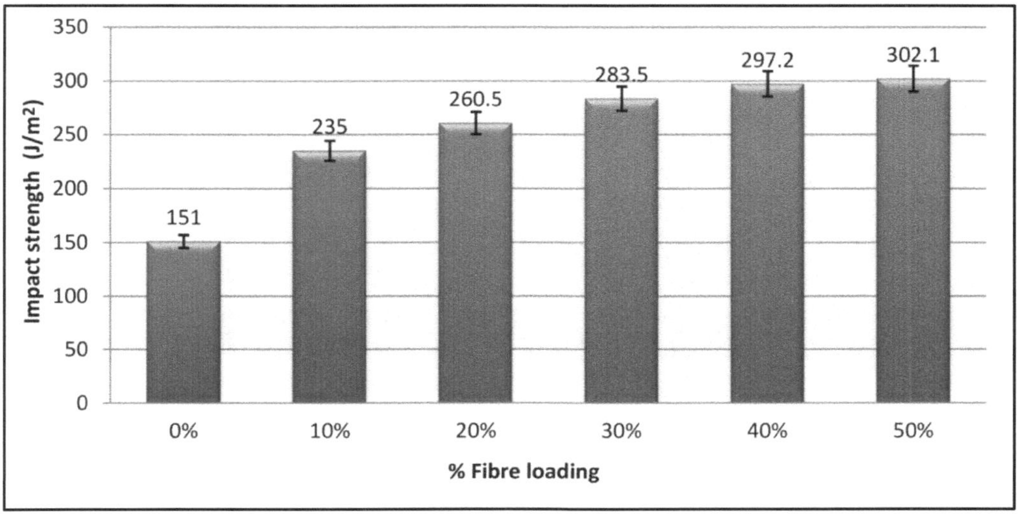

Fig. 7: Effect of Percentage Fibre Loading on Izod Impact Strength of SMF/UPR Composites

4.4 Water Absorbency

Water absorption capacities for different % fibre loaded composites were calculated and reported in Figure 9. It can be seen that the water absorbency increases as the immersion time increases for all composites but after 18 hrs the increase in water absorption was very less because of saturation of water within the composites. With increase in fibre loading water absorption also increases from the Figure 8 it can be observed that for 50% of fibre loading water absorption is higher as compared to other compositions, this was

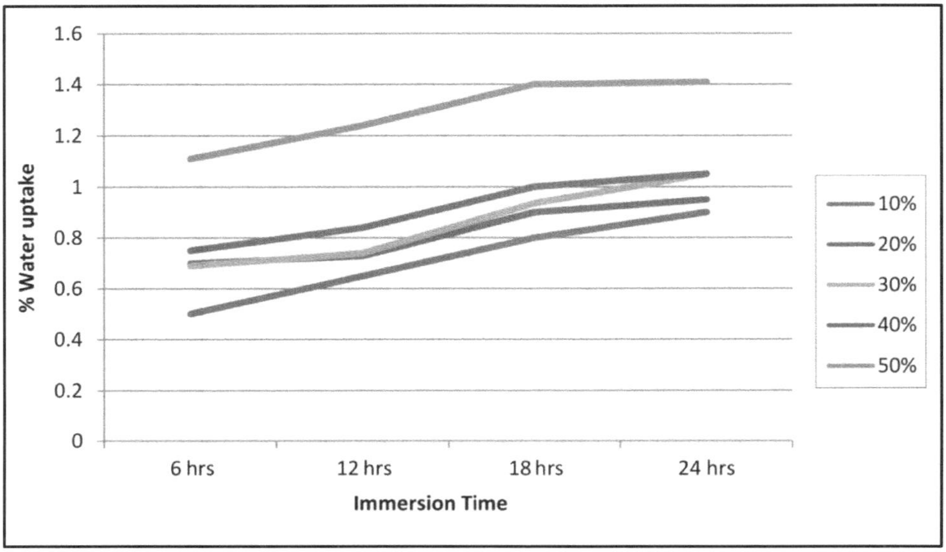

Fig. 8: Effect of Fibre Loading on Water Absorbency of SMF/UPR Composites

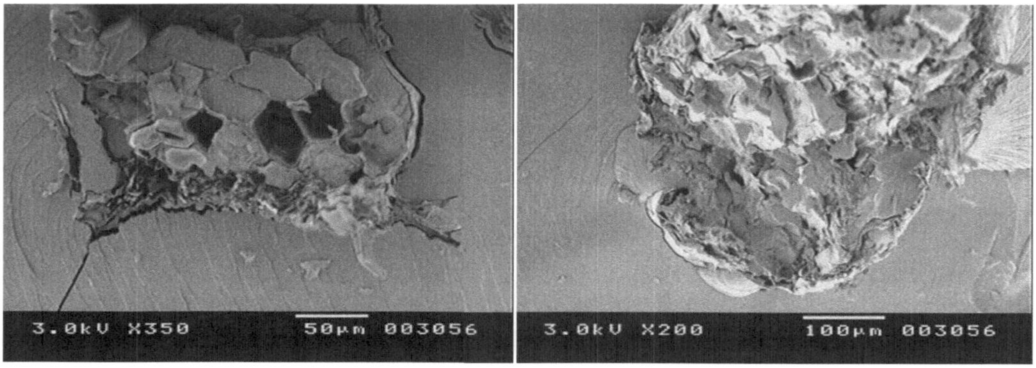

Fig. 9: SEM Image of Composites

due to increase in the fibre loading leading to increase in the amount of cellulose which contain OH group to attract water molecule. The overall water absorption of the composites was very less because UPR is hydrophobic in nature and it restricts water molecules to reach into the fibre because of better mechanical bonding between UPR and reinforcing fibres.

4.5 SEM Analysis

SEM images of fracture samples in impact strength test were taken to analyse the bonding behavior between fibre and resin. From the images it is observed that fibre and resin are properly bonded together as there is very small gap or no gap between fibre surface and resin. During impact test composite sample was broken but reinforcing fibres resist impregnation of cracks developed during testing.

5. CONCLUSION

Various nonwoven fabrics were successfully prepared from the fibres extracted from *Saccharum munja* grass using conventional jute processing machines. These fabrics further used as reinforcing material to prepare composites along with UPR matrix. The maximum tensile properties were found in composites prepared with 50% loading of the fibres. 50% Loading of fibres have processing difficulties i.e., wetting of fibres with UPR thus 40% fibre loading is optimized. From this study it can be concluded that nonwoven fabric from *Saccharum munja* has good reinforcing ability to be utilized in composites.

REFERENCES

[1] Begum, K., Islam, M.A. and Huque, M.M. (2015). Investigation on The Tensile and Flexural Properties of Coir-Fibre-Reinforced Polypropylene Composits, *J. Sci. Res.* 7, 97–111.

[2] Kalaprasad, G., Joseph, K., Thomas, S. and Pavithran, C. (1997). Theoretical Modelling of Tensile Properties of Short Sisal Fibre-Reinforced Low-Density Polyethylene Composites, *Journal of Material Science,* 32, 4261– 4267, Downloaded by [University of Sussex Library] at 10,35 07 January 2016.

[3] Koradiya, S.B., Patel, J.P. and Parsania, P.H. (2010). The Preparation and Physiochemical study of Glass, Jute and Hybrid Glass-Jute Bisphenol-C-Based Epoxy Resin Composites, *Polymer-Plast. Technol. Eng.*, 49, 1445–1449.

[4] Marques, M.F.V., Lunz, J.N., Aguiar, V.O., Grafova, I., Kemell, M., Visentin, F., Sartori, A. and Grafov, A. (2015). Thermal and Mechanical Properties of Sustainable Composites reinforced with Natural Fibres, *Journal of Polymer Environment,* 23, 251–260.

[5] Monteiro, S.N., Calado, V., Rodriguez, R.J.S. and Margem, F.M. (2012). Thermogravimetric Stability of Polymer Composites Reinforced with Less Common Lingo Cellulosic Fibres—An Overview, *Journal of Material Research and Technology.* 1, 117–126.

[6] Naveenkumar, R., Sharun, V., Dhanasakkaravarthi, B. and Rajakumar, I.P.T. (2015). Comparative Study on Jute and Kenaf Fibre Composite Material, *Int. J. Appl. Sci. Eng. Res.* 4, 250–258.

[7] Ngo, W.L., Pang, M.M., Yong, L.C. and Tshai, K.Y. (2014). Mech. Properties of Natural Fibre (Kenaf, Oil Palm Empty Fruit Bunch) Reinforced Polymer Composites, *Advance Environ. Biol.* 8, 2742–2747.

[8] Ramesh, M., Palanikumar, K. and Reddy, K.H. (2016). Influence of Fibre Orientation and Fibre Content on Properties of Sisal-Jute-Glass Fibre-Reinforced Polyester Composites, *Journal of Applied Polymers Science*, 133: DOI: 10.1002/APP.42968.

[9] Reddy, K.O., Maheswari, C.U., Reddy, K.R., Shukla, M., Muzenda, E. and Rajulu, A.V. (2015). Effect of Chemical Treatment and Fibre Loading on Mechanical Properties of Borassus (Toddy Palm) Fibre/Epoxy Composites, *International Journal of Polym. Analysis Charact*, 20, 612–626.

[10] Sekaran, A.S.J., Kumar, K.P. and Pitchandi, K. (2015). Evaluation on Mechanical Properties of Woven Aloevera and Sisal Fibre Hybrid Reinforced Epoxy Composites, *Bulletine of Material Science* 38, 1183–1193.

[11] Shibata, M., Takachiyo, K.-I., Ozawa, K., Yosomiya, R. and Takeishi, H. (2002). Biodegradable Polyester Composites Reinforced with Short Abaca Fibre, *Journal of Applied in Polymer Science* 85, 129–138.

[12] Singh, G.P., Vishwas, P.M., Jagtap, R.N. and Adivarekar, R.V. (2014). Extraction of Fibres from *Saccharum Munja Grass* and its Application in Composites, *Journal of Applied Polymer Science,* DOI: 10.1002/APP.40829, 40829 (1 of 8).

[13] Xia, G., Reddy, K.O., Maheswari, C.U., Jayaramudu, J., Zhang, J., Zhang, J. and Rajulu, A.V. (2015). Preparation and Properties of Biodegradable Spent Tea Leaf Powder/Poly (Propylene Carbonate) Composite Films, *International Journal of Polymer Analysis of Charact*eristics 20, 377–387.

SECTION-II

Textile Chemical Processing

Eco-Friendly Approaches for Improving Colour Fastness, UV Protection and Antimicrobial Properties with Natural Resource Based Functional Finishing of Natural Dyed Silk and Cotton Fabrics

A.K. Samanta[1] **and Pubalina Samanta**[2]

[1]Department of Jute and Fibre Technology, IJT, University of Calcutta,
35 BC Road, Kolkata, West Bengal, India

[2]Rani Birla Girls' College, 38, Shakespeare Sarani, Kolkata, West Bengal, India

E-mail: [1]ijtaksamanta@hotmail.com; [2]pubalina@gmail.com

ABSTRACT: *There are immense lack of scientific study on exploring effects of different pre and post-treatments with natural resource materials for pre or post treatment of natural dyed cotton, silk and other textiles. Hence, in the present study, natural bio-mordants (alum + harda) were used for subsequent dyeing of silk with Punica (pomegranate rind/anar peel) and same pre-mordanted cotton fabrics for dyeing with Tesu (palash i.e. Butea moniosperma). These two varieties of natural dyed fabrics were subjected to post dyeing after-treatments with chitosan and CTAB to improve their wash fastness and post-treated with aqueous and MeOH extract of eucalyptus leaves and few Commercial UV absorbers with or without citric acid/lemon juice as acid catalyst to enhance light fastness and UV protection factor (UPF) respectively. These fully eco-friendly bio-mordanting, bio-dyeing and bio-finishing process for silk and cotton textiles render satisfactory wash fastness improvement by both chitosan and CTAB and post treatment with MeOH extract of Eucalyptus leaves or hydroxy–sulphono-benztriazole as UV absorber as good light fastness enhancer besides improving its UPF values up to 40 and bacterial reduction up to 99%. The mechanisms of chemical interactions for these pre and post treatment for natural dyed silk and cotton textiles were discussed to understand the role of such pre or post treatment to impart enhanced colour strength, colour fastness, uv-protection, and antimicrobial action.*

Keywords: Bio-dyeing, Bio-finishing, Bio-mordanting, Chitosan, UV Protection Factor (UPF).

1. INTRODUCTION

Scanty reports are available related to the sun-light/uv-light fastness properties of coloured textiles dyed with natural dyes. A review to improve sun-light fastness of natural dyes on few important textile fibres is reported by Cook (1982) including post treatment with tannin-based natural materials to enhance washing fastness of coloured cotton dyed with natural dyes. The light fastness is the resistance of dyestuffs to the influence of UV-light energy especially for ultraviolet (UV-A) part, as UV-A (320–400 nm) light reaches easily to earth from Sunlight which amounts to approx. 5–6% only (Cristea & Vilarem, 2006). Amongst, three category of UV light e.g. UV-A (320–400 nm), UV-B (280–320 nm) and UV-C (100–280 nm), UV-C is fully absorbed by ozone layer in stratosphere and UV-B light that reaches earth is less than 1% and is not that harmful.

Factors affecting light and wash fastness of natural dyed textiles including UPF values are reported (Samanta, 2020). The relative light stability of a range of natural dyes was reviewed (Padfield & Landi, 1966). Moreover, Hofenk & Graaff (1983) examined the sunlight fading character of few natural dyed textiles having aimed to determine the chemical/physical changes of original dye/colours molecules in faded textiles. Samanta *et al.,* (2018) described important factors affecting light fastness properties and fundamentals of natural dyes. Changes in colour on exposure to Microscal MBTF lamp fade-o-meterare studied quantitatively (Duff *et al.*, 1977). Role of different chemical structural entity and substituents of natural dye molecules responsible for sunlight fading has been reported by Gupta (1999). Samanta (2020) has recently reported advantages of Bio-mordants. Bio-Dyes and Bio-Finishes for eco-friendly dyeing of cotton and jute textiles, also mentioning the role of chemical nature and type of metal/coordinating power and dosages of mordants/dyes on wash and light fastness. The role of variety of mordanting agents for wash and light fastness of few yellow dyes from natural source were reported in literature to observe that turmeric (halud), fustic and marigold (genda) natural colours have poorer colour fastness than other yellow coloured natural dyes for stannous (Tin) and aluminium (Alum) based mordant in agents than chromium, ferrous (iron) or copper based mordant in agents (Crews, 1982). Oda (2001; 2001b) reported the influence of variety of chemical/ natural additive compounds on sunlight fading of carthamin natural colour in cellulose-acetate sheet or film. Researchers have attempted many new approaches to improve light fastness of natural dyed textiles (Cristea & Vilarem, 2016).

Gulrajani *et al.,* (2003) determined physico chemical parameters of dyeing and colour fastness properties for dyeing of protein fibre like wool and polyamide fibre like nylon with extract of red sandalwood and with mixture of turmeric and madder and also turmeric and red sandalwood and mixture of harda (myrobolan) with madder and mixture of harda and red sandalwood. Turmeric showed poor wash fastness amongst all combinations used; though it gave maximum colour strength in simultaneous mordanting method. Wash fastness of such natural dyes like turmeric however can be improved with a commercial dye fixer by mixing with that dye with dyes of higher wash fastness like red sandalwood. Samanta *et al.,* (2011) studied role of various single and double mordanting agents and also studied the effect of variations of dyeing conditions for natural dyeing of jute based fabric with tesu (fire of forest or palash). A double pre-mordanting using hard a treatment followed by potash-alum or Al-sulphate mordanting and subsequent dyeing with aqueous extract of tesu showed 20% mordant concentration for myrobolan and alum or aluminium-sulphate (1:1) is most suitable for dyeing jute fabric with tesu extract, where dyeing at *pH* of 11 renders overall good colour yield and colour fastness properties and by subsequent post treatment, ½ to 1 grade of improvement in wash and light fastness were achieved. Samanta *et al.,* (2018) also reported compatibility between individual dye for use of mixture of binary pair of natural dyes for dyeing of jute and cotton fabrics with jackfruit wood and other natural dyes. Such naturally dyed jute and cotton fabrics were after-treated with a querternary ammonium compound based cationic agent for improving wash fastness and also after-treated with 1% benztriazole; an UV-absorber, where wash fastness was improved to one grade higher and light fastness ratting was improved by half to one grade higher respectively. Recently, Singhee & Samanta (2019) have studied standardization of dyeing process variables for dyeing silk fabric with tesu to derive optimum dyeing conditions. A latest report on use of bio-mordant with or without metallic mordant and their scientific analysis has been compiled covering all the above issues (Vankar, 2017).

With the above background information in mind, in the present work, cotton and silk fabrics were dyed and finished using natural-bio-mordants and selective anionic natural dyes and were post treated with natural finishing agents (applying separate natural bio-finishing agents to compare the results with performances of eco-friendly synthetic cationic dye fixer or UV absorbers etc.) to improve colour fastness to wash and to UV-Light exposure; and also to improve antimicrobial property and UV protection properties of silk fabrics dyed with Anar peel/Punica and same functional properties of cotton fabrics dyed with tesu with harada+natural potash alum as bio-mordant for both type of fabrics.

2. MATERIALS AND METHOD

2.1 Fabrics

- ***Cotton Fabric*:** Plain weave desized, scoured and bleached 100% cotton fabric with 112 ends/inch and 88 picks/inch, and area density of 128 g/m^2 and thickness 0.21 mm was obtained from NTC Ltd., Kolkata.
- ***Silk Fabric*:** Plain weave, degummed and bleached 100% silk fabric was obtained from Sujapur RK. Sangha, Malda.

2.2 General Chemicals

Commercial grade a) acetic acid (CH_3COOH) and b) sodium acetate as buffer for adjusting *pH* and c) Sodium chloride (NaCl) as electrolyte/salt added in dye bath, all obtained from E-Merck, India, were used in the present work. Rests of the chemicals were obtained from local supplier.

2.3 Mordants and Mordanting Process

Natural alum (Fatkiri i.e. Potash alum) and Harada (natural haritaki i.e. myrobolan) were used here.

- ***Natural Alum:*** Potash alum/fatkiri is frequently used by dyers using natural dyes. This alum is refined from bauxite (aluminum ore) and are freed from all impurities (eg. iron) to get $K_2SO_4.Al_2(SO_4)_3.24H_2O$.

- ***Harda:*** Harda or myrobolan i.e. *Terminalia chebula* was used as bio-mordanting assistant along with alum. This plant species is found throughout South East Asia including sub-himalayan region to east towards West Bengal and Assam. *T. chebula* contains almost 32% of tannin, which are pyro-gallol (hydrolysable tannin) containing 14 components of hydrolysable tannins and chebulinic acid.

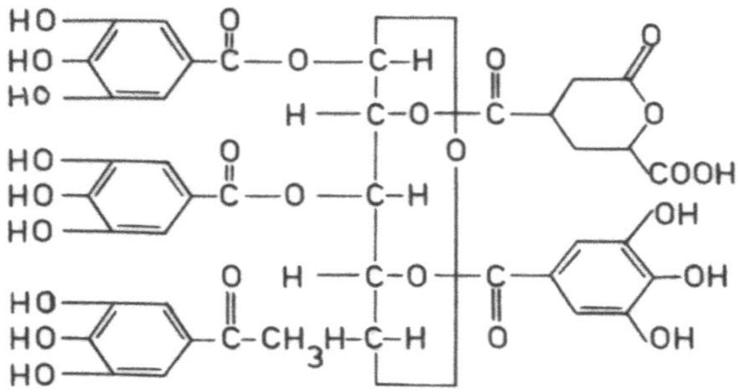

Fig. 1: Chemical Structure of Chebulinic Acid Residue Present in Harda (Myrobolan extract)

These mordants were collected from the local market and used as eco friendly bio-mordanting agent.

Sinnur *et al.,* (2004) have optimized overall 15% application of harda and potash alum (50:50) to apply in sequence as a combined bio-mordant, which renders maximum K/S value and overall good colour fastness. Hence, in this work, the same mordant concentrations and mordanting procedure was used for dyeing cotton fabric with tesu and for dyeing silk fabric with anar peel. Harda mordanted liquor was prepared by overnight soaking followed by boiling in water for 1hr before cooling and straining through nylon cloth (60 mesh) to get filtrate as harda liquor. Both the cotton and silk fabrics were immersed in harda liquor and were heated maintaining MLR 1:30 at 80°C for 30 minutes followed by same procedure of 2nd mordanting with natural alum solution applied by same procedure in sequence for both cotton and silk using this unique combination of bio-mordant and mordant assistant.

2.4 Natural Dye and Dyeing Procedure

Natural Dye 1-Punica: Purified and dried powder of Anar Peel (Punica), was obtained from Ama Herbal, UP, as purified dry powder of punica as ready natural dyeto use. Punica extract contains (A) Granatonine, the colour compound of punica and (B) Pelletierine (Type of Gallo-tannin). Chemical structures of these two components of anar peel extract are shown in Figure 2.

Purified dye solution of anar peel (Punica) was obtained by dissolving pre-weighed purified dry powder of anar peel taken into a beaker containing requisite quantity of water and was heated under a water-bath at around 50–60°C for 30 min to dissolve it followed by straining it in 60 mesh nylon cloth, to get the filtrate as final dye liquor of punica (anar peel).

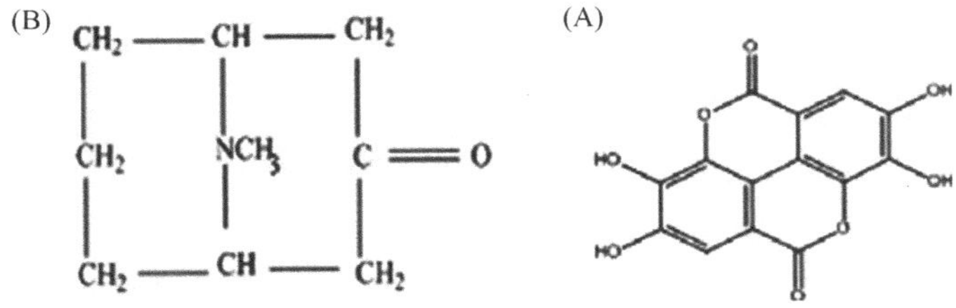

(A) (B)

Fig. 2: Chemical Structure of (A) Granatonine and (B) Pelletierine (type of Gallo-tannin) in Punica Rind (anar Peel)

***Dyeing Procedure for Dyeing Silk with Punica/Anar Peel Extract*:** Dyeing with a pre-fixed % of shade at optimized conditions of dyeing was followed to dye the said 15% overall Harda + alum (50:50) pre-mordanted silk fabric, using 1.5% dye extract solution as dye concentration, salt concentration 5 gpl, MLR 1:50, *pH* 4, dyeing temperature at 80°C for and dyeing time of 60 minutes. After dyeing, the sample were rinsed twice in running water and lightly soaped in 1 gpl soap solution using a neutral soap for soap wash at 50°C for 15 min, followed by thrice washing in tap water and final drying in air.

Natural Dye-2-Tesu: Tesu *(Butea monosperma)* i.e. Palash flower petal extracted natural dye was used in this work to dye cotton fabric. Tesu is commonly known as Flame of forest, belongs to class Fabaceae. It is locally called palash/dhak, and is found throughout India, Burma, and Ceylon. The main colouring pigment of palash flower/tesu is butein. Besides butein, there are isobutein, coropsin, isocoreopsin, sulphurein (glycoside) and flavonoids compounds in it. Studies have shown that isobutein slowly changes to butein on drying. The bright orange red colour of the flower is attributed to the presence of chalcones, ptero-carpans and aurones which are also present in traces. The Chemical structure of butein, as the main colour component of Tesu (*Butea monosperma*) is shown in Figure 3.

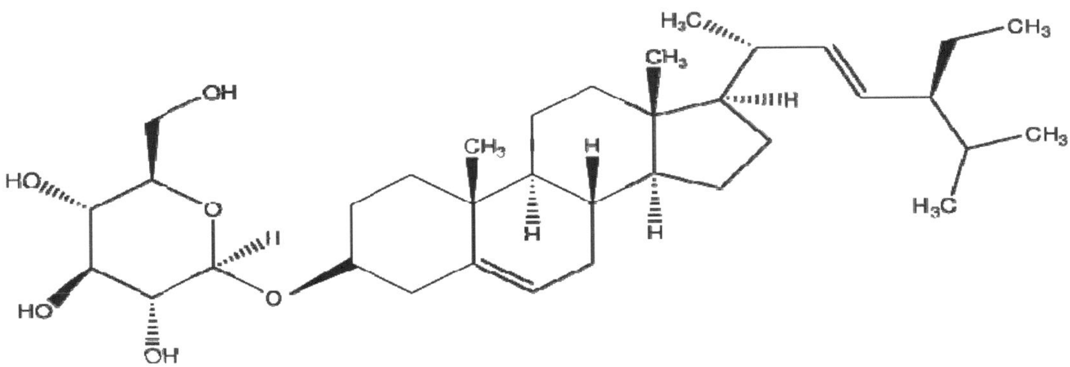

Fig. 3: Chemical Structure of Butein, Main Colour Component of Tesu (*Butea monosperma*)

Pre-fixed mass of tesu flower petals (30 g dry weight of raw tesu flower petal containing 5% colour component butein) i.e. 1.5 g dry powder of purified tesu colorant) was taken in

100 ml distilled water and then it was nearly boiled (90°C) at optimal extraction conditions (Samanta *et al.*, 2009): i.e. 60 min extraction time, 90°C temperature, MLR 1:20 and pH 11. After the complete extraction of dye-liquor, it was filtered with 60 mesh nylon cloth and the filtrate (coloured liquor) was taken as final dye liquor of tesu.

Dyeing Pre-mordanted Cotton with Tesu Extract: Scoured and bleached cotton fabric was pre mordanted with Harada+natural alum [50:50] using overall 15% mordant concentration, and was then dyed with tesu extract at optimized dyeing conditions (Samanta *et al.*, 2009) using dye concentration of 30% (percentage over dry weight of raw tesu flower petal) i.e. 1.5% of purified dry powder tesu colorant) at pH 12, MLR 1:30, time 60 min, temperature 90°C and salt concentration-10 gpl. After dyeing is over, the samples were rinsed twice in running water and then lightly soaped in 1gpl soap solution using a neutral soap for soap wash at 50°C for 15 min, followed by washing in running tap water thrice and final drying in air.

Natural Finishing Agents: The leaves and barks of Eucalyptus have about 11% polyphenol based natural tannins and flavanoids. The major compounds found in the Eucalyptus bark and leaves are (a) Gallic acid (b) Rutin (c) Ellagic acid (d) Quercetin and (e) Rhamnetin, besides Eriodictyol, Naringenin, Rhamnazin, and Toxifolin in traces. Figure 4 shows the chemical structures of its major constituents.

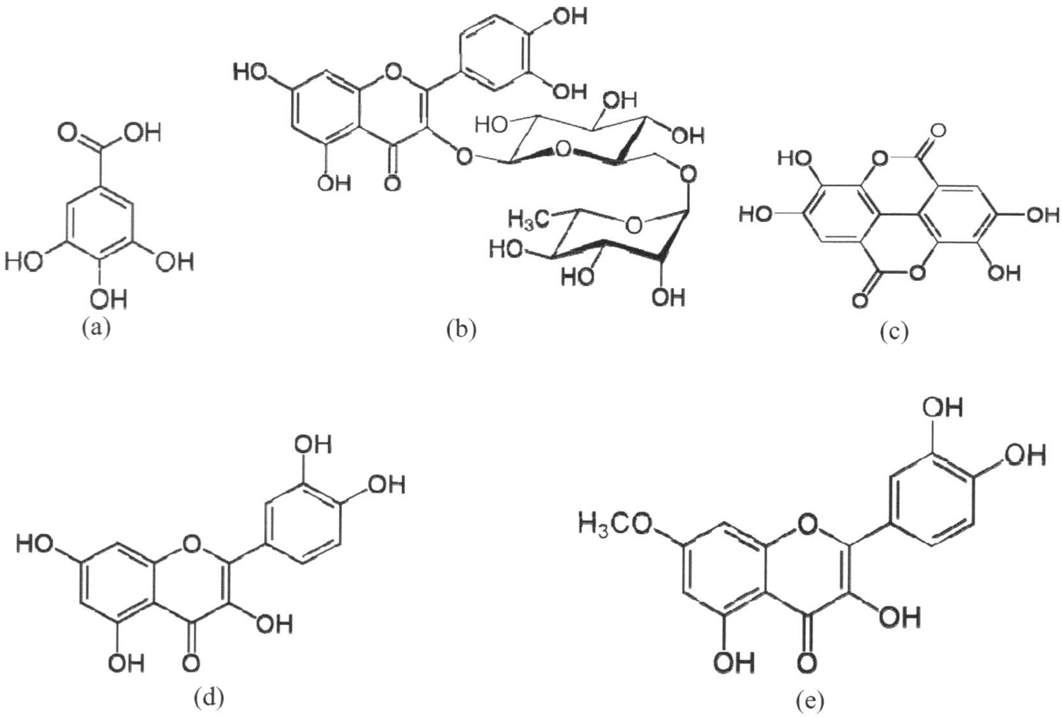

Fig. 4: Chemical Structure of (a) Gallic acid (b) Rutin (c) Ellagic acid (d) Quercetin and (e) Rhamnetin, Present in Eucalyptus Leaves Extract

Quercetin present in eucalyptus leaves/bark is the main colour compound (besides 11% tannins) and it also acts as an antioxidant/uv-protector agent. Eucalyptus leaves also contain 11% tannin (gallic acid and ellagic acid) and some flavonoids and light yellow in colour. So it can be used as both natural dye cum finishing agent having both antimicrobial and UV-protective criteria.

Extraction and Post treatment with eucalyptus leaves extract for improving Light Fastness/UPF etc. for extraction: 100 grams of eucalyptus dried leaves (solid dry mass) was weighed and mixed in distilled water of 1 litre for aqueous extraction and 50:50 water: MeOH mixture for alcoholic MeOH extract. For pH 5 to 6, dilute acetic acid was added during extraction of eucalyptus leaves. This solution was then heated at about 60°C for 60 min for aqueous extraction at pH 5–6 and finally strained in 60 mesh nylon cloth to obtain the filtrate as 10% solution of eucalyptus leaves in two modes (i) Aqueous extract of Eucalyptus leaves and (ii) MeOH extract of Eucalyptus leaves separately.

***For post finishing treatment with eucalyptus extract*:** The said tesu dyed cotton and punica dyed silk fabrics were immersed in 10% eucalyptus extract using both aqueous extract and MeOH extract separately, maintaining MLR 1:10 adding 4 to 5 gpl (0.4 to 0.5%) citric acid/lemon juice (as catalyst) and then the eucalyptus extract pre-soaked fabric was padded and squeezed by 2-dip-2-nip process in lab padding mangle maintaining 100% wet pick up. Padded fabric samples were then dried at 95±5°C (Nr 100°C) for 10 min followed by further heating (curing) at 130±5°C for 3 minutes followed by soaping in 1 gpl neutral soap at 50°C for 15 min and then washing thrice in water before drying in air.

2.5 Post Treatment with Chitosan/CTAB and UV Absorbers with/without Catalyst

2–4% Chitosan (hot water soluble) post-treatment on the said pre-mordanted and dyed silk and cotton fabrics was done with 0.4% citric acid or equivalent lemon juice by simple pad-dry-cure process.

2–4% CTAB based dye fixing agent post-treatment on the said pre mordanted and dyed silk fabric was done in alkaline pH (without any catalyst) followed by Pad-dry process.

Post-treatment with prefixed percentage of UV Absorbers compounds like Benztriazole, Hydroxy Benztriazole and Super-UV Fix (commercial grade UV Absorber having Suphono-Hydroxy Benztriazole) with or without any catalyst at specific conditions of treatment were carried out for improving colour fastness to Light and UV-Protective property of dyed silk fabric.

2.6 Testing and Evaluation

2.6.1 *Measurement of Surface Colour Strength Values and Related Colour Parameters*

Reflectance and surface colour strength values (K/S): Reflectance (R) and Surface colour strength (K/S) values of cotton and silk fabrics were measured using computer aided reflectance spectrophotometer (Make-Premier Colour Scan Instrument Ltd.; Model SC 5100A) and associated software calculating K/S values of fabric samples as per Kubelka–Munk equation (Singh *et al.*, 2005) shown below:

$$K/S = \alpha C_D \frac{(1-R_{\lambda\max})^2}{2R_{\lambda\max}}$$

Where, S represent scattering coefficient, $R_{\lambda max}$ represent reflectance value at wavelength having maximum absorbance value (λmax), C_D is the dye concentration and α is the constant.

Total Colour Difference (ΔE): The total colour difference values were obtained by measuring corresponding lightness/darkness (ΔL^*), blueness/yellowness (Δb^*) and redness/greenness (Δa^*) values and employing following CIE-Lab equations against control bleached/pre-mordanted sample.

$$\Delta E = [(\Delta L^*)^2 + (\Delta a^*)^2 + (\Delta b^*)^2]^{½}$$

Changes in Chroma (ΔC*): Chroma (psychometric chroma) values according to CIELAB (Shah & Gandhi, 1990) color space equations was determined by the following formula:

$$\textit{Chroma Value: } C_{(ab)}^* = (a^{*2} + b^{*2})^{1/2},$$

$$\text{Chroma difference} = \Delta C^* = C^*_{1(ab)} - C^*_{2(ab)} \text{ as per CIE lab.}$$

Where, $C^*_{1(ab)}$ and $C^*_{2(ab)}$ are the chroma values for sample-1 (Given std.) and produced sample-2.

Changes in Hue (ΔH): CIELA – (CIE 1976) Hue-Difference (ΔH) was calculated as follows (Shah & Gandhi, 1990):

$$\Delta H_{ab} = [(\Delta E_{ab}^*)^2 - (\Delta L^*)^2 - (C_{ab}^*)^2)]^{1/2}$$

Metamarism Index (MI): Metameric index (MI) was determined employing following formula (Shah & Gandhi, 1990):

$$MI = \frac{\Sigma(\Delta RX_0)^2}{X^2} + \frac{\Sigma(\Delta RY_0)^2}{Y^2} + \frac{\Sigma(\Delta RZ_0)^2}{Z^2}$$

Where, X_o, Y_o and Z_o represent the Tristimulus values of a standard white calibrating tile (WI-100) and X, Y and Z are the Tristimulus values of the produced sample to test.

Brightness Index (BI): Brightness Index (BI) was investigated as per ISO-2470-1977 method (Shah & Gandhi, 1990):

$$BI = \frac{R_{457\text{ nm}} \text{ value of sample to test}}{R_{457\text{ nm}} \text{ value standard white calibrating tile of WI-100}} \times 100$$

Colour Difference Index (CDI): CDI values represents the combined effects of all major colour difference parameters between any two set of samples when dyed with varying conditions of dyeing, indicating dispersion of colour difference values to understand total effects by a single index. Colour Difference Index (CDI values) was calculated by following empirical relationship (Samanta *et al.*, 2009).

$$\text{Colour Difference Index (CDI)} = \frac{\Delta E \text{ X } \Delta H}{\Delta C \text{ X } MI}$$

2.6.2 Evaluation of Colour Fastness Properties

Colour Fastness to Washing: Washing fastness of dyed cotton and silk fabric samples (10 cm × 4 cm) were determined as per ISO-II (IS: 3361–1979) method (ISI (BIS), 1981/SP-15-Part-4, 1988), with neutral soap solution (5 gpl) at 50°C for 45 minutes, keeping material to liquor ratio at 1:50 using SASMIRA Launder-O-Meter.

Colour Fastness to Rubbing: Dry and wet rubbing fastness of dyed cotton and silk fabrics were assessed as per IS: 766–1984 method (ISI (BIS), 1981/SP-15-Part-4, 1988). A dry piece of dyed fabric specimen (14 cm × 5 cm) was fastened tightly to the base of Mag-Solvic make Electronic Crockmeter, allowing to rub the dyed specimen to and fro for 10 cm length for ten cycles by abraider white cloth mounted on the abraider finger. The degree of staining on abraider white cloth was assessed by 'Grey Scale' for rub-fastness rating. Wet rub fastness was determined in the same procedure after soaking the sample in water maintaining 100% moisture.

Colour Fastness to Light: Light fastness of dyed cotton fabrics were evaluated as per IS: 2454-1967 method (ISI (BIS), 1981/SP-15-Part-4, 1988). Dyed cotton samples (10 cm × 2 cm) were exposed to UV-light for fading in a Shirley MBTF-micro-scal fade-o-meter along with 8 blue wool standards (BS1006: BOI: 1978). After exposure for 10 h, 20 h, 40 h, 80 h, 160 h, the sunlight simulated light fastness was determined by comparing the fading behavior of 1–8 grade blue wool standard fabrics, after different hours of exposure to UV light.

2.7 Characterization of Natural Dye through UV-Visible Scan

Tesu: UV-VIS spectra of 0.1% aqueous extract of the purified Tesu (Palash flower petal) dye powder was separately obtained by wavelength scan in a Hitachi-U-2000 UV-VIS absorbance spectrophotometer for 200 nm to 1100 nm range (UV-visible range- IR zone), which is shown in Figure 5(a) showing small humps at 350 nm (UV Zone) and a large peak considered as λmax 490 nm.

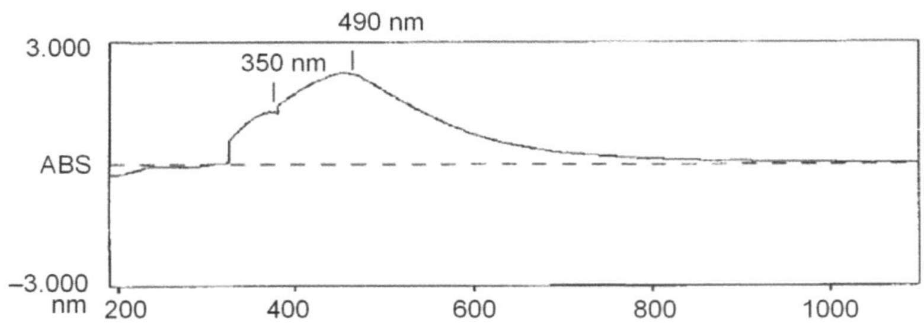

Fig. 5(a): UV VIS Spectral Scan of Aqueous Extract of Tesu Powder (*Butea monosperma*)

Anar Peel Extract: UV-VIS spectra of 0.1% aqueous solution of the purified punica (anar peel) dye powder was separately obtained by wavelength scan in a Hitachi-U-2000 UV-VIS absorbance spectrophotometer for 200 nm to 900 nm range (UV-visible range), which is shown in Figure 5(b). Showing a hump considered as λmax–460 nm and sharp peak at 216 nm and 278 nm for tannins.

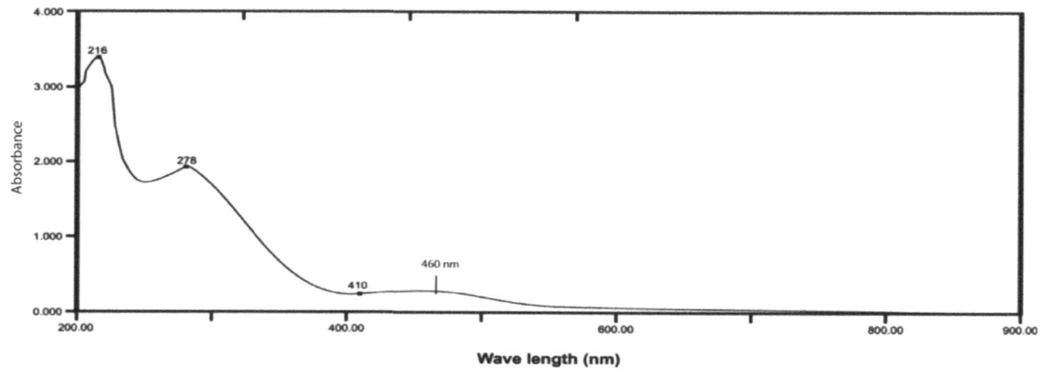

Fig. 5(b): Wavelength Scan of Aqueous Extract of Anar Peel

Eucalyptus Leaves Extract: UV-VIS spectra of 0.1% aqueous solution of the purified Eucalyptus aqueous extract is shown in Figure 5(c).

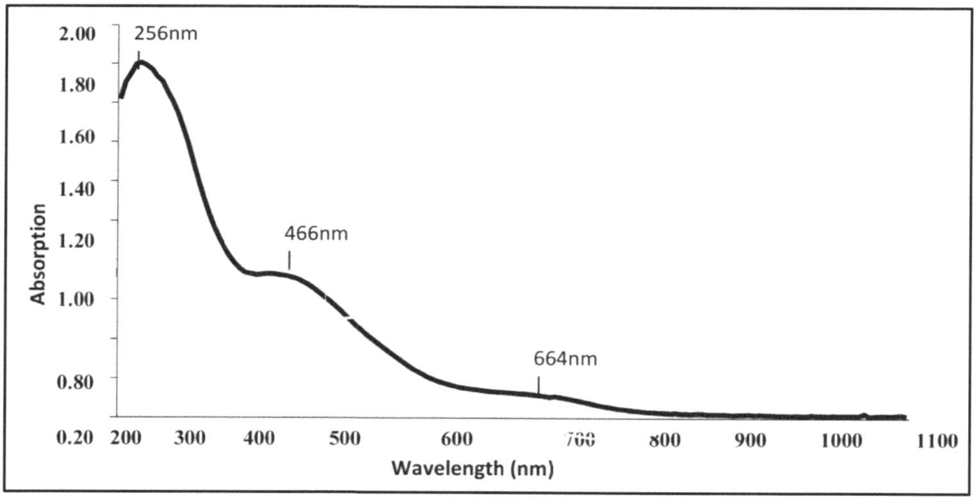

Fig. 5(c): UV-VIS Spectral Scan of Aqueous and MeOH extract of Eucalyptus Leaves

UV-VIS spectra of extract of Eucalyptus leaves were obtained using Hitachi-U-2000 UV-VIS absorbance spectrophotometer for 200 to 1100 nm. It shows a small hump considered as λmax-420 nm and a sharp peak at 256 nm in UV region for tannins present in it, besides a small hump at 664 nm.

2.8 Test of UV Protection Factor (UPF)

Untreated and dyed and/or finished cotton fabrics were tested for determining UV protection factor (UPF) as per AATCC: 183: 2004/2010/2014 (revised) method. The Ultraviolet Protection Factor (UPF) value i.e. ultraviolet radiation blocked or transmitted by control fabric, Harda+Alum pre-mordanted fabrics and punica dyed silk and tesu dyed cotton fabrics without or with post treatments was measured at 200–380 nm using Labsphere Inc. (USA) UV-Transmittance Analyzer to calculate UPF values.

2.9 Test of Anti-Microbial Property

Antimicrobial property of control, pre-treated, dyed, and finished cotton and silk fabrics were tested in terms of bacterial reduction % as per AATCC 100–2012 standard method using Gram positive bacteria—*Staphylococcus aureus* and Gram negative bacteria *Klebsiella pneumoniae*. In this method, percentage reduction of bacterial growth was assessed, and picture of Petri plates were taken allowing the sample to incubate at 37°C for 24 h for measuring rate of inhibition of microbial growth.

3. RESULTS AND DISCUSSION

3.1 Effect of Selective Post Treatments for Improving Wash Fastness of Anar Peel Dyed Silk Fabric

Effect of selective post-treatment of natural dyed silk fabric dyed with anar peel/punica (Pomogrenate Rind) extract, post treated with Chitosan (Natural agent) and CTAB (Cetyl tri-methyl ammonium bromide as an eco-friendly chemical cationic agent) with or without suitable catalyst for improving wash fastness of punica dyed silk fabric after pre-mordanted with 15% overall dosages of harda + natural alum (50:50) as bio-mordants applied in sequence is studied and discussed.

Post-treatment of 1.5% purified Anar Peel dyed silk fabric (after pre-mordanted with 15% overall application of Harda + Alum 50:50 applied in sequence) was carried out by application of pre-fixed percentage of chitosan (from natural source) with predetermined percentages of suitable catalyst (like citric acid (CA) or lemon juice containing citric acid, as catalyst, maintaining 1/5th of the catalyst dosages with respect to weight of main post treating chemicals (i.e. chitosan or CTAB) by pad – dry (100°C for 10 min) cure (130°C for 3 min) process, for subsequent assessment of their changes in surface colour strength and colour fastness to wash, light and rubbing after these post treatments.

Relevant results for changes in surface colour strength (K/S value) and other related colour parameters before and after the said post treatments are tabulated and shown in Table 1. Relevant data in Table 1 indicates that a blank post-treatment on 1.5% purified Anar Peel Dyed Silk fabric (premordanted with 15% overall application of Harda + Alum –50:50 applied in sequence) with 2% Chitosan with 0.4% Citric Acid or lemon juice treatment shows that there is a small reduction in K/S value in both the cases. Post-Treatment with 2% Chitosan in presence of 0.4% citric acid show a higher reduction of K/S values than use of 0.4% lemon juice as catalyst instead of citric acid. Presence of citric acid for post treatment with chitosan thus show a higher leaching out effect causing a bit higher reduction in surface colour strength on punica dyed silk fabric, which is much lesser in case of lemon juice.

Table 1: Effect of Post Treatments with Chitosan and CTAB on pre Mordanted [15% Harda + Alum (50:50) Premordanted] and Dyed [1.5% purified Anar peel extract dyed] Silk Fabric on its Colour Strength and Related Other Colour Parameters

Type of Post Treatment	*K/S*	*DE*	*DL*	*Da*	*Db*	*DC*	*DH*	*MI*	*BI*	*CDI*
Controlled Degummed Silk (No Dyeing)	0.22	–	–	–	–	–	–	2.83	56.85	–
Harda + Alum (50:50) 15% overall application (No Dyeing) premordanted silk fabric	4.40	2.65	2.45	–2.39	2.35	2.39	2.33	3.23	11.97	0.82
15% overall application of Harda + Alum (50:50) and 1.5% punica (pure)dyed	10.48	3.86	3.79	–0.56	1.82	1.84	–3.76	3.34	9.49	2.36
Post treatment on 15% (Harada + Alum–50:50) pre mordanted and 1.5% punica dyed silk										
2% Chitosan + 0.4% Citric Acid	8.96	3.28	1.43	0.54	–2.89	–2.91	–0.48	2.64	11.90	0.20
2% Chitosan + 0.4% Lemon Juice	9.43	3.26	1.58	0.37	–2.82	–2.83	–0.30	2.46	12.83	0.14
2% Chitosan (No catalyst)	8.49	3.08	0.91	0.71	–3.18	–3.19	–0.66	2.66	13.67	0.24
2% CTAB (at Alkaline pH)	11.10	7.81	6.03	–2.96	1.976	1.97	–2.00	3.39	10.45	2.33
Blank treatment with two types of catalyst only										
0.4% Citric acid (No Chitosan)	8.98	5.54	4.78	–1.41	2.42	2.40	1.44	2.78	13.10	1.19
0.4% Lemon Juice (No Chitosan)	10.48	6.43	5.47	–2.34	2.44	2.48	2.29	3.10	11.32	1.91

Chitosan treatment with citric acid as catalyst makes the fabric somewhat stiffer than lemon juice used as catalyst at comparable conditions of treatment, but for chitosan treatment, after the colour change due to such post-treatment, the CDI values are found to become much lower, but higher CDI values is obtained for lemon juice used as catalyst, indicating more colour uniformity in case of chitosan treatment than lemon juice used as catalyst for chitosan post-treatment. The main objectives of these two types (chitosan treatment with two types of catalysts) of treatments are to improve washing fastness. Colour fastness to wash, light and rubbing for these post treatments are shown in Table 2. For comparative purpose, the same anar peel dyed silk fabric is treated with a common commercial grade dye fixer based on querternary ammonium compound i.e. CTAB (Cetyl Tri Methyl Ammonium Bromide) in alkaline pH on the above said pre-mordanted and 1.5% anar peel dyed silk fabric to enhance its colour fastness to washing and their colour fastness results are also shown in Table 2.

Table 2: Colour Fastness to Washing, Light and Rubbing after Post Treatment with Chitosan with or Without Catalyst Like Citric Acid and Lemon Juice as Compared to Commercial Cationic Dye Fixer

Type of After-Treatment	*Washing Fastness*			*Crocking/Rubbing Fastness*		*UV-Light Fastness*
	Loss of Depth	*Silk (Staining scale)*	*Cotton (Staining scale)*	*Dry rubbing*	*Wet rubbing*	*MBTL fado-meter*
15% overall—Harda + Alum (50:50) followed by 1.5% purified punica dyed sample	3/4	3/4	3/4	4	4	3/4
After-treatment on 15% (Harada + Alum –50:50) pre mordanted and 1.5% punica dyed silk						
2% Chitosan + 0.4% Citric Acid	4/5	4	4	5	4/5	4
2% Chitosan + 0.4% Lemon Juice	4/5	4/5	4/5	4/5	4/5	4
2% Chitosan (No Catalyst)	3/4	3/4	4	5	5	3/4
2% CTAB (In Alkaline pH)	4/5	4/5	4/5	4	4	4
Blank treatment with two types of catalyst only						
04% Citric Acid (No Chitosan)	3	3/4	3/4	4	3/4	3/4
0.4% Lemon Juice (No Chitosan)	3/4	3/4	4	4	4	4

Use of a commercial grade cationic dye fixer based on CTAB, is when applied in alkaline pH, it show allround very good colour fastness including wash fastness. Chitosan post-treatment with lemon juice or citric acid as catalyst instead of CTAB render at par improvement in wash fastness for punica dyed silk fabric (Table 2), whereas chitosan post-treatment without catalyst or blank treatment with only lemon juice or only citric acid (without chitosan) cannot give this improvement in wash fastness.

Chitosan may be readily reactive to silk or cotton by utilizing its primary amino group and the primary and secondary hydroxyl groups. Glycolated-chitin/Acetylated chitin, a partially o-hydroxy-ethylated/acetylated chitin, after occurring many reactions with chitin form chitosan. The primary amino group of chitosan in acidic pH/acidic catalyst form cationic derivative generating $-NH_3^+$ functionality in acidic medium (in presence of CA or lemon juice) which can therefore act as cationic agent to attach to anionic natural dyes to enhance its washing fastness grade of anionic mordant able natural dyes. Therefore, use of chitosan instead of CTAB is recommended with acidic catalyst (citric acid or lemon juice), where lemon juice is preferred, as it shows less reduction in colour value and improves colour fastness to light also due to known UV-absorber criteria of lemon juice besides its content of citrates.

Thus, chitosan itself chelates (forms giant big coordinating complex insoluble in water) with metal ions used in mordants (aluminium from natural alum) applied along with harda as bio-mordant) and creates an insoluble matrix for immobilization of such mordantable natural dyes fixed by forming Fibre-Mordant-Dye-Chitosan giant co-ordinating complex, improving wash fastness for punica dyes.

Even without any post treatment, punica (containing granatonine as colour component and pelletierine as tannin in it) dyed silk fabric after pre-mordanting with metallic mordant (natural alum i.e. potash alum) with or without harda (containing additional chebulinic acid based tannin) is able to form a complex with Fibre-metal–tannin–dye component, as shown in Figure 6 and show moderate wash fastness 3 or 3–4 for punica dyed silk fabric after pre-mordanted with alum with or without harda.

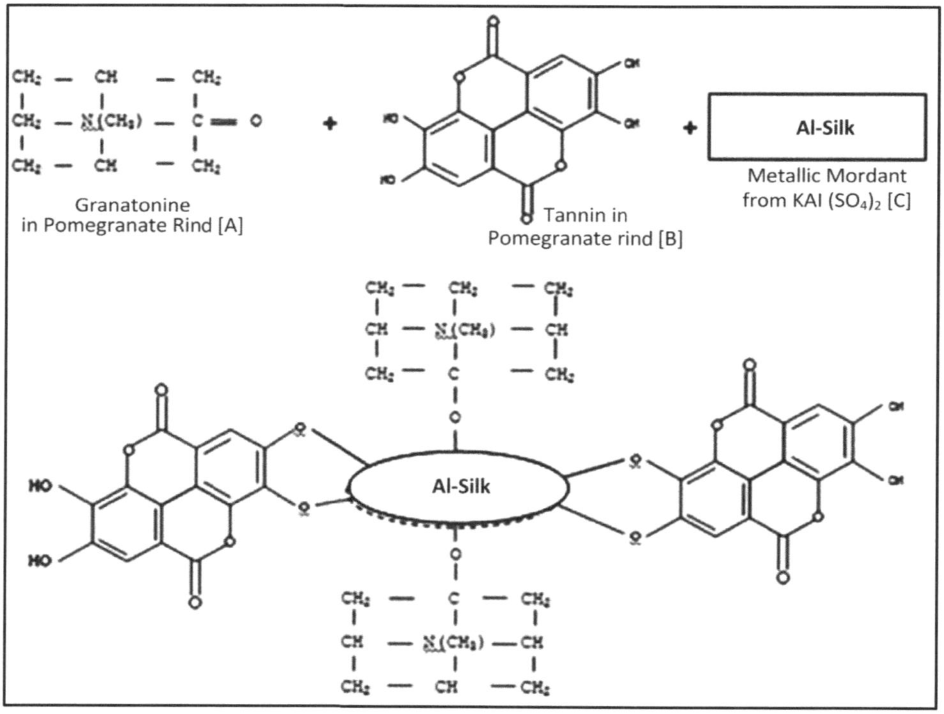

Fig. 6: Mechanism of Complex Formation amongst Fibre-Metallic Mordant (Aluminium in natural alum), Pelletierineas Tannin and Granatonine as Colour Component in Punica Extract

Data in Table 2 thus show all round better colour fastness for Chitosan post treated anar peel dyed silk fabric in presence of citric acid or lemon juice, as an acidic catalyst. Chitosan itself is believed to form chelates (forms coordinating complex) with metal ions, especially with the transition metals, and also a thin film matrix for immobilization of natural dyes pre-fixed on fibre by mordanting with metallic mordant like aluminium from alum or by H bonding to fibre, forming Fibre-metallic mordant–punica tannins-punica colour component (Dye)–chitosan as much bigger chelated complex after post treatment with chitosan-NH_3^+ (in presence of lemon juice) and dye anions linkages for better fixation of such anionic natural dyes like punica and renders it improved wash fastness.

The Overall good washing fastness rendered by CTAB (a synthetic but eco-friendly cationic dye fixing agent as a querternary compound) in presence of alkali (alkaline pH) may be explained by the following mechanism, as shown in Figure 7.

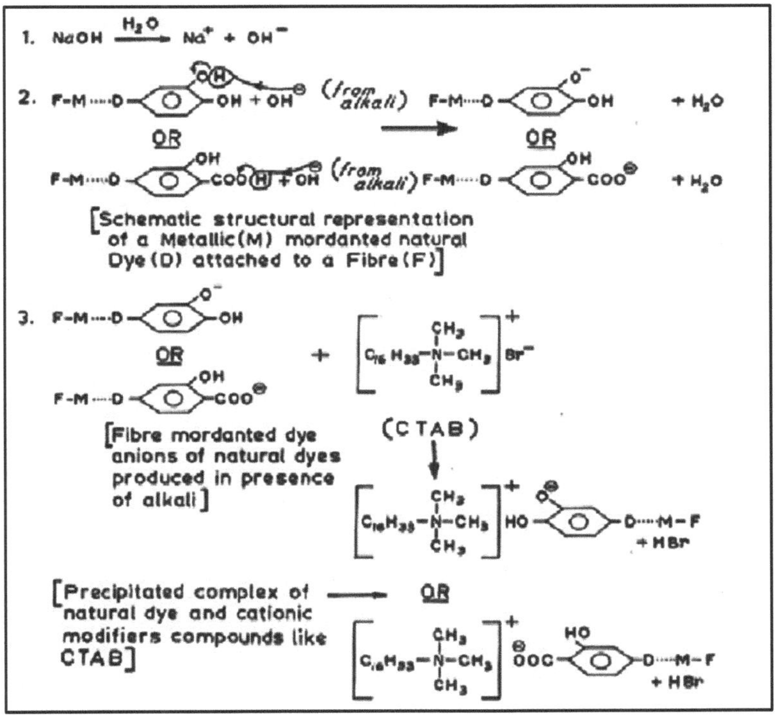

Fig. 7: Mechanism of Complex Formation of Punica Dye Anion (In Alkali Medium) with CTAB (cationic dye fixer) for Improving Washing Fastness of Punica Dyed Silk

3.2 Effect of Different Post Treatments on Improving Light Fastness of Anar Peel Dyed Silk Fabric

As mentioned earlier, amongst three kinds of UV bands (UV-A, UV-B and UV-C) of sunlight, fading of dyed textiles occur mainly by UV-A: 320 to 380 nm, 94% of which reaches earth and is mostly responsible for sunlight fading of dyed textiles. To protect/ delaying such UV fading, after-treatment with suitable UV absorbers is necessary which can absorb UV-A. Few reports on natural source based UV absorbers are available in literature (Pargai *et al.*, 2020), which include UV absorption criteria of extracts of anar peel, orange peel and eucalyptus leaves and wall nut wood etc. as natural dyes cum uv-protecting agents. In this part of the work, for improving light fastness of anar peel/punica extract dyed silk fabric, the effect of various post-treatments on anar peel dyed silk fabric with one natural UV Protecting agent (Eucalyptus Leave extract) and few eco-safe synthetic UV-absorbers like benztriazole, hydroxy-benztriazole and sulphono-hydroxy-benztriazole (Supra UV Fix—a commercial UV absorber) were carried out, for comparison of their performances with performance of eucalyptus leaves extract. Relevant results of surface colour strength and other colour parameter of after-treated samples with above said UV-absorbers on punica dyed silk are shown in Table 3. Relevant data indicate that there are hardly any noticeable changes in K/S value (surface colour strength) and other colour parameters due to the said after-treatments on said pre-mordanted and 1.5% punica dyed silk fabrics.

Table 3: Effects of Different UV Absorber Post Treatments on Colour Strength and Other Colour Interaction Parameters

Type of Post Treatments	*K/S*	*DE*	*DL*	*Da*	*Db*	*DC*	*DH*	*MI*	*BI*	*CDI*
Controlled Degummed Silk Fabric (No Dyeing)	0.22	–	–	–	–	–	–	2.83	56.8	–
Harda + Alum (50:50) 15% overall application pre-mordanted silk fabric (No Dyeing)	4.40	2.65	2.5	–2.39	2.35	2.39	2.33	3.23	11.9	0.82
15% overall application of Harda + Alum (50:5) and 1.5% purified punica dyed	10.48	3.86	3.8	–0.56	1.82	1.84	–3.76	3.34	9.49	2.36
15% Harda + Alum (50:50) + 1.5% Punica + Benztriazole–2%	10.28	3.78	3.1	5.09	2.03	2.55	–4.07	3.49	10.1	1.73
15% Harda + Alum (50:50) + 1.5% Punica + Hydroxy-Benztriazole-2%	10.12	3.87	3.2	5.93	2.09	2.23	–4.61	3.45	9.60	2.31
15% Harda + Alum (50:50) + 1.5% Punica + Supra – UV – Fix- 2% (Commercial grade Sulphono-Hydroxy Benztriazole)	10.24	3.84	3.3	5.22	1.87	1.90	–4.19	3.25	9.70	2.60
15% Harda + Alum (50:50) + 1.5% Punica + Post treated with 10% Eucalyptus leave extract + citric acid (5 gpl)	10.49	3.54	3.2	5.73	2.31	2.37	–4.80	3.60	9.44	1.99

Results of colour fastness to wash, rubbing and light, with or without the above said post treatments with the selective natural and synthetic UV absorbers applied on 1.5% (purified) punica dyed silk fabric are shown in Table 4. Relevant results in Table 4 indicate that light fastness of 15% harda + potash alum (50:50) pre-mordanted silk fabric dyed with 1.5% purified Punica as natural dye without any post treatment show light fastness rating of

3–4, which however is improved to ½ to 1 grade higher by post treatment with 2% of each of Benztriazole, Hydroxy-Benztriazole and Supra-UV-Fix (a commercial grade Sulphono-Hydroxy Benztriazole). However, post-treatment with 10% MeOH extract of eucalyptus leaves and post treatment with 2% Supra-UV-Fix (Sulphono-Hydroxy benztriazole) show somewhat higher improvement than that for only Benztriazole or Hydroxy-Benztriazole post-treatment in terms of improvement in light fastness grade, vide results in Table 4.

Table 4: Relevant Data for Washing, Rubbing and Light Fastness for UV Absorber and Eucalyptus Leaves Extract Post Treated Punica Dyed Silk Fabric

Type of Post-Treatments	*Washing Fastness*			*Rubbing Fastness*		*Light Fastness under MBTL fade-o-meter*
	Loss of Depth	*ST (Silk)*	*ST (Cotton)*	*Dry rub*	*Wet rub*	
15% application of Harda + Alum (50:50) + 1.5% Punica [i.e. only dyed silk fabric (No post treatment)]	3–4	3–4	3–4	4	4	3–4
15% Harda + Alum (50:50) + 1.5% Punica + Benztriazole-2%	4	4	3–4	4	4	4
15% Harda + Alum (50:50) + 1.5% Punica + Hydroxy – Benztriazole -2%	4–5	4	4	4–5	4	4–5
15% Harda + Alum (50:50) + 1.5% Punica + Supra – UV – Finish-2% (A commercial grade of Sulphono-Hydroxy Benztriazole)	4–5	4	4	4–5	4	4–5
15% Harda + Alum (50:50) + 1.5% Punica + post treated with 10% Eucalyptus leave extract + citric acid (5 gpl)	4–5	4	4	4–5	4	4–5

Thus, it is established from this work that using post treatment with Eucalyptus leaves extract (in MeOH) on anar peel natural dyed silk textiles after the said pre-bio-mordanting, light fastness can be maintained/enhanced at the same level as obtained by commercial UV absorber compound like Supra UV Fix (a commercial grade Sulphono-Hydroxy Benztriazole) and sometimes it often shows higher improvement in light fastness rating than the performance of the said commercial UV absorbers.

3.3 Effect of Functional-Finishing with Natural Materials to Improve Antibacterial Property and UPF Values of Tesu Natural Dyed Cotton Fabric

Few studies on UV resistance (Xiuliang *et al.*, 2013) as well as antibacterial property (Suping *et al.*, 2017) are sparingly reported earlier. An attempt for use of such natural dyes cum finish compounds or a separate natural dye and a separate functional finish from natural source are investigated in this work for tesu dyed cotton fabric to improve its antibacterial and UPF values by post treatment with separate natural resource based finishing agents over and above natural bio-mordanting and natural dyeing are reported.

3.3.1 *Antibacterial Finish on Tesu Natural Dyed Cotton Fabric with Natural Finishing Agents*

Many researchers (Suping *et al.*, 2017) have earlier reported antibacterial finish by natural dye itself like catechu and rheumemodi or by separate natural source based functional finishing materials like extract of eucalyptus leaves and barks, curcumin, peel of banana etc. Current status in studies on natural dye based functional finishing or separate functional finishing applied on natural dyed textiles have been compiled and reported recently (Pargai *et al.*, 2020). In the present work, both aqueous and MeOH extract of eucalyptus leaves are used as natural source based UV protecting as well as bacterial resisting agent for tesu natural colour dyed cotton fabric. Influences of application of 10% aqueous extract and MeOH extract of eucalyptus leaves by 'padding (2-dip-2-nip) oven-drying–hot air curing method for multifunctional after-treatment on 30% tesu dyed (based on dry weight of natural source) cotton fabric or by concurrent mordanting, dyeing and finishing in a single bath with or without required catalyst (citric acid-5 gpl i.e. 0.5%) were studied to improve microbial resistance character of such treatment with eucalyptus extract on tesu/palash flower petal extract dyed cotton fabric [(after bio-mordanting with harda + potash alum (overall 15% application in 50:50 ratio)].

Simultaneous mordanting-dyeing and finishing were carried out with the said two bio-mordants in combination, pre-fixed dosages of aqueous extract of tesu as natural colourant and 10% aqueous extract of eucalyptus leaves with required $1/5^{th}$ dosages of citric acid/lemon juice were added in a single bath and was heated at 90°C for 90 minutes adding all ingredients in a single bath in exhaust process. Observed results for changes in surface colour strength and other colour parameters and corresponding different types of colour fastness criteria (light/wash/rubbing) are tabulated in Table 5 and Table 6.

In Table 5, relevant data indicates that 10% eucalyptus extract post-treated (by padding–drying–curing process) on Tesu dyed cotton fabric after the said harda plus alum double premordanted sample, its colour strength value is found to enhance better as compared to simultaneous dyeing and finishing single bath process at comparable conditions using the same amount of all agents used. Also for the said tesu dyed cotton fabric samples is when sequentially after-treated with same amount of both aqueous extract and MeOH extract of eucalyptus leaves by same padding–drying–curing process, MeOH extract of eucalyptus is found to be better showing higher K/S values and much yellower and darker as compared to similar fabric produced otherwise and is even found better than that obtained by single bath simultaneous natural dyeing and finishing process.

Darker shades obtained after sequential post treatment with 10% eucalyptus extract as a separate finishing treatment by pad–dry–cure process on tesu dyed cotton fabric, is considered due to the additional effect of more and more attachment of active coloured constituents of eucalyptus containing tannins having poly-phenols and carboxylic acid groups (in gallic acid) and querectin, a colour component of eucalyptus, able to attach/react in more quantity with cotton cellulosic fibre, while the attachment of coloured compounds

Table 5: Surface colour strength and other colour parameters showing influences of finishing-treatment with eucalyptus leaves extract using citric acid catalyst by pad-dry-cure process and single bath simultaneous dyeing and finishing process for 15% Harada + Alum (50:50) bio-mordanting and tesu flower petal dyeing of cotton fabric

Treatments	*Surface Colour Strength*	*ΔE*	*ΔL*	*Δa*	*Δb*	*ΔC*	*ΔH*	*BI*	*MI*	*CDI*
Bleached cotton control fabric	0.02 ±0.01							92.1 ±0.9	0.91 ±0.04	–
15% application of Harda + Alum (50:50) (No dyeing)	0.16 ±0.01	6.1±0.90	–1.1	–0.91	5.85	–0.04	–5.90	44.6 ±1.0	3.43 ±0.02	–
15% application of Harda + alum (50:50) and 30% tesu extract dyed	2.65 ±0.05	17.8 ±1.2	–4.58	–1.06	17.1	11.20	–13.1	15.9± 1.1	7.26 ±0.14	2.8
After-treatment with 10% aqueous and MeOH extract of eucalyptus leaves on 30% tesu dyed cotton fabric after pre mordanted in sequence with 15% Harda + alum (50:50)										
(10%) aqueous extract of eucalyptus leaves + citric acid (5gpl)	5.49 ±0.10	15.4± 1.1	3.18	–0.73	4.31	4.37	11.0	18.6± 1.4	6.44 ±0.06	5.9
10% Eucalyptus leaves extract (in MeOH) + Citric Acid (5 gpl)	7.49 ±0.11	20.4 ±1.0	3.19	–0.82	4.11	5.33	12.0	19.6± 1.1	6.75 ±0.25	6.8
After-treatment with 10% MeOH extract of eucalyptus leaves on 30% tesu dyed cotton fabric after pre-mordanted with 15% alum only (No Harda)										
10% Eucalyptus leaves extract (in MeOH) + Citric Acid (5 gpl) after 15% Alum pre-mordanted	4.49 ±0.09	14.6 ±1.3	3.00	–0.68	5.36	5.71	11.7	16.3 ±1.2	6.02 ± 0.04	4.97
Concurrent Single bath dyeing and finishing of 15% harda + alum (50:50) pre-bio-mordanted cotton fabric with 30% extract of tesu flower petal and 10% eucalyptus leaves with citric acid										
Tesu (30%) extract + Eucalyptus leaves aqueous extract (10%) + citric acid (5 gpl)	4.85 ± 0.07	14.4 ± 1.2	2.38	–0.52	3.68	3.72	12.01	17.5 ± 1.0	5.61 ± 0.03	8.29

Table 6: Wash Fastness, Rubbing/Crocking Fastness and UV-light Fastness for 15% harda + alum Pre-bio-mordanted and 30% tesu extract dyed cotton fabric after-treated with 10% aqueous and MeOH extract of eucalyptus leaves by pad-dry-cure process and also in single bath exhaust process of concurrent/simultaneous dyeing-finishing process

Treatments	*Wash Fastness*		*Rubbing Fastness*	*Light Fastness under MBTL Microscal Fade-o-Meter*
Variation in Mordanting and Dyeing Conditions and Post Dyeing Finishing Treatments	*Loss of Depth of Colour*	*Cotton (Staining scale)*	*Dry Rubbing*	
15% Harda + alum (50:50) and 30% tesu aqueous extract dyed	4	3/4	4–5	3
After-treatment with 10% extract of eucalyptus leaves on 30% tesu aqueous extract dyed cotton after pre mordanted with 15% harda + alum (50:50) in sequence				
(10%) Eucalyptus leaves Aqueous extract + Citric acid (5 gpl)	4/5	4–5	4	4
10% Eucalyptus leaves MeOH Extract + Citric Acid (5 gpl)	4/5	4	4	4
After-treatment with 10% extract of eucalyptus leaves on 30% tesu aqueous extract dyed cotton after pre mordanted with 15% alum only				
10% Eucalyptus leaves MeOH Extract + Citric Acid (5 gpl)	3/4	3/4	3	3/4
Concurrent or simultaneous mordanting-dyeing and finishing of cotton fabric using 15% harda + alum (50:50), 30% tesu as dye and 10% eucalyptus extract as finishing agent				
Tesu (30%) extract + 10% Eucalyptus leaves aqueous extract + citric acid (5 gpl)	4/5	4/5	4	4/5

and tannins of eucalyptus is much less due to dilution effect in exhaust process in single bath simultaneous dyeing and finishing technique. MeOH extract is darker than aqueous extract which may be viewed as an effect of additional extract of colour component of eucalyptus by MeOH compare to water extract. Hence, for post finishing treatment with eucalyptus extract after dyeing by pad dry cure technique is recommended, where surface colour strength value is found to be superior to that obtained by single bath simultaneous dyeing–finishing process. Higher K/S values for eucalyptus post treatment may be viewed as an effect of the formation of giant (bigger) molecular adduct in large size of complex formed amongst 'Cotton fibre-Harda + alum bio-mordant-tesu (natural dye)-Eucalyptus leaves extracted substances" in presence of citric acid used as a catalyst during post treatment process during pad-dry-cure two step sequential process. While, in case of single bath simultaneous dyeing–finishing (exhaust process) technique, chances of such giant complex formation is reduced due to dilution effect which causes less strike rate and less absorption and less diffusion of eucalyptus ingredients producing less formation as a giant molecular adduct of "Cotton fibre-Harda + Alum bio-mordant-tesu (natural dye)—Eucalyptus leaves extracted substances" for the tesu dyed cotton fabric, resulting less colour yield in simultaneous—dyeing and finishing exhaust process.

However, taking the colour fastness results into consideration, data in Table 6 shows that fastness to UV-light observed for concurrent natural dyeing and finishing process using 10% eucalyptus extract is at par or sometimes to a little extent better than the similar sample produced by tesu natural dyeing followed by separate eucalyptus leaves extract finishing process, using 10% aqueous or MeOH extract of eucalyptus leaves separately. The wash fastness and rub fastness obtained have almost no differences amongst the two processes and are found to be much similar. Moderate wash fastness for 30% tesu natural dyed cotton can be viewed as an effect of complex formation amongst the fibre-metal-tannin in tesu-dye molecules of tesu (containing butein and some flavonoids/tannins) with natural potash alum as metallic bio-mordant along with harda (i.e. myrobolan, containing chebulinic acid with huge coordinating power). Antimicrobial tests were carried out for both types of samples i.e. (i) cotton fabric after-treated/finished with aqueous and MeOH extract of 10% Eucalyptus leaves (with 5 gpl citric acid catalyst) by pad–dry-cure process applied on 30% tesu dyed cotton fabric after pre-bio-mordanted with 15% application of harda + alum (50:50) and (ii) cotton fabrics undergone simultaneous mordanting-dyeing and finishing process adding 15% harda + alum (50:50) and 30% tesu extract and 10% MeOH extract of Eucalyptus leaves and 5 gpl citric acid in a single bath and heated at 90°C for 90 min in exhaust process with MLR 1:20 and those results are shown in Table 7.

Table 7: Bacterial Reduction % as per AATCC-100–2012/2014 for 15% harda + alum (50:50) pre-mordanted sample and subsequent 30% tesu dyed sample and aqueous and MeOH extract of 10% eucalyptus leaves (with 0.50% citric acid) finished cotton fabrics

Treatments for Bio-Mordanting/ Natural Dyeing and Natural Finishing	*Reduction (%) in Bacterial Growth as per AATCC-100–2012/2014*	
	K. Pneumoniae: AATCC 4352 (Gram –ve)	*S. Aureus: AATCC 6538 (Gram +ve)*
Bleached Control Cotton Fabric	NIL	NIL
15% application of Harda + Alum (50:50) in sequence (No dyeing with any dye or tesu)	99.98	98.43
15% [Harda + Alum (50:50)] bio-pre mordanting and 30% tesu aqueous extract dyeing of cotton	NIL	68.88
After-treatment with aqueous and MeOH extract of 10% eucalyptus leaves applied on [15% (harda + alum) in 50:50 ratio] pre-mordanted and subsequent 30% tesu extract dyed samples		
10% Eucalyptus leaves Aqueous extract + Citric acid (5 gpl) on said premordanted and dyed fabric	NIL	93.01
10% Eucalyptus leaves MeOH Extract + Citric Acid (5 gpl) on said pre-mordanted and tesu dyed fabric	99.97	99.78
After-treatment with MeOH extract of 10% eucalyptus leaves on 30% tesu aqueous extract dyed cotton fabric after pre mordanting with 15% alum only cotton fabric		
10% Eucalyptus MeOH Extract + Citric Acid (5 gpl) on 15% alum pre-mordanted and 30% tesu dyed fabric	99.89	99.72
Concurrent mordanting-dyeing-finishing of cotton fabric in single bath using 15% harda + alum (50:50), 30% tesu and 10% eucalyptus leaves extract (in MeOH) with 5 gpl citric acid		
15% harda + alum (50:50), and subsequently Tesu (30%) extract + Eucalyptus leaves aqueous extract (10%) + citric acid (5 gpl) all applied together in a single bath for simultaneous dyeing/finishing.	NIL	54.01

Thus, as per the results shown in Table 7, *K pneumoniae* as gram –ve bacteria and *S. aureus* as gram +ve bacteria, there is 99.98 and 98.43% reduction of bacterial growth on cotton fabric sample for said double pre-bio-mordanting with sequential application of 15% application of Harda + Alum (in 50:50 ratio) without any dyeing (No Dyeing) vide corresponding Petri-plates S code 2Q and S Code 2R as compared to S Code 1Q and S Code 1R in Figure 8. This can be explained by the influence of the natural alum (known as Fatkiri) as bio-mordant along with harda as bio mordanting assistant, which is known to be antiseptic and antibacterial, working here as alum antibacterial function of fitkari (alum) for prevention of bacterial growth on pre-mordanted cotton fabric sample. However, on dyeing with tesu, this antibacterial property imparted by fitkari is lost fully for gram –ve bacteria and lost partly for gram +ve bacteria (vide Petri-plate S Code 3Q and 3R), as the fitkari/alum are partly/fully consumed in complex formation amongst [mordant + fibre + tannins present in tesu + tesu as dye] after dyeing of pre-mordanted cotton fabric, showing remarkable increase in bacterial growth for gram –ve bacteria with nil bacterial reduction and also increase of growth of gram +ve bacteria with only 68% reduction in bacterial growth, when harda + alum pre-mordanted cotton fabrics are dyed with 30% tesu.

Sample code 4P, 4Q and 4R in Figure 8 and corresponding data in Table 7 shows that harda + alum pre-bio-mordanted and subsequently 30% tesu extract dyed cotton fabric after treated with aqueous extract of 10% eucalyptus leaves in presence of citric acid catalyst, there is no bacterial reduction against gram –ve bacterium (showing NIL bacterial reduction %) and huge growth of bacterium in Petri-plates S code 4Q, Figure No-8, while the same sample shows 93.01 bacterial reduction% against gram +ve bacterium very low growth of bacterium in Petri-plate S code 4R, Figure 8. However, the same pre-mordanted and tesu dyed sample of cotton fabric is when post treated with 10% MeOH extract (instead of aqueous extract) of eucalyptus leaves, it shows highest bacterial reduction for both Gram negative bacteria (bacterial reduction of 99.97%) and Gram positive bacteria (bacterial reduction of 99.78%) vide corresponding Petri plate S Code 5Q and 5R in Figure 8. However, for comparable samples undergoing concurrent simultaneous mordanting dyeing and finishing show antibacterial test results in much similar fashion or quite similar to the 30% tesu dyed cotton fabric samples after 15% harda + alum (50:50) pre-bio-mordanting, showing no reduction of bacterial growth for gram –ve bacteria (NIL bacterial reduction) and poor antibacterial response for gram +ve bacteria (54.02% bacterial reduction only) vide corresponding Petri plates S code 7Q and S code 7R respectively in Figure No 8. But only 15% alum pre-mordanted (without use of harda) and 30% tesu extract dyed cotton fabric sample is when after treated with MeOH extract of 10% eucalyptus leaves it become highly antibacterial in nature showing Petri plates S code 6Q and S code 6R against gram –ve and gram +ve bacterium giving 99.89% bacterial reduction for gram –ve bacterium and 99.72% bacterial reduction for gram +ve bacterium, vide data in Table 7, which are almost at par or similar level of bacterial reduction performance that obtained for 15% application of Harda + Alum (50:50) pre-mordanted and tesu dyed sample after treatment with MeOH extract of 10% eucalyptus leaves by pad-dry-cure process, for gram –ve bacteria (bacterial reduction of 99.97%) and for gram +ve bacteria (bacterial reduction of 99.78%), vide data in Table 7 and Petri-plates S code 5Q and 5R (Figure 8).

<table>
<tr><th>(P) Initial Sample to Test (Exposed to Bacterial Species for 0 hour)</th><th>K Pneumoniae-ATCC 4352 (Q) Gram –ve bacteria, after exposure of 24 hours</th><th>S. Aureus-ATCC 6538, (R) Gram +ve bacteria after exposure of 24 hours</th></tr>
<tr><td>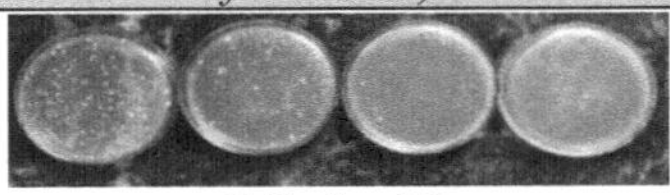
S Code -1P</td><td>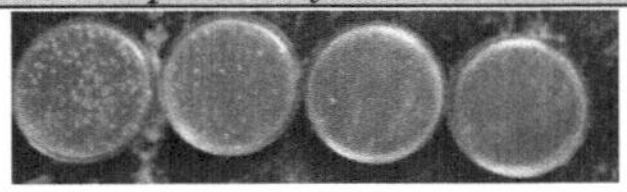
S Code -1Q</td><td>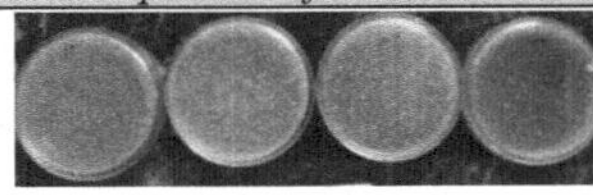
S Code -1R</td></tr>
<tr><td colspan="3">(1): Control bleached cotton fabric sample: un-mordanted and undyed i.e. no mordant and no dye</td></tr>
<tr><td>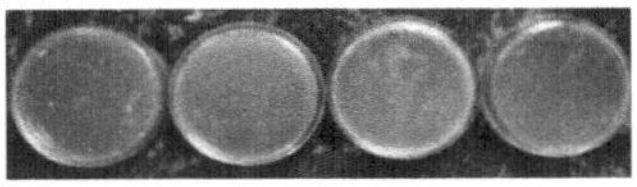
S Code -2P</td><td>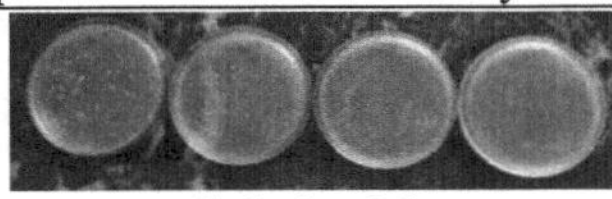
S Code -2Q</td><td>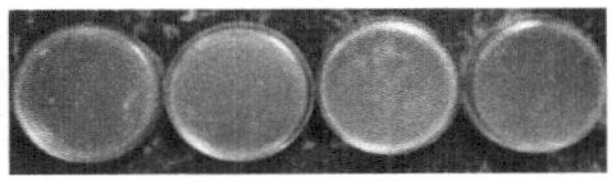
S Code -2R</td></tr>
<tr><td colspan="3">(2): 15% overall application of Harda+Alum [50:50] treated cotton fabric: (No dyeing)</td></tr>
<tr><td>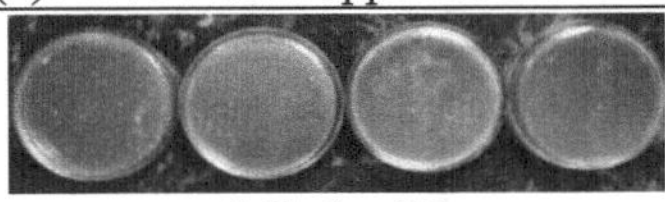
S Code -3P</td><td>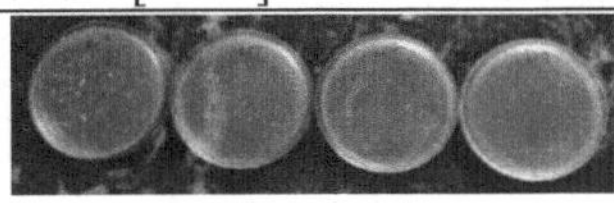
S Code -3Q</td><td>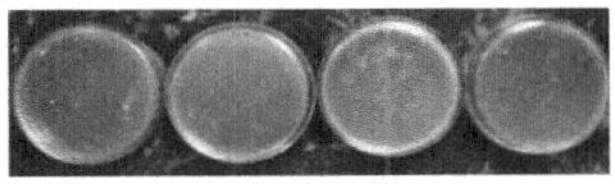
S Code -3R</td></tr>
<tr><td colspan="3">(3): 15% Harda+Alum [50:50] pre-mordanted +30% Tesu Extract dyed cotton fabric</td></tr>
<tr><td colspan="3">After-treatment with Aqueous and MeOH extract of 10% eucalyptus leaves on 30% tesu dyed cotton fabric pre-mordanted with 15% harda + alum (50:50) or with only 15% alum.</td></tr>
<tr><td>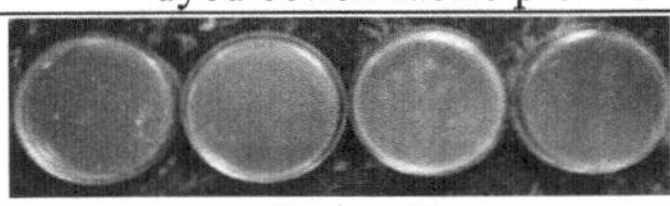
S Code -4P</td><td>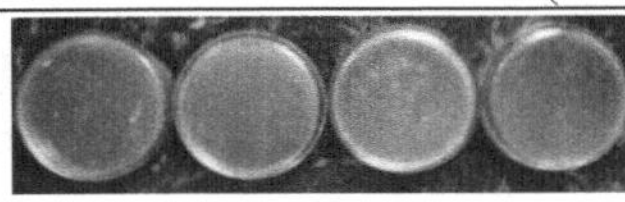
S Code -4Q</td><td>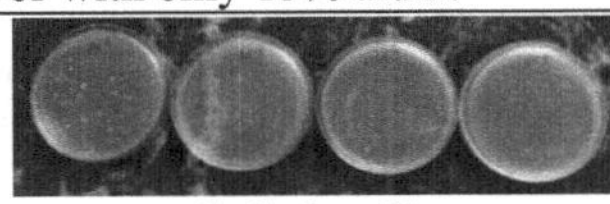
S Code -4R</td></tr>
<tr><td colspan="3">(4): After-treatment with 10% Eucalyptus extract (Aqueous) + citric acid (5 gpl) on 30% Tesu dyed after 15% total harda + Alum (50:50) Pre-bio-mordanted cotton fabric</td></tr>
<tr><td>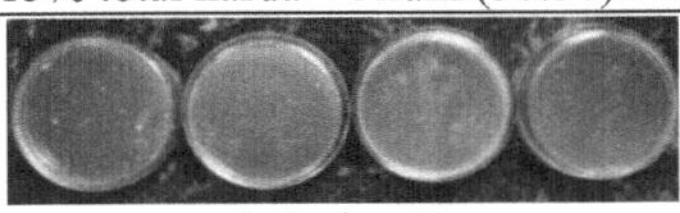
S Code -5P</td><td>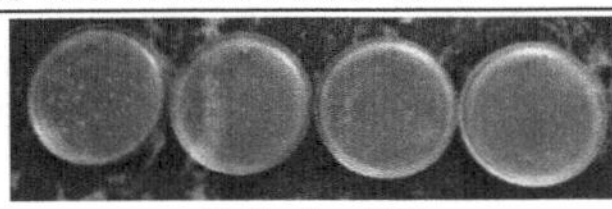
S Code -5Q</td><td>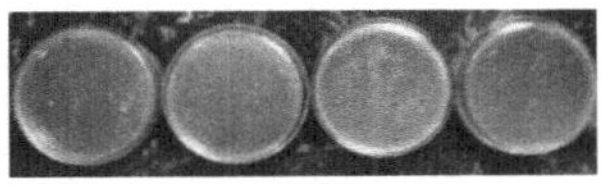
S Code -5R</td></tr>
<tr><td colspan="3">(5): After-Treatment with 10% Eucalyptus extract (in MeOH) + citric acid (5gpl) on 30% Tesu dyed after 15% total harda + Alum (50:50) Pre-bio-mordanted cotton fabric</td></tr>
<tr><td>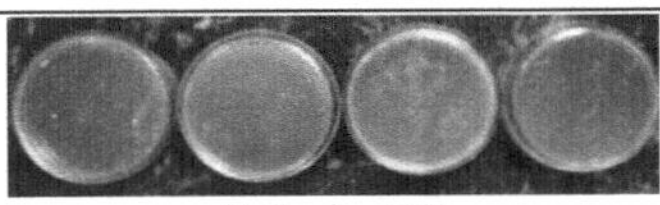
S Code -6P</td><td>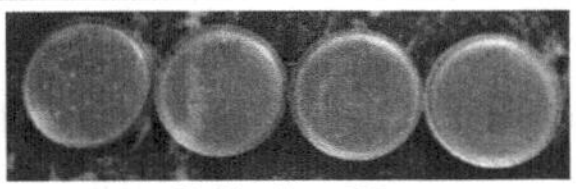
S Code -6Q</td><td>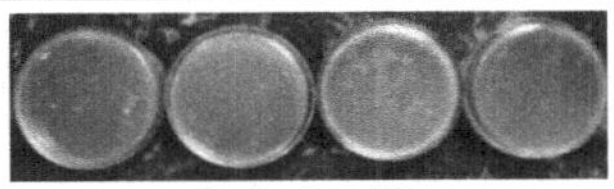
S Code -6R</td></tr>
<tr><td colspan="3">(6): After-treatment with 10% Eucalyptus extract (in MeOH) + citric acid (5gpl) on 30% Tesu dyed after 15% Alum pre-mordanted cotton fabric</td></tr>
<tr><td colspan="3">Simultaneous dyeing-finishing of cotton fabric adding 15% harda + alum (50:50), 30% tesu extract and 10% eucalyptus leaves extract (in MeOH) + 5 gpl citric acid in single bath</td></tr>
<tr><td>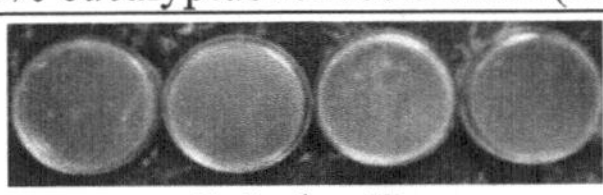
S Code -7P</td><td>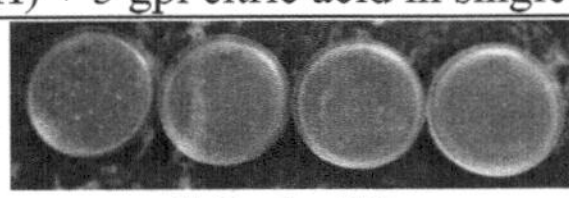
S Code -7Q</td><td>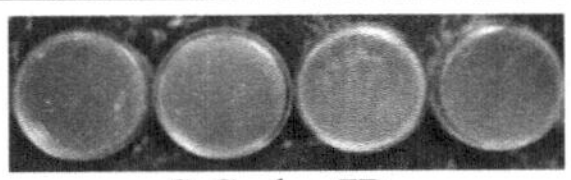
S Code -7R</td></tr>
<tr><td colspan="3">(7): Concurrent/simultaneous dyeing-finishing of cotton fabric adding 15% harda + alum (50:50), 30% tesu and 10% eucalyptus leaves extract (in MeOH) + 5 gpl citric acid in single bath.</td></tr>
</table>

Fig. 8: Photo of Petri Plates of Control, Pre-Mordanted, Dyed and Finished Cotton with Sample Code

However, unexpectedly, concurrent simultaneous mordanting, dyeing and finishing process in single exhaust bath process, the bacterial reduction % observed are not at all satisfactory and useful, as compared to said pre-mordanted and tesu dyed and post-dyeing after-treatment/finishing with aqueous extract or MeOH extract of 10% eucalyptus leaves, where the later i.e. when 10% MeOH extracted eucalyptus leaves is applied instead of its aqueous extract, antibacterial effect is enhanced significantly. So, in order to improve the bacterial reduction % to fullest/highest degree with eucalyptus extract as natural multi-functional finishing agent, the said 15% harda + alum (50:50) pre mordanted cotton fabric subsequently dyed with 30% tesu extract are to be post dyeing after-treated/finished separately by padding- drying-curing process using MeOH extract of 10% eucalyptus leaves in presence of citric acid as catalyst for good results in reduction of growth of both gram positive and gram negative bacterium as well as it also gives a darker colour for addition of its own inherent colour component with very good to excellent antimicrobial property along with reasonable UV protection reported in next sub-section of this paper, to consider eucalyptus as a multifunctional natural finishing material.

3.3.2 *UV-Resistance Action of Eucalyptus and Coconut Oil on Natural Tesu Dyed Cotton Fabric*

Nowadays, UV protection in textiles has become equally important for protection of human skin. Considering this in view, reddening of skin under any barrier like UV protective textiles or other such materials can be judged by UV Protection factor (UPF) or sun protection factor (SPF) values. Few reports are available in literature on some natural UV protective finishing materials (Xiuliang *et al.*, 2013) which are when applied on silk, cotton and other natural textiles, there is medium to good enhancement in UPF values. Considering that, Tesu dyed cotton fabrics area after treated with aqueous extract and MeOH extract of eucalyptus leaves and also with coconut oil emulsion with or without acidic catalyst. Hence, such natural UV absorbers (MeOH extract of eucalyptus leaves and coconut oil emulsion) are applied by suitable post treatment on harda + Alum pre-bio-mordanted and Tesu dyed cotton fabric by pad-dry-cure process for improving both light fastness and UPF values of tesu natural dyed cotton fabrics. A similar work for silk fabric dyed with punica is presented in Table 4. In the present work, Ranges of UV protection factor (UPF value) are classified based on its performance criteria for textiles, where it is considered that UPF values 15 to 24 is considered as 'Good' performance, UPF values 25 to 40 is considered as 'Very Good' performance and UPF values 40–50^{+} is considered as 'Excellent' performance for UV protection of textile and related other materials to protect human skin.

Hence, both aqueous and MeOH extract of 10% Eucalyptus leaves and emulsified coconut oil with or without 5 gpl citric acid (catalyst) are applied on control cotton fabric and 15% Harda + Alum (50:50) pre-bio-mordanted and 30% tesu dyed cotton fabric, for imparting a good UV protective finish on it.

Relevant data of observed results of UV protection factor (UPF) values before and after the said different pre and post treatments are measured following AATCC: 183: 2010/2014 method and are shown in Table 8. Control bleached cotton fabric shows UPF

Table 8: UV Protection Factor (UPF) Test as per AATCC-183-2010/2014 for Post Treatment with Aqueous and MeOH Extract of 10% Eucalyptus Leaves and Coconut Oil

Type of Treatments and Sample Description	*UV Protection Factor (UPF)*	*UV-A (%)*	*UV-B (%)*
Bleached Control Cotton Fabric (un-mordanted and un-dyed cotton fabric)	5 ± 1	29.6	22.1
15% overall application of Harda + Alum (50:50) (No dyeing)	10 ± 1	19.5	11.0
15% [Harda + Alum (50:50)] pre-bio-mordanted and 30% tesu aqueous extract dyed cotton fabric	12 ± 1	10.1	7.8
After treatment with aqueous and MeOH extract of 10% eucalyptus leaves and 10% emulsified coconut oil with 2 gpl citric acid by pad-dry-cure process on 30% tesu dyed cotton fabric after pre mordanting with 15% harda + alum (50:50) in sequence			
10% Eucalyptus leaves (Aqueous Extract) + citric acid (2gpl)	25 ± 2	4.1	3.6
10% Eucalyptus leaves [alcoholic (MeOH) Extract] + citric acid (2 gpl)	40 ± 2	2.0	1.8
10% Coconut oil emulsion + citric acid (2 gpl)	30 ± 2	2.7	2.0
After-treatment with MeOH extract of 10% eucalyptus leaves and 10% emulsified coconut oil on 30% tesu extract dyed cotton after pre mordanting with only 15% alum			
10% Eucalyptus leaves alcoholic (MeOH) extract + Citric Acid (2 gpl)	35 ± 2	2.6	2.8
10% Coconut oil emulsion + citric acid (2 gpl)	25 ± 2	5.3	8.2
Concurrent Simultaneous mordanting-dyeing and finishing of cotton fabric adding 15% Harda + Alum (50:50),30% tesu extract and 10% aqueous extract of eucalyptus leaves +2gpl citric acid, all applied together in a single bath heated at 90°C for 90 minutes			
Tesu (30%) extract + Eucalyptus aqueous extract (10%) + citric acid (2 gpl)	25 ± 2	4.5	4.5
Tesu (30%) + Coconut oil (10%) + citric acid (2 gpl)	20 ± 1	3.9	6.1

value of 5 only, while 15% overall Harda+Alum (50:50) double pre-mordanted (No dyeing) cotton fabric samples shows UPF value of 10 and the said pre bio-mordanted cotton fabric is when subsequently dyed with 30% tesu (aqueous) extract shows UPF value of 12 only. But the same cotton fabric sample is when selectively pre-bio-mordanted with 15% application of harda + alum (50:50) and subsequently natural dyed with aqueous extract of 30% tesu and is post dying after-treated with both aqueous or MeOH extract of 10% eucalyptus leaves or with 10% emulsified coconut oil in presence of 1/5th of citric acid (2 gpl) as catalyst, UV protection performance are found to improve upto 25, 40 and 35 respectively, vide observed UPF values in Table 8. UPF value of 40 is observed as best result for 15% harda+alum pre-bio-mordanted cotton fabrics is when dyed with 30% tesu extract and is post treated with 10% extract (in MeOH) of eucalyptus leaves with 2 gpl citric acid as catalyst by much easier and simple padding-drying-curing process instead of cumbersome micro encapsulation process, as compared to UPF value 30 found for after-treatment with same method and dosages of 10% coconut oil emulsion.

Concurrent simultaneous mordanting-dyeing and finishing of cotton fabric adding 15% Harda+Alum (50:50), 30% tesu extract and 10% aqueous extract of eucalyptus leaves + 2 gpl citric acid, all applied in a single bath exhaust process, using the said two types of UV resisting natural agents (Eucalyptus leaves extract and emulsified coconut oil) the enhancement of UPF values for concurrent dyeing and finishing process are found to be much less as UPF value of 20 for 10% emulsified coconut oil and UPF value is 25 for 10% eucalyptus leaves aqueous extract. Thus, UPF values in case of simultaneous dyeing and finishing treatment are not so high rather much below the expectation as compared to sequential post-treatment process by pad-dry-cure process in both the cases respectively. UPF value 15–24/25 is considered as moderate to good UV protection and hence observed UPF values 20 and 25 for concurrent mordanting-dyeing and finishing process using emulsified coconut oil and Eucalyptus leaves extract, may be considered to render it a moderate to good UV protection only, but not as good as the UPF value of 40 (Nr Excellent) as obtained for post-treatment with MeOH extracted Eucalyptus leaves by pad dry cure process on said 15% application of Harda + Alum (50:50) pre-bio-mordanted and 30% tesu extract (aqueous) dyed cotton fabric. For emulsified coconut oil treatment in presence of citric acid, it is generally hydrolyzed forming lauric acid (approx. 45–48%) and some α-tocopherol content of it participate in reaction with cotton. So during after-treatment of cotton fabric with emulsified coconut oil in presence of citric acid, the hydrolyzed lauric acid moieties are possibly attached to hydroxyl group of cotton cellulose by reaction of –COOH group of lauric acid with hydroxyl group of cellulose forming ester linkage. Attachment of α-tocopherol to cotton forming an ether linkage by reaction between hydroxyl group of α-tocopherol and hydroxyl group of cotton is another possibility. UV light fading is protected of coconut oil and eucalyptus leaves extract is due to their inherent composition having UV resistant components in both of them preferentially absorbing uv light protecting dye molecules causing UV filtering effect generated in treated cotton fabric improving UPF rating good (UPF-25) to near excellent (UPF-40) for after-treatment with emulsified coconut oil and MeOH extract of eucalyptus leaves respectively. Moreover, along with UV protection criteria, the presence of essential oil present in the extract of eucalyptus leaves makes the tesu dyed and eucalyptus finished cotton fabrics unaffected by microbes showing high degree of antimicrobial effect too.

From compositional analysis of eucalyptus extract, it is known that essential oil part of eucalyptus contain 1,8-cinecole (eucalyptol), benzene, nerolidol, limonene, alpha-pinene and beta-pinene, alpha-phellandrene, gamma-terpinene, caffeic acid, linaol, geraniol and thymol etc. as chemical constituents in MeOH extract of eucalyptus leaves and it is believed that some of which are strongly responsible for reducing and blocking the UV rays and protect bacterial growth by attaching with cellulosic –OH groups in presence of citric acid as catalyst showing better result for both protection against UV rays and microbial growth, which need further research to explore and identify which constituent is more and mainly responsible for this. Eucalyptol (1,8-cineole) is the most important constituent obtained in MeOH extract of eucalyptus leaves, known to demonstrate good antimicrobial activities. Thus, 10% eucalyptus extract (in MeOH) can be used as natural resource based multifunctional finishing agent for post treatment of cotton textiles as a natural bio-finishing agent to improve UPF value up to 40 with high bacterial reduction (99%), for both Gram positive and Gram negative bacterial attack to cotton.

4. CONCLUSION

From the present work, the mechanism for giant insoluble complex formation amongst "Cotton (harda + Alum) – natural dye – finishing agents" with understanding of the roles of tannin/chebulinic acid based harda as mordanting assistant along with natural alum as bio mordant (pre-mordanting), punica or tesu as mordant able anionic dye and chitosan or CTAB as dye fixer, eucalyptus extract as natural multifunctional UV protector and antimicrobial finishing agent and coconut oil as another UV protecting agent were established for use in promoting colour yield (K/S values), colour fastness to wash and light as well as enhanced antimicrobial protection and UV Protection factor, by applying these for both silk dyed with Punica (anar peel) and cotton dyed with tesu, after pre mordanting both with 15% application of harda + natural potash alum in a fully eco-friendly (oeko-tech) approach.

Chitosan is proved to be efficient as a natural resource based cationic dye fixer for natural dyed silk textiles for improving colour fastness to washing at par with Cetyl trimethyl ammonium bromide (CTAB) based synthetic dye fixer, establishing the given mechanism of action of CTAB for this.

Similarly, both aqueous and methanol based extract of Eucalyptus leaves is proved to act as good light fastness improver cum natural UV absorbers rendering better light fastness and high UV protection against sun-light exposure to human skin showing very good to excellent UV protection factor (UPF) up to 40, rendering almost at par or even better performance than use of commercial UV absorber like Supra–UV–Finish-2% (Sulphono-Hydroxy Benztriazole—a commercial UV absorber). MeOH extract of eucalyptus leaves shows much better results than use of aqueous extract of eucalyptus leaves showing at par or even better UPF values than selective commercial UV absorber.

Post treatment with eucalyptus extract on tesu dyed cotton fabric is found to be equal performer being able to improve excellent level of protection against microbial attack for both gram positive and gram negative bacterial reduction up to 99%, establishing its high antibacterial performance on cotton.

REFERENCES

[1] Cook, C.C. (1982). After Treatments for Improving the Fastness of Dyes on Textile Fibre, *Rev.Prog. Colouration*, 12(1), 78–89.

[2] Crews, P.C. (1982). The Influence of Mordant on the Light Fastness of Yellow Natural Dyes, *J. Amer Instt. Cons*, 21, 43–58.

[3] Cristea, D. and Vilarem, G. (2006). Improving Light Fastness of Natural Dyes on Cotton Yarn, *Dyes and Pigments,* 70, 238–244.

[4] Cristea, D. and Vilarem, G. (2006). Improving Light Fastness of Natural Dyes on Cotton Yarn, *Dyes & Pigments*, 70, 238.

[5] Duff, D.G., Sinclair, R.S. and Stiriling, D. (1977). The Fastness to Washing of Some Natural Dyestuffs on Wool, *Studies in Conservation,* 22, 161–16.

[6] Fadhel, B.B., Aissi, A., Ladhari, N., Deghrigue, M., Chemli, R. and Joly, J.P. (2012). Antibacterial Effects of Two Tunisian Eucalyptus Leaf Extracts on Wool and Cotton Fabrics, *The Journal of the Textile Institute*, 103(11), 1197–1204.

[7] Gulrajani, M.L., Bhaumik, S., Oppermann, W. and Hardtmann, G. (2003). Dyeing of Red Sandalwood on Wool and Nylon, *Indian J. Fibre Text. Res.*, 28(6), 221.

[8] Gupta, D. (1999). Dyeing of Ratanjot Dye on Nylon & Polyester, *Colourage,* 46(7), 35.

[9] Hofenk, D. and Graaff, J.H. (1983). Conservation—Restoration of Church Textiles and Painted Flags, *4th Int. Restorer Seminar*, Hungary, 2, 219.

[10] ISI (BIS), 1981/SP-15-Part-4, 1988. *ISI (BIS) Handbook of Textile Testing.* 1st ed.-SP-15-1981 and first Revision -SP-15-part-4, 1988), Bureau of Indian Standards, Delhi, 221–280.

[11] Joshi, M., Ali, S.W. and Purwar, R. (2009). Ecofriendly Antimicrobial Finishing of Textiles Using Bioactive Agents Based on Natural Products, *Indian Journal of Fibre and Textile Research,* 34, 295–304.

[12] Khan, M.I., Ahmad, A., Khan, S.A., Mohammad, Y., Mohammad, S., Nikhat, M. and Mohammad, F. (2011). Assessment of Antimicrobial Activity of Catechu and its Dyed Substrate, *Journal of Cleaner Production*, 19, 1385–1394.

[13] Khan, S.A., Ahmad, A., Khan, M.I., Mohammad, Y., Mohammad, S., Nikhat, M and Mohammad, F. (2012). Antimicrobial Activity of Wool Yarn Dyed with Rheumemodi L. (Indian Rhubarb), *Dyes and Pigments,* 95, 206–214.

[14] Oda, H. (2001b). Improvement of Light Fastness of Natural Dyes on Cotton Yarn, *Colouration Technology,* 117(5), 254.

[15] Oda, H. (2001). Improvement of Light Fastness of Natural Dyes, *Colouration Technology*, 117(4), 204.

[16] Pargai, D., Shahnaz, J. and Gahlot, M. (2020). Functional Properties of Natural Dyes, In *Chemistry and Technology of Natural and Synthetic Dyes and Pigments*, Edited By: A.K. Samanta, N.S. Awwad and H.M. Algarni, 207–225. (In Tech Open; UK).

[17] Samanta, A.K. (2020). Bio-Dyes, Bio-Mordants and Bio-Finishes: Scientific Analysis for Their Application on Textiles, in book on *'Chemistry and Technology of Natural and Synthetic Dyes and Pigments'* Ed by Samanta A.K. *et al.* (InTech Open publisher, London, UK), Ch-1, 3–39.

[18] Samanta, A.K., Agarwal, P., Singhee, D. and Dutta, S. (2009). Application of Single and Mixtures of Red Sandalwood and Other Natural Dyes for Dyeing of Jute Fabric: Studies on Colour Parameters/Colour Fastness and Compatibility. *Journal of Textile Institute,* 100(7), 565–587.

[19] Samanta, A.K., Konar, A., Chakrabarti (Mukherjee) and Datta, S. (2011). Dyeing of Jute Fabric with Tesu Extract: Part-1: Effects of Different Mordants and Dyeing Process Variables on Colour Yield and Colour Fastness Properties, *Indian Journal Fibre & Text. Res.,* 36, 63–73.

[20] Samanta, P. (2020). A Review on Application of Natural Dyes on Textile Fabrics and Its Revival Strategy, *In the Book on Chemistry and Technology of Natural and Synthetic Dyes and Pigments, Ed. by A.K. Samanta, N.S. Awwad & H.M. Algarni,* InTech Open, London UK, Ch-3, 75–98.

[21] Samanta, P., Singhee, D. and Samanta, A.K. (2018). Fundamentals of Natural Dyeing of Textiles-Pros. and Cons., *Current Trends in Fashion Tech. and Textile Eng. (CTFTTE),* 2(4), 01–07.

[22] Shah, H.S. and Gandhi, R.S. (1990). *Instrumental Colour Measurements and Computer Aided Colour Matching for Textiles* (Mahajan Book Distributors, Ahmedabad, India), 76–116.

[23] Singh, R., Jain, A., Panwar, S., Gupta, D. and Khare, S.K. (2005). Antimicrobial Activity of Some Natural Dyes, *Dyes and Pigments*, 66, 99–102.

[24] Singhee, D. and Samanta, P. (2019). Studies on Dyeing Process Variables for Application of Tesu (*Butea Monosperma*) as Natural Dye on Silk Fabric, *Journal of Natural Fibres,* DOI: 10.1080/15440478.2018.1453430, Published on line 30 Mar 2018, 16(8); 1098–1112.

[25] Suping, H., Li, W. and Huayin, W. (2017). Dyeing and Antibacterial Properties of Liriope Platyphylla Fruit Extracts on Silk Fabrics, *Fibres and Polymers* 2017, 18, No. 4, 758–766.

[26] Xiuliang, H., Xinzi-Chen, Y.C., Helan, X., Lifen, C. and Yiqi, Y. (2013). Dyeing and UV-Protection Properties of Water Extracts from Orange Peel, *Journal of Cleaner Production*, 52, 410–419.

[27] Young-Hee, L., Seul-Gi, L., Eun-Kyung, H., Young-Mee, B., Sunja, C. and Han-Do, K. (2016). Dyeing Properties and Deodorizing/Antibacterial Performance of Cotton/Silk/Wool Fabrics Dyed with Myrrha (*Commiphora Myrrha*) Extract, *Textile Research Journal*, 87(8), 973–983.

Novel Approach of Dyeing Polyester with Acid Dyes

D.V. Patel[1], P.K. Patel[2], Rajveer Singh Ataliya[3], Utsav Shah[3], Tanushree Pathak[3] and Jayant G. Patil[3]

[1,2]Department of Textile Chemistry, Faculty of Tech. & Eng.,
The M.S. University, Baroda, Gujarat, India

[3]Students of BE-IV Textile Processing (2019–20)

E-mail: [1]disant.v.patel-tchem@msubaroda.ac.in;
[2]priyank.patel-tchem@msubaroda.ac.in

ABSTRACT: *Polyester fibre is characterized by high strength, good crease recovery, excellent biological resistance, however due to the hydrophobic and oleophilic nature; lack of moisture absorbency, harsh feel, development of static charge, etc. is observed. Polyester being hydrophobic is dyed basically with non-ionic disperse dyes. The study emphasises on modifying the surface morphology of polyester by introducing some protein micro particles on the surface to enhance its dyeability with protein dyes. Change in the properties (absorbency, strength, dyeability, etc.) of the polyester fabric with the applications of the solubilized keratin extracted from waste wool fibre has also been studied. The major advantage of this modification is that polyester fabric can also be dyed with dyes applicable on protein fibres.*

Keywords: Disperse Dyes, Etching, Keratin Extraction, Polyester, Wool.

1. INTRODUCTION

Polyesters are polymers that contain ester linkage groups along their main chains. Polyester has been one of the most popular fibres because of its high strength and good stability against most of the chemicals but poor resistance to alkali (Vasantha, 1991). Polyester fibres have many desirable properties including relatively high tenacity, low creep, good resistance to stain and deformation, high glass transition temperature and good resistance to acids and oxidizing agents (Chinta & Singh, 2012). Polyester fibre has conquered the leading position among the synthetics because of its excellent properties such as high strength, abrasion resistance and wrinkle-free characteristics. However, polyester retains little moisture and does not transport aqueous fluids, the hydrophobic nature of polyester makes it difficult to dye (they require a carrier) and to finish in aqueous media (Tavanai, 2009). To overcome some of these problems and to improve the property of polyester, it requires some modification. The modification can be done in two ways: a) polymer modification i.e. by introducing acidic groups in the polymer chain or b) fabric modification i.e. with the help of chemicals like sodium hydroxide (Dave *et al.*, 1987).

2. MATERIALS AND EXPERIMENTAL METHOD

In this study, 100% polyester fabric plain weave was used to know the effect of NaOH on the structural modifications. Polyester fabric (previously heat set) and having a surface mass of 119 g/m^2 was used in this study; Picks per inch (PPI) = 220 and Ends per inch (EPI) = 102. The fabrics are subjected to washing to eliminate the presence of impurities.

Raw wool fibre (in staple form) was purchased from sheep breeder of Surat, having a staple length of 3.5–4.5 cm and 20–25 micron thickness.

2.1 Alkaline Modification of Polyester Fabric

The polyester fabric is treated with different concentrations of NaOH i.e. 1%, 3% and 5% at 90°C temperature for 30 minutes with material to liquor ratio of 1:30. All the chemicals used in this work were laboratory grade reagents.

2.2 Modification of Polyester Fabrics (by pre-treatment with keratin extract)

The raw wool fibres were degreased and cleaned by scouring treatment with 4% Detergent (Non-ionic) and 2% Soda ash at 60–70°C temperature for 30–45 minutes. Alkaline hydrolysis of degreased wool fibre was carried out in a solution containing 5% NaOH at boiling temperature for 20–30 minutes, the mixture was then filtered by 125 mm Ø Whatman filter paper (grade-1) and neutralized with acetic acid (Gupta & Sharma, 2016).

This solution was evaporated by boiling until the volume of solution was reduced to half the amount, cooled, and used for the pre-treatment of polyester fabric (Rajabinejad *et al.*, 2016).

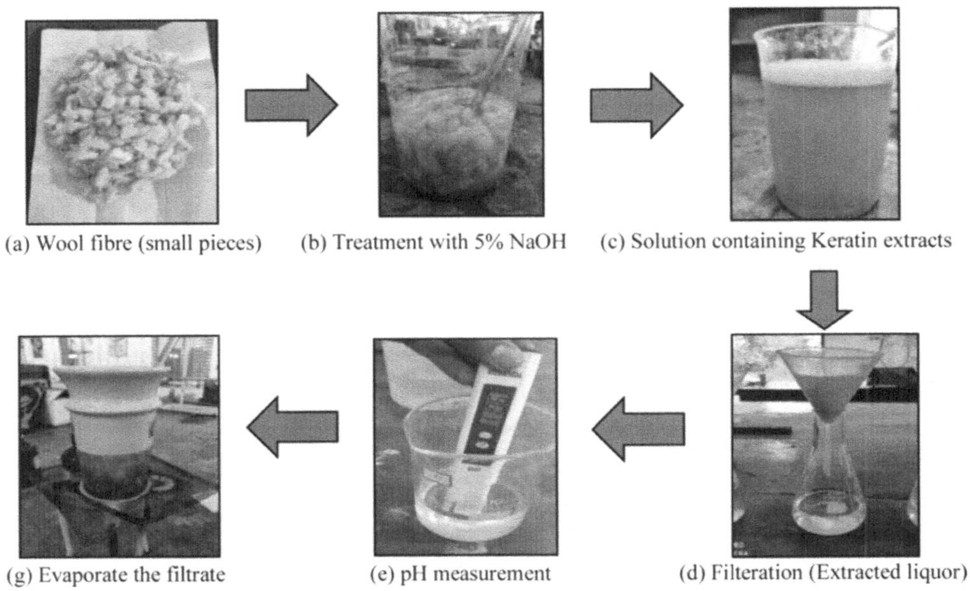

Fig. 1: Sequence of Keratin Extraction Process from Wool Fibre

Pre-treatment procedure

Process sequence for application of keratin extract on polyester fabric

Polyester fabric (modified with NaOH)

↓

Immerse in wool filtrate solution (For 10–15 minutes)

↓

Padding (3-dip.3-nip)

↓

Dry the fabric at Room Temperature

↓

Curing at 180°C temperature for 90 seconds

2.3 Dyeing

Alkali-modified as well as keratin extract pre-treated polyester fabrics were dyed with Disperse Navy Blue S-3G (C.I. Disperse blue 301) with a percent shade of 5% on weight of fabric, 1% Dispersing agent at 130°C temperature. The pH of dye bath was maintained at 5.0–5.5 using 1% Acetic acid in a high-temperature high-pressure beaker dyeing machine for 30–45 minutes (Mowafi *et al.*, 2018).

The major advantage of polyester fabric modified by second method (i.e. by using keratin extract) is that dyeing with acid dyes is also possible (Kumawat *et al.*, 2018). Dyeing of these samples was carried out using Colocid Blue BAR (C.I. Acid Blue 324) and 2% Sulphuric acid at boiling temperature for 30–45 minutes using exhaust dyeing method.

2.4 Determination of Weight Loss

The effect of NaOH concentration on the polyester fabric can be studied by finding the weight loss of fabric after modification. The weight loss (WL) is expressed as relative WL according to the equation:

$$WL = ((W1 - W2)/W1) \times 100$$

Where, W1 and W2 = Weights of the samples before and after treatment, respectively.

2.5 Absorbency Measurement

The water absorbency of treated fabrics was tested as per AATCC Test Method 79 to know the time taken by a drop of water placed on the fabric surface to completely absorb.

2.6. Analysis of Particle Size of Liquor Containing Keratin Extract

Dynamic Light Scattering (DLS) method was used to measure particle size of the extracted keratin particles. The flow mode option enables the system to be connected to an SEC (Size exclusion chromatography) system to use as a detector for the size of proteins micro particles.

2.7 Color Measurements

The relative color strength (K/S) of dyed fabrics was measured using SS5100A premier color scan spectrophotometer at a wavelength 400–700 nm. Reflectance values were measured, and relative color strength (K/S) was calculated using Kubelka-Munk equation.

$$K/S = (1-R)^2/2R$$

Where, K = absorption and S = scattering characteristics, respectively.

2.8 Determination of Washing Fastness of Dyed Fabric

ISO standard test no. III (IS: 764: 1979) was used for the determination of washing fastness of dyed fabrics. Launder-O-meter was used for this purpose.

3. RESULTS AND DISCUSSION

3.1 Determination of Weight Loss of Polyester Fabric Modified with NaOH

The effect of NaOH concentration on the polyester fabric can be studied by finding the weight loss of fabric after modification. Here 1%, 3% & 5% concentrations of NaOH was used for the treatment, which gives 2.45%, 4.32% and 9.33% weight loss respectively.

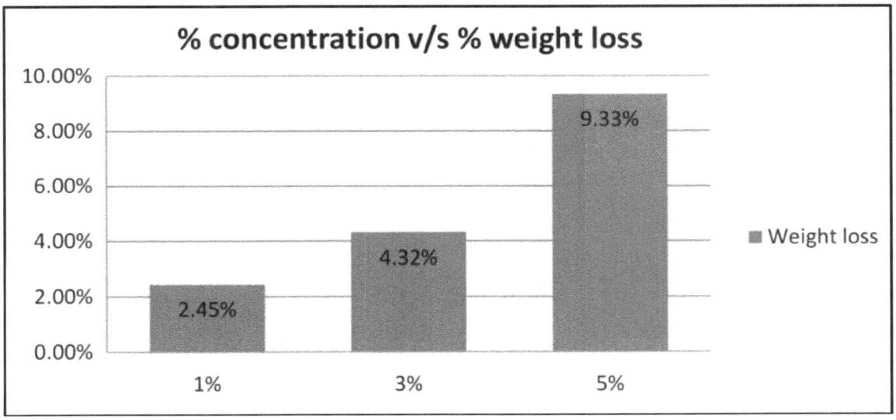

Fig. 2: Effect of NaOH Concentration on % Weight Loss of Polyester Fabric

From the above graph it is observed that, as the concentration of NaOH increases, the weight loss also increases almost linearly.

3.2 Absorbency Measurement (AATCC Test Method 79)

The absorbency for different polyester fabrics were carried out and it was observed that, untreated polyester fabric shows very less absorbency, but after the treatment with NaOH absorbency of fabric improved significantly. The data for absorbency test is shown in the Table 1.

Table 1: Absorbency of Different Samples

Sr. No.	*Fabric Type*	*Drop Test (Seconds)*
1.	Control polyester fabric	13–15
2.	Polyester fabric modified with Alkali	3.63
3.	Polyester fabric modified with keratin extract	1.98

The polyester fabric modified further with keratin extract showed very good absorbency (compared to the sample treated only with alkali), because of the inherent moisture regain behavior/capability of protein fibres.

3.3 Determination of Particle Size

Here particle size of different concentrations of Sodium hydroxide (1%, 3% and 5%) was checked for choosing the appropriate concentration for wool extraction process.

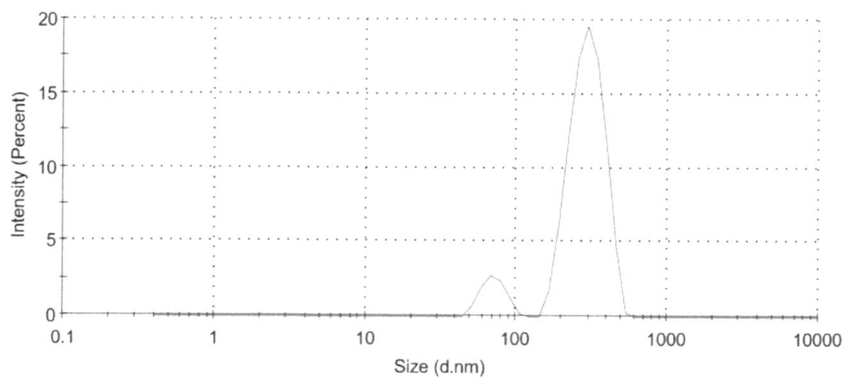

Fig. 3: Graph of Size Distribution by Intensity (for 1% NaOH solution)

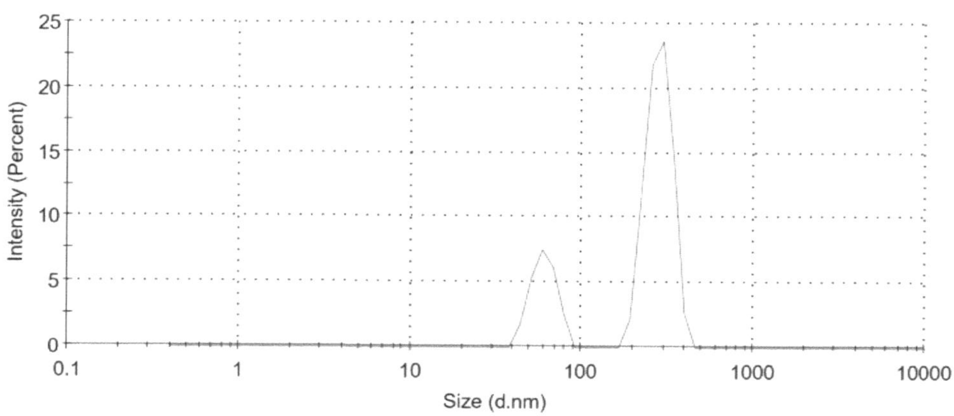

Fig. 4: Graph of Size Distribution by Intensity (for 3% NaOH solution)

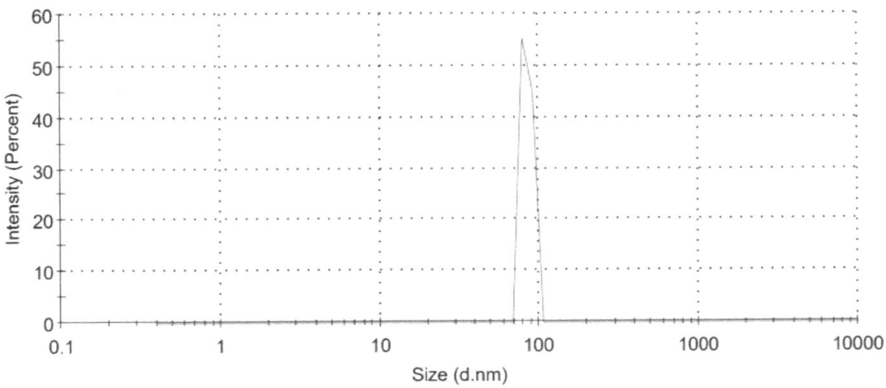

Fig. 5: Graph of Size Distribution by Intensity (for 5% NaOH solution)

By studying the above results, we observed that in 5% concentration constant particle size of keratin particles were obtained, while in 1% & 3% concentration particles of varying sizes are obtained. Hence 5% concentration was chosen for the extraction process.

3.4 Measurement of K/S Value and Other Colour Co-ordinates

The control polyester fabric, NaOH modified fabric and polyester modified with keratin extracts were dyed with disperses dyes separately and spectrophotometer was used to evaluate their colour value by K/S values. From the Figure 6 it can be observed that NaOH

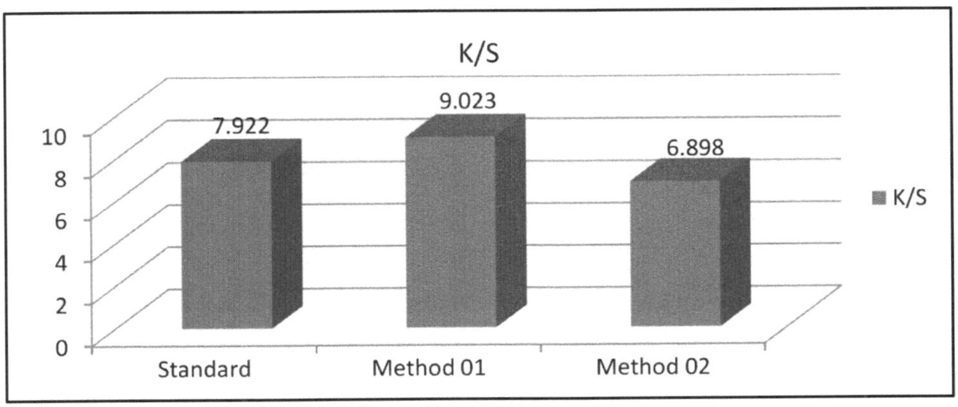

Fig. 6: K/S Values for Disperse Dyes on Modified Polyester Fabric

Sample Name	*Description*	*K/S Value*
Standard	Control sample/Untreated polyester	7.922
Method 1	Dyed polyester sample (modified with NaOH)	9.023
Method 2	Dyed polyester sample (modified using keratin extracts)	6.898

modified sample has higher K/S value compared to control sample and polyester modified with keratin extracts. Modification with NaOH will help improve the dyeability of polyester with disperse dyes.

Control polyester fabric, NaOH modified fabric and polyester modified with keratin extracts were dyed with acid dyes separately and tested by spectrophotometer to know the K/S values. From the Figure 7 it can be observed that control polyester sample shows very low K/S value while NaOH modified sample showed some improvement dyeing and K/S was also increased.

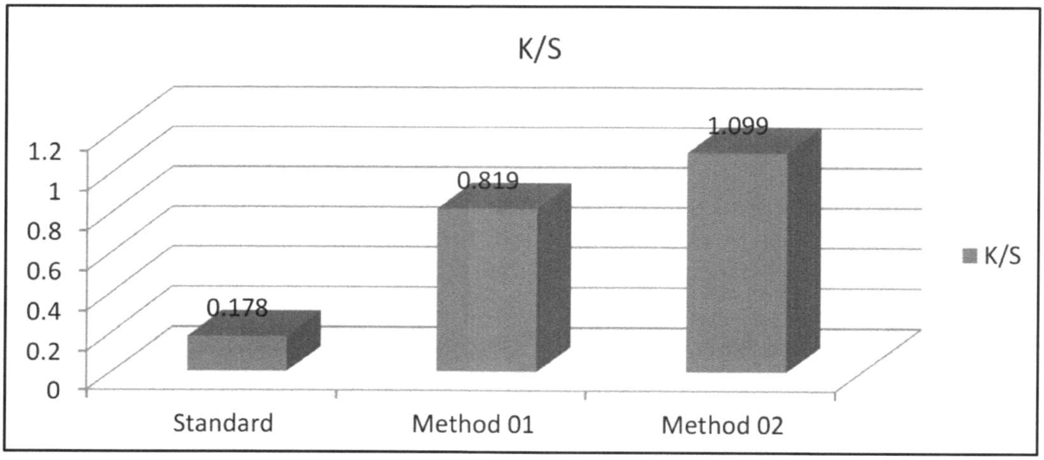

Fig. 7: K/S Values for Acid Dyes on Modified Polyester Fabric

Name of Sample	*Description*	*K/S Value*
Standard	Control sample/Normal polyester	0.178
Method 1	Dyed polyester sample (modified with NaOH	0.819
Method 2	Dyed polyester sample (modified using keratin extracts)	1.099

The polyester modified with keratin extracts gave highest K/S value compared to other two samples because of improved dyeability of fabric with acid dyes and hence good dyeing results were obtained.

3.5 Evaluation of Washing Fastness of Dyed Fabric

The fabric to be tested having standard size of 10 × 4 cm was placed between two adjacent undyed fabrics of same size, sewed along four sides to form a composite specimen.

The composite specimen was placed in a glass jar containing 5 g/l soap solution and 2 g/l soda ash solution, keeping material to liquor ratio 1:50. Jars were then closed and placed in Launder-O-meter.

Table 2: Washing Fastness Rating of Different Samples

Sr. No.	*Treatment*	*Rating*
1.	Untreated polyester fabric (dyed with acid dyes)	1
2.	Polyester modified with NaOH (dyed with acid dyes)	2–2.5
3.	Polyester modified with extracts of keratin (dyed with acid dyes)	3–3.5
4.	Untreated polyester fabric (dyed with disperse dyes)	3.5
5.	Polyester modified with NaOH (dyed with disperse dyes)	4
6.	Polyester modified with extracts of keratin (dyed with disperse dyes)	3

4. CONCLUSION

Treatment of polyester fabric with sodium hydroxide imparts changes in the properties of fabric expressed by changes in the weight loss, absorbency, etc. The modification of polyester with keratin extracts from wool showed further improvements in the polyester fabric with respect to dyeability especially with acid dyes. Untreated polyester fabric does not dye with acid dyes (only staining of fabric) and large amount of colour will be removed during washing. On the contrary, polyester fabric modified with keratin extract dyed with acid dyes gave good washing fastness and better colour yield with acid dyes as reflected from the K/S values of dyed samples. This modification leads to further scope of improvement and lay down foundation to improvise such methods of surface modification by which the unthinkable can be thought of in future.

REFERENCES

[1] Chinta, S.K. and Singh, R. (2012). Processing Problems of Polyester and its Remedies, *International Journal of Engineering Research & Technology*, 1(7), 1–3.

[2] Dave, J., Kumar, R. and Srivastava, H. (1987). Studies on Modification of Polyester Fabric I: Alkaline Hydrolysis, *Journal of Applied Polymer Science,* 33, 455–477.

[3] Gupta, A. and Sharma, S. (2016). Sustainable Management of Keratin Waste Biomass: Application and Future Perspectives, *Brazilian Archives of Biology and Technology*, 59, 2–4.

[4] Kardas, I., Lipp-Symonowicz, B. and Sztajnowski, S. (2009). Comparison of the Effect of Polyester Fibre Surface Modification Using Enzymes and Chemical Substances with Respect to Changes in Mechanical Properties, *Fibres and Textiles in Eastern Europe*, 4(75), 93–94.

[5] Kumawat, T., Sharma, A., Chandra, S. and Sharma, V. (2018). Keratin Waste: The Biodegradable Polymers, *Asian Journal of Applied Sciences*, 9(3), 151–155.

[6] Mowafi, S., Taleb, M. and El-sayed, H. (2018). Utilization of Proteinic Biopolymers: Current Status and Future Prospects, *Textile colour and Polymer Science*, 15(1), 16–22.

[7] Rajabinejad, H., Buciscanu, I. and Maier, S. (2016). Practical Ways of Extracting Keratin from Keratinous Wastes and by-Products: A Review, *Environmental Engineering and Management Journal,* 15(5), 1132–1136.

[8] Tavanai, H. (2009). A New Look at Modification of Polyethylene Terephthalate by Sodium Hydroxide, *The Journal of Textile Institute*, 10, 633–639.

[9] Vasantha, M. (1991). A Study on the Surface Modification of Polyester Fabrics by Alkaline Hydrolysis (Unpublished Ph.D. Thesis), *The Faculty of Science and Humanities, Anna University (Madras)*, 19–21.

Effect of Metallic Salts in Dyeing of Wool with *Acacia Arabica Wild* Bark Extract

Praveen Jha[1] and Bharat Patel[2]
Department of Textile Chemistry, Faculty of Tech. Eng, The M.S. University of Baroda, Gujarat
E-mail: [1]praveen05txch41@gmail.com; [2]bharatextile@gmail.com

ABSTRACT: *The word mordant (the Latin word) means 'to bite'. Most of the dyes obtained from natural resources possess a low affinity towards textile fibre. After mordanting fibre's pores open and dye absorption takes place. Mordant is having the ability to fix the dyes. Mordants may be natural, metallic salts, or tannins. Tannins are polyphenolic compound in dyes which is responsible for fixing onto the fibre. After mordanting, a Dye-mordant complex is formed and the hue of fabric change according to this complex. Our study aims to know the effect of mordanting on the dyeing of wool fabric. A natural dye extracted from the skin of Acacia Arabica Wild bark has been applied on the woolen substrate in the presence and absence of metallic salts (Mordant's). The application of the dye liquor is done using the pad-dry-steam technique. Meta-mordanting technique has been employed for the application of mordant. The remarkable effect of mordant has been visualized on the colour strength and fastness properties of the dyed sample. The application of Acacia Arabica Wild on woolen fabric by the pad-dry-steam technique of dyeing can be considered as an effective eco-option because it gives extremely good results with substantial minimization of the effluent load.*

Keywords: *Acacia Arabica Wild*, Metallic Mordant, Natural Dye, Padding Technique, Wool Fabric.

1. INTRODUCTION

In the last few years, people have become more aware of health and hygiene. Due to unbalance pollution our eco system gets sun control day by day. After some time, man may starve for even a drop of pure water and pure food (Ali *et al.*, 2010). The word pollution is defined by an imbalance between nature and environmental life cycle, due to living in habitant's activity. Water, Air & Land became polluted due to synthetic compounds used in the different manufacturing processes. In dyeing with synthetic dyes people used non-biodegradable, carcinogenic chemicals (Bhattacharyya, 2011). Wastewater disposal is another problem. Several times animals die by drinking such water and soil also get unfertile. Natural dye is a reasonable solution for these problems. But the limitation of natural dyes is limited sources, shade repeatability and cost. So, we must increase sources of natural dyes. Natural dyes can be obtained from any part of the plant viz., leaves, fruits, seeds, flowers, barks, and roots. Now a day's benefit of using natural dyes is additional UV-protection and antimicrobial properties.

Pros of natural dyes are that it can be dyed to cellulosic (cotton, viscose), synthetic (polyester) as well as wool-like protein fibre. Wool is an animal fibre having properties of air permeability, so it is used for the winter season. If renewable resources as fibres like wool and dyes like natural dyes will be used then the ecosystem will be balanced (Billmeyer & Saltzman, 1981). In this experiment, we have used natural dyes (*Acacia Arabica wild*) on wool with metallic salt ($CuSO_4$, $FeSO_4$ and $SnCl_2$) as mordant's which have less or no harsh effect. Mordants are responsible for opening up the pores in fibre and improve dye absorption (Cardon, 2007). Mordant react with fibre and dye and able to fix the dye (Gupta, 1999). After mordanting with different metal, dye-metal complexes are formed, which influence washing, rubbing, or light fastness of the fabric (Needles, 1986). Metallic salts and tannins are a different class of mordants (Ratnapandian *et al.*, 2011). In this experiment, we aim to assess the change in depth, tone, and fastness properties after the application of metallic mordents on wool fabric.

2. EXPERIMENTAL

Acacia Arabica wild is found all over India with the common name 'babool'. 12–20% tannin is found in babool bark. Babool bark *(Acacia arabica)* was collected from the Baroda region (India). Also found in Burma, Sri Lanka, Egypt, and Bangladesh. Padding mangle was used for padding wool fabric. Hot air oven was used for drying and curing (Manufacturer-Paramount). For washing fastness testing gyro wash machine (Manufacturer-James H. Heal & Co. Ltd, Halifax, England). Crockmeter was used for rubbing fastness testing (Manufacturer-James H. Heal & Co. Ltd, Halifax, England). For light fastness testing mega solo light fastness tester was used (Manufacturer-James H. Heal & Co. Ltd, Halifax, England). Color-eye software was used for color value testing.

3. EXTRACTION OF DYE

Collected bark of *Acacia Arabica*. Remove the outer skin of babool bark & cut into small chips. Dry & Grind this bark up to powder form. Soak these small, dried chips into cold water for 24 hours. Again, it boiled for 2 hrs yield a brown colour solution. Afterward take this powder with solvent Methanol. Then boil with solvents and filtered the solution with Whatman paper. Then keep it in a dark place for cooling. We got the brown colour of dye which is used for sample padding.

4. MATERIALS

Wool fabric has purchased from Surat textile market. The structure of the fabric is plane weave and G.S.M of fabric is 198 g/m^2, 62 ends/inch, 55 picks/inch. The fabric was washed with soda ash 2% and non-ionic soap 1% at 95°C for 30 minutes. After giving one hot wash at 70°C followed by cold wash and neutralization with green acid.

5. DYEING OF THE CONTROL SAMPLE—WITHOUT MORDANT

Now for energy conservation purposes, the preferred dyeing technique is a continuous manner (Patel *et al.*, 2003). In Exhaust dyeing only a small part of dyes retain or exhaust with substrate and remaining dyes will drain or waste. In continuous dyeing, we are applying the Pad-Dry-Cure technique. We are using *Acacia arabia wild* extracted dye on scoured wool fabric. Firstly 15 ml of dye extract. Dip the wool fabric for 10 minutes in dye liquor at 70°C. For proper level dyeing, 3 dip-3 nip padding process and certain concentrations of wetting agent require. Wet pickup is 70% for wool normal composition (Billmeyer & Saltzman, 1981). Following are the dyeing liquor recipe.

Dye	: 15% (Dissolved *Acacia arabia* Dye)
Wetting agent	: 2 gpl
Anti-migrating agent (sodium alginate)	: 2 gpl
Resist salt	: 1 gpl
Urea (hygroscopic agent)	: 5 gpl
Green acid	: 2–5 gpl (For maintain dye bath pH: 3.5–4.5)

Make the volume 100 ml by adding water.

After the 3 dip-3 nip padding, wool fabric is dried at 80°C & cured at 160°C for 90 seconds at hot air oven (Manufacturer – Paramount).

6. MORDANTING

Mordanting is required to fix the dye or improve fastness properties. There are three different techniques used for application of mordents (Billmeyer & Saltzman, 1981). Pre-mordanting, meta-mordanting, Post-mordanting. Among these three techniques, post mordanting gives better colour yield and fastness properties. So, we are using metallic salt ($CuSO_4$, $FeSO_4$ and $SnCl_2$) as a mordant by post mordanting technique.

6.1 Mordanting with Metallic Salts

(a) *Post-Mordanting with Copper Sulphate ($CuSO_4$)*

In this technique scoured wool sample is firstly padded with the same recipe as the control sample 3 dip - 3 nip and 70% expression afterward without dry, again pad with 10% $CuSO_4$ (10 gm in 100 ml) solution 2 dip – 2 nip and 70% expression. The wool fabric is dried at 80°C and cured at 160°C for 90 seconds in a hot air oven.

(b) *Post-mordanting with Ferrous Sulphate ($FeSO_4$)*

The scoured wool sample is firstly padded with the same recipe as the control sample 3 dip - 3 nip and 70% expression afterward without dry, again pad with 10% $FeSO_4$ (10 gm in 100 ml) solution 2 dip - 2 nip and 70% expression. The wool fabric is dried at 80°C and cured at 160°C for 90 seconds in a hot air oven.

(c) *Post-mordanting with Stannous Chloride ($SnCl_2$)*

Here also scoured wool sample is firstly padded with the same recipe as the control sample 3 dip-3 nip and 70% expression afterward without dry, again pad with 10% $SnCl_2$

(10 gm in 100 ml) solution 2 dip-2 nip and 70% expression. The wool fabric is dried at 80°C and cure at 160°C for 90 seconds in a hot air oven.

7. TESTING AND ANALYSIS

7.1 Colour Difference Measurement

For colour, the difference measurement of dyed samples is tested on method CIE *Lab*. The light source is D65-10. The equipment used for testing the dyed sample was a spectrophotometer. Color eye software was used for color value testing. The colour value i.e. L^* (–ve indicates darker and +ve indicates lighter), a^* (–ve indicates greener and +ve indicates redder), b^* (–ve indicates bluer and +ve indicates yellower), C* (–ve indicates duller and +ve indicates brighter) and H* indicates hue of substrate.

7.2 Testing of Colour Fastness to Washing

For assessment of colour fastness to washing of the dyed samples ISO 105-C10: 2006 (C2) method was performed at 60°C for 30 minutes with 5 gpl an ionic soap and 2 gpl soda. For washing fastness testing Gyro wash machine, manufacturer-James H. Heal & Co. Ltd, Halifax, England our uses. Dry the fabric. Evaluate the samples at the AATCC grey-scale.

Two types of greyscale available

(a) Change in colour,
(b) Colour stain on adjacent fabric.

7.3 Testing of Colour Fastness to Rubbing

Crock meter used for rubbing fastness testing, manufacturer-James H. Heal & Co. Ltd, Halifax, England. ISO 105-X12-2001 method was preferred. Generally, ten cycles rubbed on dry and wet condition with adjacent fabric. Afterward, the rating was given on the greyscale.

7.4 Testing of Light Fastness

For light fastness testing Megasol light fastness tester used, manufacturer-James H Heal & Co. Ltd, Halifax, England. ISO 105 BO2 method was applicable. Testing time was 20 hrs. Than given the rating on the greyscale.

8. RESULTS AND DISCUSSION

8.1 Effect of *Acacia Arabica wild* Bark Extract on Wool Fabric Applied by Pad-Dry-Cure Technique

The extract of *Acacia Arabica Wild* bark extract was applied on wool fabric by a continuous Pad-Dry-Cure technique. Figure 1 represents the photographic images of the wool samples dyed with and without the use of metallic mordant.

Control Sample
(Wool dyed without mordant)

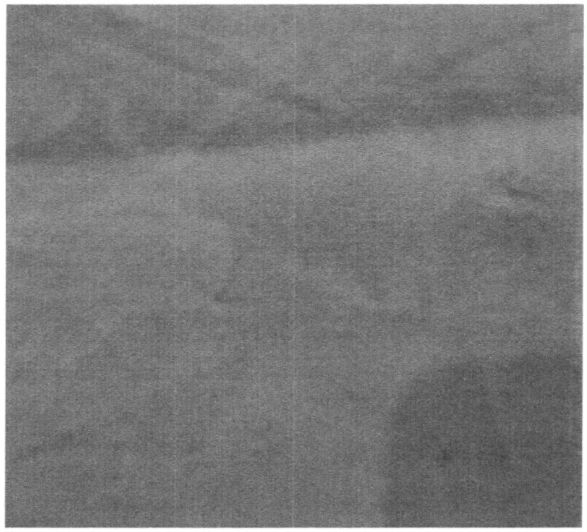

Wool dyed with $FeSO_4$ as mordant

Wool dyed with $CuSO_4$ mordant

Wool dyed with $SnCl_2$ mordant

Fig. 1: Wool Fabric Samples Dyed with the Skin Extract of *Acacia arabica wild* using a Pad-dry-cure Technique

It can be seen from the actual photos, that the skin extract of the plant can directly be applied to wool fabric. With the use of metallic salts alter the colour of the dyed samples. Different shades viz., beige, dark brown, copper colour, light brown can be produced on wool fabric respectively with iron, copper, and tin salts.

8.2 Effect of Different Metallic Salt on Color Value of Shade

By observing the results in Table 1, it may be seen that wool fabric can be dyed with *Acacia Arabica wild* without the use of any mordant. But the depth of the samples was found to be improved by using different metal salt. Ferrous sulphate gives the highest depth.

Table 1: Color Value of Wool Fabric Dyed by *Acacia arbica wild* Bark with or Without Metallic Mordant

Sample	*DL**	*Da**	*Db**	*DC**	*DH**	*Ecmc*	*Remark*
Without mordanted sample	–	–	–	–	–	–	Taken as standard
Mordanted with $CuSO_4$	–11.65D	–8.2G	–1.28B	–2.14D	8.02G	13.09	Darker, greener, duller
Mordanted with $FeSO_4$	–16.71D	0.14R	7.19Y	6.86B	2.16G	8.72	Darker, yellower, brighter
Mordanted with $SnCl_2$	3.42D	1.31R	5.9Y	5.98B	0.87B	4.56	Darker, yellower, brighter

Standard sample description: L* = 69.70, a* = 5.68, b* = 13.85, c* = 14.97, h* = 67.7

8.3 Effect of Different Metallic Salt on Fastness Properties of Dyed Wool by *Acacia Arabica Wild*

From Table 2, it is observed by using metallic salt mordant on wool dyed by *Acacia arabia*, light fastness improved. Washing fastness increases up to ½ rating. In case of wet rubbing fastness decrease slightly up to ½ rating. Without mordant sample also fastness is good due to tannin present in *Acacia Arabica wild.*

Table 2: Fastness Properties of Wool Fabric Dyed by *Acacia arabica wild* Bark with or Without Metallic Mordant

Sample	*Wash Fastness*			*Rubbing Fastness*		*Light Fastness*
	Colour Change	*Staining on*		*Dry Rubbing*	*Wet Rubbing*	
		Wool	*Cotton*			
Without mordanted sample	4	4–5	4	4–5	4	4
Mordanted with $CuSO_4$	4	4	3–4	4	3–4	5–6
Mordanted with $FeSO_4$	4–5	4	4–5	4	3–4	5–6
Mordanted with $SnCl_2$	4–5	4	4–5	4	3–4	5

9. CONCLUSION

Wool fabric can easily be dyed with natural dye extracted from *Acacia Arabica wild*, with and without mordanting by using metallic salt ferrous sulphate, copper sulphate and stannous chloride. Overall depth increases with metallic salt mordant. In the case of $FeSO_4$ and $SnCl_2$ mordant depth and brightness both increase. But in case of $CuSO_4$ mordant depth increases but brightness decrease. Mordanting with all metallic salt improves in color fastness to washing. Mordanting with ferrous sulphate gives best light fastness and colour depth.

REFERENCES

[1] Ali, A., Ali, S., Saleem, H. and Hussain, T. (2010). *Asian Journal of Chemistry,* 22(9), 7065–7069.

[2] Bhattacharyya, N. (2011). Natural Dyes for Textiles and Their Eco-Friendly Applications, *Journal of Fibre Bioengineering and Informatics*, 4(3), 267–76.

[3] Billmeyer, F.W. and Saltzman, M. (1981). *Principles of Color Technology, Second Edition, A Willey Interscience Publication.*

[4] Cardon, D. (2007). *Natural dyes,* Archetype London.

[5] Gupta, D. (1999). *Colourage*, 46, 35.

[6] Needles, H.L. (1986). Textile Fibres, Dyes, Finishes, and Processes, *A Concise Guide, Noyes Publications,* Park Ridge, New Jersey, U.S.A., 58–63.

[7] Patel, B.H., Agarwal, B.J. and Patel, H.M. (2003). Novel Padding Techniques for Dyeing Babool Dye on Cotton, *Colourage*.

[8] Rajeswari, V. (2020). Application of Natural Dye from Babool Bark on Cotton Fabric Using Mordants, *IJSDR*, 5, Issue 2.

[9] Ratnapandian, S., Fergusson, S.M., Wang, L. and Naebe, M. (2011). Effect of Atmospheric Plasma on Pad-Dyeing Natural Dyes on Wool, *Journal of Fibre Bioengineering and Informatics*, 4(3), 267–76.

Antimicrobial Properties of Cotton Finished with Natural Biocide and Chitosan

J.N. Shah[1] **and Rashmi D. Pachauri**[2]

Department of Textile Chemistry, Faculty of Tech. and Eng.,
The M.S. University of Baroda, Gujarat, India
E-mail: [1]jayendra.n.shah-tchem@msubaroda.ac.in; [2]2468rashmi@gmail.com

ABSTRACT: *By imparting novel antimicrobial treatments onto cotton textiles using natural products is a breakthrough for textile users to compete in the tough global marketplace. This will add values to textile in the direction of protection and safety. This research paper is directed towards suitable finishing formulation of Natural Biocide and Chitosan on cotton fabrics by exhaust application method to evaluate its use as a durable antimicrobial finish to meet functional and environmental demands to extend their high impart applicable properties. The morphology by surfacing electron microscopy (SEM) shows few fibrillation and surface coating on fabric surface. The antimicrobial activity of our formulation has also been evaluated qualitatively by Parallel Streak Method and quantitatively by AATCC 100 test method. Treatments showed washing durability up to 25 washes.*

Keywords: Antimicrobial Activity, Chitosan, Cotton Fabric, Morphology, Natural Biocide.

1. INTRODUCTION

Growth of microbes on cotton textiles results into bad odour, deterioration in quality and pigmentation; this affects both product and the wearer. Antimicrobial finishes on textiles not only avoid and control infection but also reduces odour, pigmentation and bio deterioration (Kavitha *et al.*, 2006). There are two types of antimicrobials available such as non-leaching antimicrobials and leaching antimicrobials. Nowadays, synthetic agents like quaternary ammonium compounds and nano materials are available for antimicrobial finishing of textiles (Reshma *et al.*, 2018). These synthetic agents cause environmental issues and synthesis of nano particles is a very expensive process (Kolli *et al.*, 2018). Use of chitosan and natural biocides for antimicrobial finishes can be used for clothing and medical supplies. Natural biocides are very effective products against bacteria and fungi (both gram positive and negative). As high amount of moisture, oxygen, and nutrient transport are the superior mediums for the growth of microorganisms on the fibres. The natural bioactive agents used in antimicrobial finishing are very effective against undesirable organisms as they are non-toxic, safe, and biodegradable but should also have long durability (Hooda *et al.*, 2013). The natural biocides have no adverse consequences on environment and are being used widely to overcome the problems caused by harmful synthetic agents on environment.

1.1 Biocides

The perfect technique of preventing microbial degeneration is to use natural materials, which are naturally resistant to attack. Another method is to apply antimicrobial chemicals known as "biocides which are generally included into the finished textile product (Malpani, 2013). So far, no additive agent has been unearthed, which provides neither complete safety nor is without some drawback. Perfect biocides include following requirements (Sinta *et al.*, 2011).

- Efficient against a large range of microorganisms, especially bacteria and microfungi.
- Operative during the life of the product.
- Of low mammalian toxicity and non-toxic to humans at the concentrations used.
- Lacking colour and odour.
- Influential at low concentrations.
- Not expensive and easy to use.
- Resistant to sunlight and percolating from the fabric.
- Fabric handle and health and unaffected.
- Adaptable with water-repelling and flame-reducing agents, dyes, and other textile accessories.
- Does not intensify the fabric to destruction by light or other effects.

1.2 Natural Biocide: *Azardirachta indica* (Neem)

In various countries Neem (*Azardirachta indica*) has gained a lot of importance due to its natural antibacterial properties and used as healthcare formulations worldwide. Neem belongs to mahogany family Meliaceae (Dey *et al.*, 2010). The major property includes its non-toxicity to human body. Neem tree has more than 300 active compounds obtained from its various parts such as leaves, seeds and bark such as limonoids. Since in ancient India, Neem has been used as an herbal medicine for various human requirements, also has been found in Ayurveda as traditional medicine (Gupta & Bhaunik, 2007). The Neem leaves extract has pest repellent properties and as herbal pesticides is very effective against bacteria and fungi (both gram positive and negative).

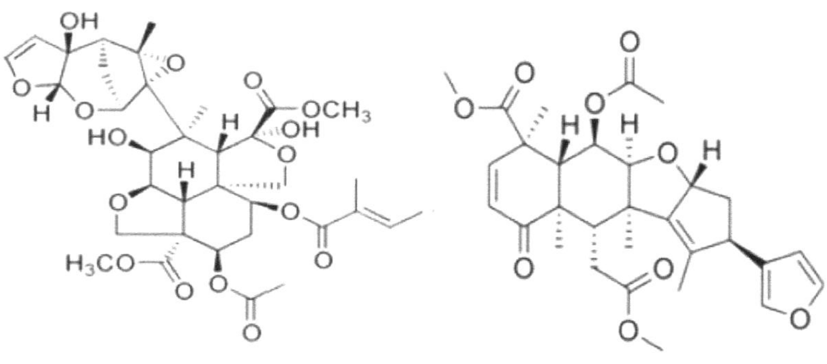

Fig. 1: Azadirachtin Nimbinin-Active Limonoids in Neem

1.3 Chitosan

In nature, Chitin is 2nd polysaccharides made of linear chain of acetyl glucosamine groups (Dewick, 2009). Chitosan is obtained by removing acetyl glucosamine group (CH_3–CO) and by leaving the active amine group (NH_2) as shown in figure below. Chitosan has natural antimicrobial properties as it deforms and shrinks cell membrane of bacteria which leads to death of the microbes. Chitosan at the surface of bacteria binds with negatively charged residues of molecules, which further controls the growth of bacteria (Choudhury *et al.*, 2012). Chitosan is biodegradable, polycationic in nature, has no toxicity and excellent wound healing properties (Landage & Wasif, 2012).

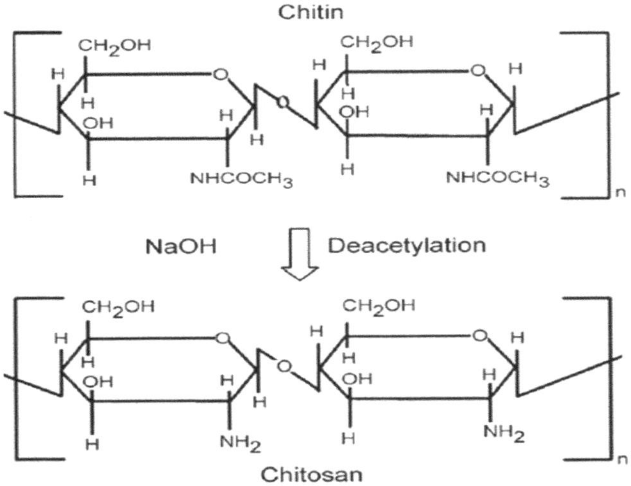

Fig. 2: Chemical Structure of Chitin and Chitosan

2. MATERIALS AND METHOD

2.1 Processing of Natural Biocide (NB)

The natural biocide (medicinal herb) collected for the study was *Azadirachta indica*. This herb was selectively collected from nearby areas of my office campus at Ahmedabad. Firstly, the collected neem leaves were thoroughly washed under tap water and kept for drying under sunlight for three days to reduce moisture. These dried leaves were then grinded to a very fine powder and sieved. 20 grams of neem leaves powder was suspended in 100 ml methanol and incubated overnight at room temperature. The supernatant was then filtered by using Whatman No. 1 filter paper and was further used as neem extract for the treatment of cotton fabrics.

2.2 Procedure for Chitosan Solution (CS) Preparation

Chitosan was collected from Meghmani Dyestuffs Pvt. Ltd., Ahmedabad. Chitosan solution (1%) was prepared by adding 1 g of chitosan solution was dissolved first with 1% acetic acid (1 ml of acetic acid in 99 ml of distilled water) and stirred at room temperature for 24 h till a fine homogeneous suspension was formed. The solution was then kept overnight at stand-still condition to remove air bubbles formed during stirring.

2.3 Procedure for Neem Chitosan Complex (NCC) Preparation

The complex was prepared at room temperature with constant stirring of 15 ml extracted Neem solution along with prepared chitosan solution (added dropwise). A neem chitosan complex (NCC) of blackish-brown colour was obtained which precipitated at the bottom of the glass beaker.

2.4 Fabric Treatment

A fine-medium weight 100% bleached cotton fabric was collected from the local market of Ahmadabad and used for the application purpose. Fabric was also washed well in 0.2% Na_2CO_3 solution at 80ºC for 30 minutes with material to liquor ratio of 1:50, Rinsed thoroughly under tap water and air dried before use.

2.5 Application Process

Properly cleaned and dried cotton fabric was firstly prepared for further procedures by cutting it to a size of 25 × 25 cm. It was then immersed in different solutions with 5% conc. of Natural Biocide and 5% conc. of Chitosan solution in separate beakers. The similar size of sample was then treated with the prepared solution of (NCC) having same conc. of 5%. By using Exhaust method with laboratory water heating bath, these processes were carried out at the temperature of 90°C with pH 5 for the time period of 60 mins., maintaining MLR of 1:30. The finished samples were then air dried.

2.6 Antimicrobial Assessment Methods for Finished Cotton Fabrics

These finished samples were then further assessed for antibacterial activity by AATCC 147–2004 method and for durability of antibacterial finish by AATCC 100–2004 method using a launder-O-meter (Test organisms were *Staphylococcus aureus, Escherichia coli*).

2.7 Qualitative Test (Parallel streak method)

Antibacterial activity by the AATCC 147–2004 method is a qualitative test. This method is generally used to detect antibacterial activity of diffusible antimicrobial agents on treated textile materials. This method is used for initial screening of bacteriostatic activity shown by different antimicrobial agents applied on cotton materials.

2.8 Quantitative Bacterial Reduction Test (AATCC test method 100–2004)

The washed samples were further subjected to AATCC 100 antimicrobial test for the evaluation of the degree of antibacterial activity of antibacterial finishes on cotton finished material.

2.9 Test for Washing Durability

After several washed cycles, these finished fabrics were than further tested for anti-microbial activity. The durability of antibacterial finish to washing by using launderometer was carried out as per ISO Test No. 1. Treatment time was of 30 minutes with 5 gpl conc. Non-ionic detergent was used as soaping agent at 70°C; with MLR of 1:30. After proper washing and drying these samples were then further sent for durability of antibacterial activity tested by AATCC 100–2004 method for up to 25 laundering wash cycles.

2.10 Test for Scanning Electron Microscopy

Testing of Scanning Electron Microscopic (SEM) allows visualization of surface features of a sample by scanning through an electron beam. SEM has better resolution capability and depth of field than a light microscope. Therefore, good quality three-dimensional like images was obtained. The morphology of surface of the control sample, NB, CS, NCC treated fabrics was studied using SEM (JEOL JSM IT 200) machine, magnification 10,000X and resolution view of about 10 nm. The surface view and cross-sectional view of the samples can be easily observed.

3. RESULTS AND DISCUSSION

The antibacterial activity of finished fabric was evaluated qualitatively using parallel streak AATCC 147–2004 method and quantitatively using AATCC 100 test method in which the percent reduction in number of bacterial colonies in different treated sample as compared to the control sample gives the antibacterial activity of the fabric.

The unwashed samples were subjected to parallel streak method to evaluate the antibacterial activity. In general, it was found that as the NCC treated samples have maximum zone of inhibition as compared to NB and CS treated samples as shown in Figure 3. As both chitosan and neem themselves are very good antimicrobial agents, they result in an increased zone of inhibition, when used together in a single complex (NCC).

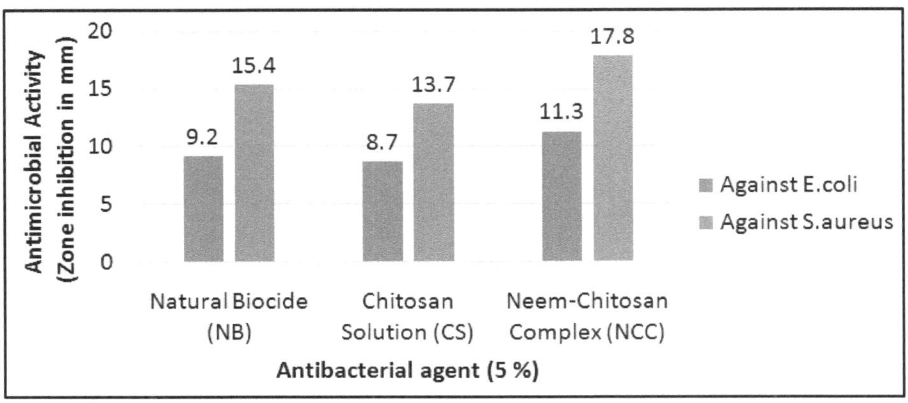

Fig. 3: Comparison of Treated Cotton Samples Showing Qualitative Test Antibacterial Activity (AATCC147)

Hence on comparison, the complex of both neem-chitosan shows an increased zone of inhibition when compared to neem and chitosan treated fabrics treated separately.

The washed samples were subjected further to AATCC 100 method and it was found that NCC treated samples have highest Bacterial reduction (%) against both *E. coli* and *S. aureus* as shown in Figure 4. On the other hand NB and CS treated samples showed high >85% antibacterial activity as compared to antibacterial activity for control sample which was almost Nil against both *E. coli* and *S. aureus*.

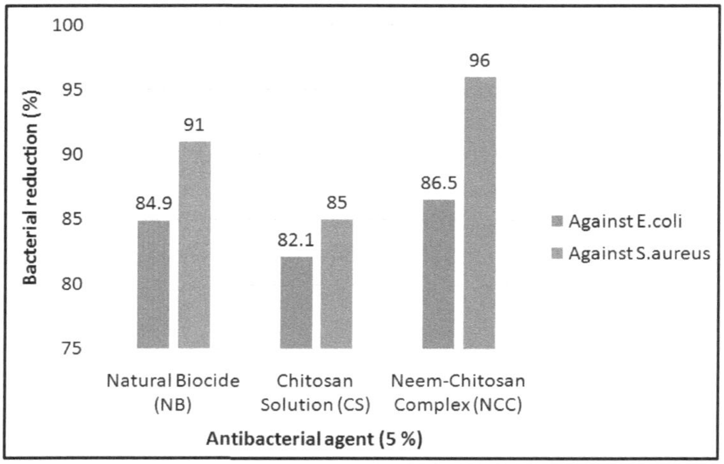

Fig. 4: Comparison of Treated Cotton Samples Showing Quantitative Bacterial Reduction (AATCC 100)

3.1 Test Results of Washing Durability

All the cotton antimicrobial finished samples were subjected to laundering procedures. The test results of durability to washing for the cotton treated samples showed reduction in percentage of antibacterial activity after 25 laundering cycles as shown in Figure 5.

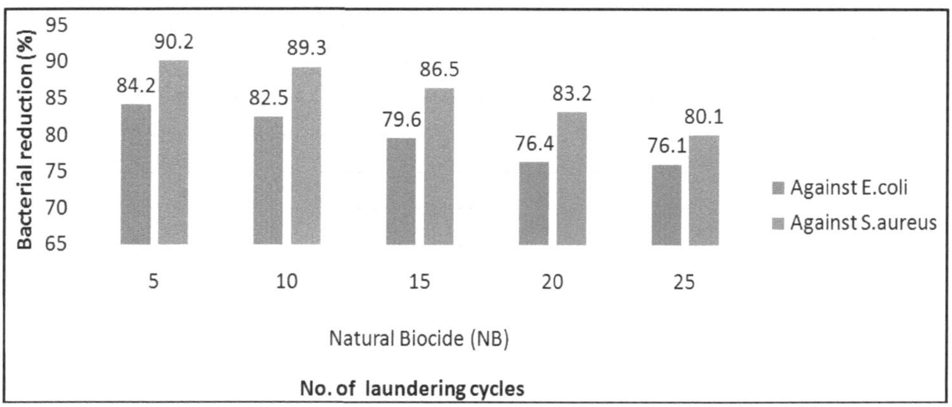

Fig. 5(A)

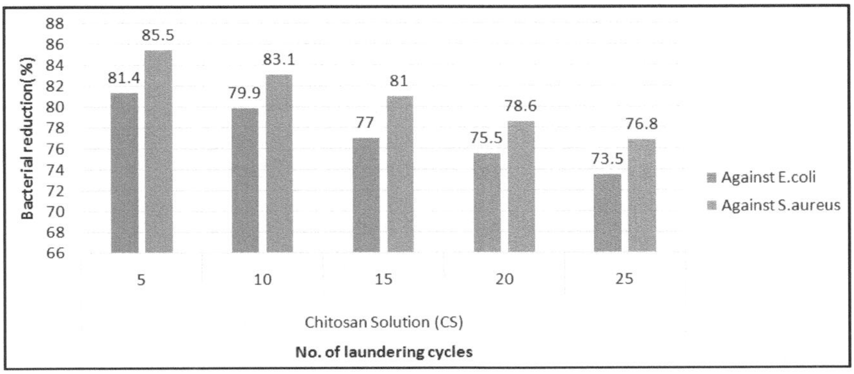

Fig. 5(B)

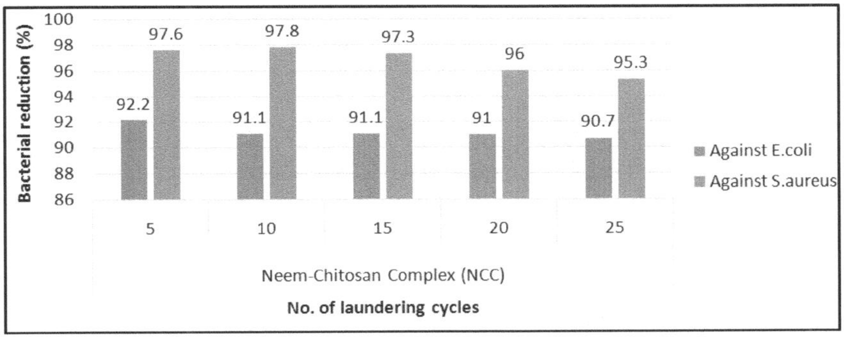

Fig. 5(C)

Fig. 5: Wash Durability of the Treated Fabrics (A, B, C)

3.2 SEM Analysis

Antibacterial properties of NCC, NB and CS fabrics and the bacterial growth on these treated fabrics were further examined by scanning electron microscopy (SEM). SEM micrographs clearly indicates that the NCC act as an effective bactericidal agent on the cotton fabrics.

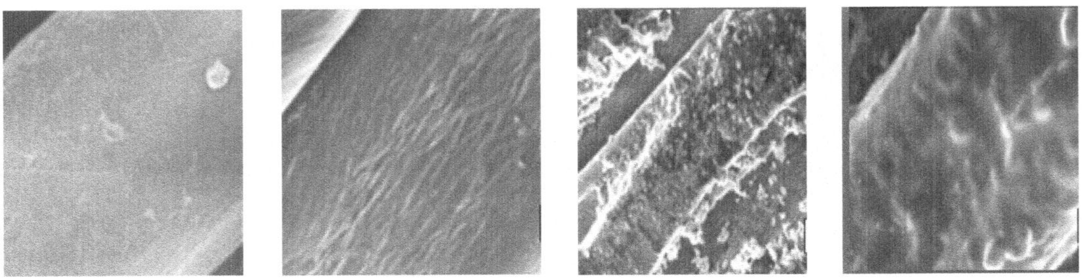

(a) (b) (c) (d)

Fig. 6: SEM of (a) Control, (b) Natural Biocide (NB) (c) Chitosan Solution (CS) and (d) Neem-Chitosan Complex (NCC) Finished Cotton Fabrics

4. CONCLUSION

The above results have been compared in order to conclude the best optimized recipe. The sample treated with the complex; 5% of Natural biocide and Chitosan gave the best results as compared to samples treated separately. They showed excellent antimicrobial activity when tested quantitatively and qualitatively. Durability to washing for anti-microbial cotton finished with the complex of NCC was 95% after 25 washes. Also, samples treated with Natural Biocide and Chitosan Solution separately gave high antibacterial property when compared to control sample which was almost Nil against both *E. coli* and *S. aureus.*

REFERENCES

[1] Choudhary, S., Sharan, L. and Sinha, M.P. (2012). Phytochemical and Antimicrobial Screening of Psidiumguajava L. Leaf Extracts against Clinically Important Gastrointestinal Pathogens, *Journal of Natural Prod. Plant Resource,* 2(4), 524–529.

[2] Dewick, P.M. (2009). *Medicinal Natural Products: A Biosynthetic Approach*, 3rd Edition, West Sussex, John Wiley & Sons ltd.

[3] Dey, S.K., Banerjee, D., Chattopadhyay, S and Karmakar, K.B. (2010). Antimicrobial Activities of Some Medicinal Plants of West Bengal, *International Journal of Pharma and Bio Sciences,* 1(3), 1–10.

[4] Gupta, D. and Bhaunik, S. (2007). Review Article: Antimicrobial Treatments for Textiles, *Indian Journal of Fibre & Textile Research*, 32, 254–263.

[5] Hooda, S., Khambra, K., Yadav, N. and Sikka, V.K. (2013). Eco-friendly Antibacterial Finish for Wool Fabric, *Journal of Life Sciences*, 5(1), 11–16.

[6] Kavitha, T., Padmashwini, R., Giridev, R. and Neelankandan, R. (2006). Antimicrobial Finishes for Textiles from Plants, *Indian Textile Journal*, 36, 4–14.

[7] Kolli, V., Ginjupalli, G., Ghosh, M., Manne, N., Hambuchen, M., Kevin, M. and Eric, R. (2018). Review of Antimicrobial Textile Finishes, *International Journal of Textile Science and Engineering,* DOI: 10.29011/IJTSE-124/100024.

[8] Landage, S.M. and Wasif, A.I. (2012). Nanosilver—An Effective Antimicrobial Agent for Finishing of Textiles, *International Journal of Engineering Sciences & Emerging Technologies*, 4 (1), 66–78.

[9] Malpani, S. (2013). Antibacterial Treatment on Cotton Fabric from Neem Oil, Aloevera and Tulsi, *International Journal of Advance Research in Science & Engg.,* 2 (Issue No. 7).

[10] Reshma, A., Brindha, V. and Amutha, K. (2018). Sustainable Antimicrobial Finishing of Fabrics Using Natural Bioactive Agents, *Int. H. Life Sci. Pharma Research,* Oct; 8(4), (L) 10–20.

[11] Sinta, T., Agrawal, Y. and Kapoor, R. (2011). Antimicrobial Finish—A Review, *Indian textile Journal*, January, 69–74.

Section-III

Garment and Fashion

A Study on Impact of COVID-19 on Purchase of Fashion Apparel

Pradeep Joshi
Amity University, Uttar Pradesh, India
E-mail: pjoshi2@amity.edu.in.

ABSTRACT: *Apparel Industry is one of the largest net foreign exchange earners for India. It provides employment to second largest number in country and contributes 4% of India's gross domestic product (GDP) and constitutes 15% of the country's export earnings. Textile and Apparel sector is most dynamic manufacturing sector where eternal demand for change keeps Fashion alive. Food, clothing, and home will always constitute basic and foremost requirement of society however Purchase pattern of Fashion Apparel changes with propensity to buy which is dependent upon state of economy. Size of Indian Fashion and Apparel market has considerably increased in the last decade and witnessed increase in demand to fulfill changing requirement of customers. India has large a number of retail outlets of various formats besides presence of leading International fashion and apparel brands. Easy access to international trends of Fashion and Apparel due to world being global village and increasing media penetration has contributed in increasing the demand in the year 2019–2020; which can also be called as Covid era has affected purchase pattern of Fashion Apparel items primarily due to work from home becoming a buzz word for some and loss of income, jobs etc. in market has impacted disposable income of target customers. Purchase pattern has changed as fashion apparel became less priority item in current time. This study is to explore the impact of COVID-19 on purchase pattern for fashion apparel and to understand current purchase behavior for fashion and apparel. This understanding will be useful for Fashion Apparel manufacturers, brands and retailers for making their market planning.*

Keywords: Covid-19, Impact, Preparedness, Purchase of Fashion and Apparel, Retailers.

1. INTRODUCTION

Indian apparel and textile sector is very critical to the Indian economy, being the highest net foreign exchange earner for the country. The competitiveness of fashion industry is determined by combination of various factors i.e. cost, supply chain, vertical integration, timely delivery, product adaptation and development as per requirements of target customers.Though India has emerged as one of the main textile and clothing exporting countries, low and stagnant market share in the world trade, comparatively lower unit value realization, lack of presence in high value segment and aggressive performance of other Asian countries are concerns of Indian Industry (Joshi, 2006).

India has a large and growing domestic market for fashion and apparel. With increase in disposal income, changing lifestyle and increased influence and access to media has influenced purchase pattern of customers. Role of media is important as it provides easy access to latest trends in fashion which makes customers buy more or increase consumption of fashion and apparel. Brands/Retailers of fashion apparel have encashed upon it with concept of speed and changing offering quite frequently which helped industry to grow in the past.

Latest trends, design, fit; brand is key criterion for purchase of fashion apparel. Price has always played an interesting role in purchase of fashion apparel as customers on one hand compare prices for utility or basic necessity items but for fashion apparel price and brand value determine status. Luxury brands, retailers, designers' brands have grown up due to exclusivity in design. There is correlation between price and perceived image of brand, label, retailer and designer. Fashion apparel purchase decision has been influenced by celebrity, stars, icons, sports person. Perceived quality is also associated with the price for customers. There is not much role of input cost in the price of premium fashion apparel products.

There is increasing global competition in fashion market; making industry to work out strategies for sustaining or increasing market share. It is important to understand how Indian fashion industry is responding to requirements of Industry 4.0 and what action plan is being made by them to remain competitive globally.

COVID-19 has put up stiff challenge for fashion market as slower economy, closed marketplaces and changed lifestyle of customers due to work from home has becoming new normal and lesser outings/social gatherings etc. are fast changing demand pattern. Various aspects including Production, retail of fashion industry is affected and hence it's important to understand impact of COVID-19 on purchase of fashion apparel. This will enable stakeholders for optimum planning in difficult and changed time.

2. RESEARCH METHODOLOGY

Research with sample size of 100 was undertaken to study impact of covid'19 on purchase pattern of target customers for buying fashion apparel. Further to it, Research with focus on ascertaining Preparedness of Indian Fashion Industry to face challenges of Industry 4.0 are studied by extensive desk research and primary research with interview of sample representatives 50 industry members comprising of apparel manufacturers, retailers, designers etc.

(A) Findings of Target Customer Survey
(B) Sample description—personal information.

2.1 Age

61.6% of respondents are from 18–25 year age group followed by 27.3% respondents from 25–35 yr age group. Hence this study primarily covers 18–35 yrs age group with approximately 89% representation in study.

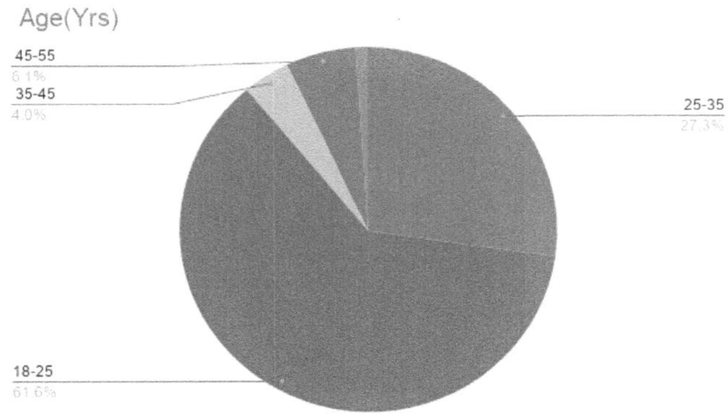

Fig. 2.1: Age

2.2 Gender

80% of respondents are female followed by 20% being male.

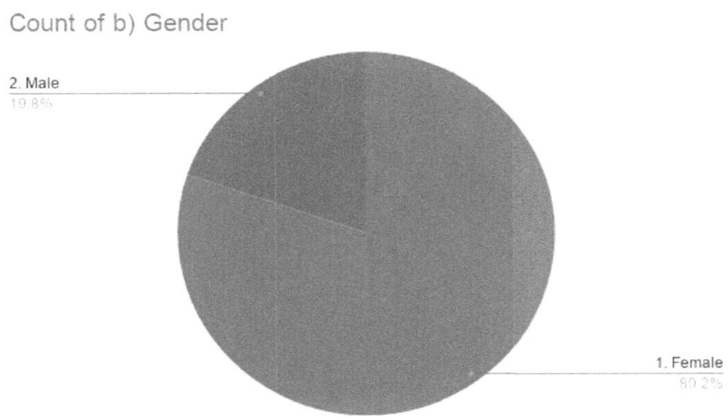

Fig. 2.2: Gender

2.4 Family Income (per month in Rupees)

There is almost equal representation of response from all income (family) groups above ₹ 50000/pm.

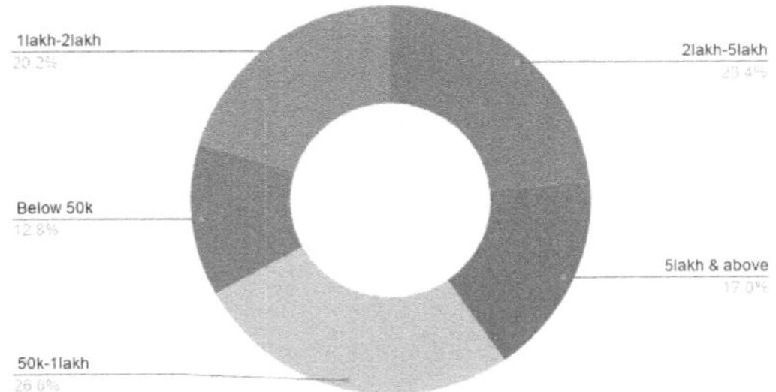

Fig. 2.3: Family Income (per month in Rupees)

2.5 Change in Purchase of Fashion Apparel

More than 85% respondents have responded that their purchase pattern has been affected by covid'19.

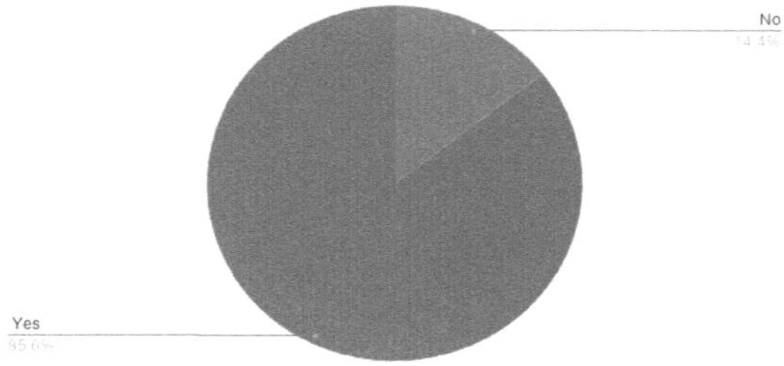

Fig. 2.4: Change in Purchase of Fashion Apparel

2.5 Fashion Apparel (pieces) Purchased in Last Year (Jan.–Dec. 2019)

39.4% respondents purchased 6–12 pcs of fashion apparel last year followed by 23.2% respondent's purchasing 12–24 pcs.

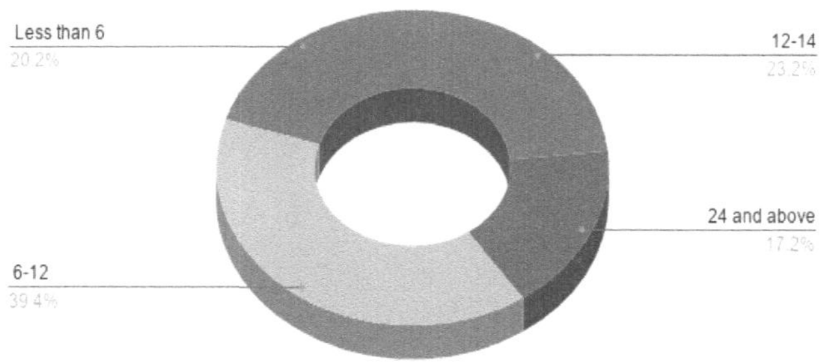

Fig. 2.5: Fashion Apparel (pieces) Purchased in Last Year (Jan.–Dec. 2019)

2.6 Fashion Apparel Purchased in Last One Year (Jan.–Dec. 2020)

44.9% respondents purchased less than 6 pieces of fashion apparel this year followed by 33.7% respondents purchasing 6–12 pieces of fashion apparel this year.

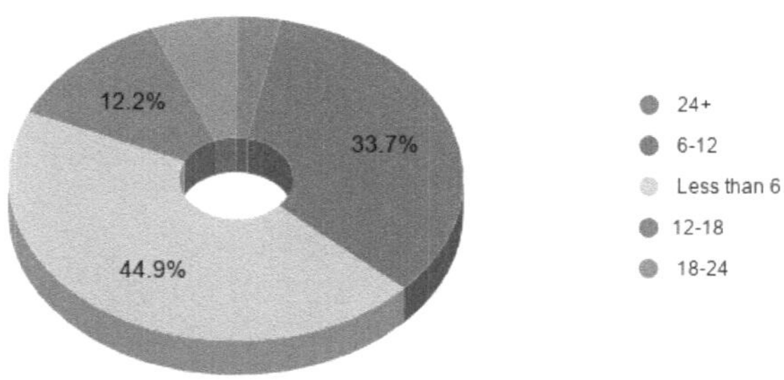

Fig. 2.6: Fashion Apparel Purchased in Last One Year (Jan.–Dec. 2020)

2.7 Preferred Mode of Purchase of Fashion Apparel Prior to Covid

Online as well as offline were equally preferred mode of purchase prior to covid'19. Majority of respondents (52.5%) preferred both as per convenience.

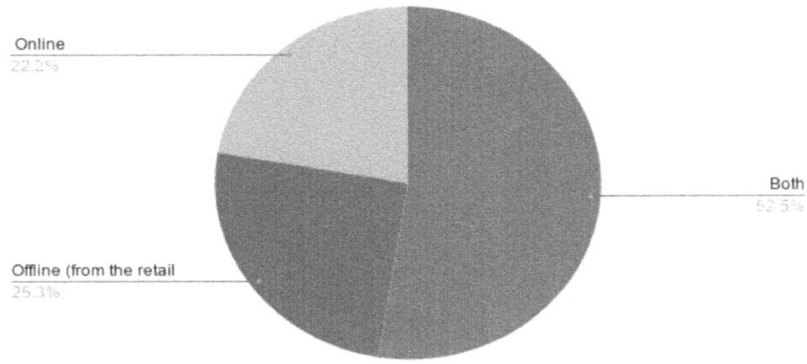

Fig. 2.7: Preferred Mode of Purchase of Fashion Apparel Prior to Covid-19

2.8 Preferred Mode of Purchase during Covid-19

Online is preferred as mode of purchase of fashion apparel by 72.7% respondents. Only 8% respondents preferred to purchase offline.

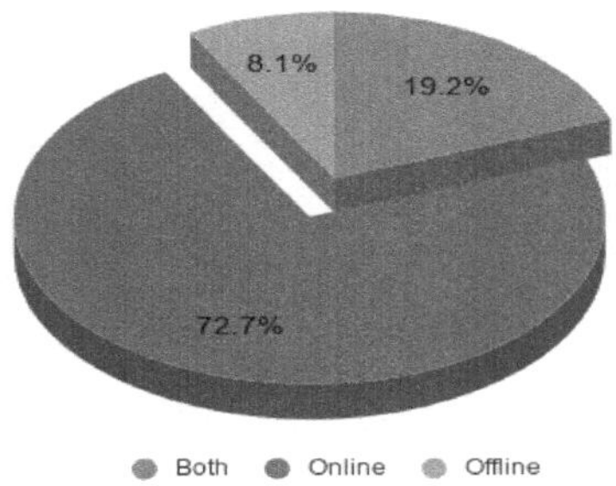

Fig. 2.8: Preferred Mode of Purchase during Covid-19

2.9 Willingness to Spend for Purchase of Fashion Items

Majority (more than 60%) of respondent's willingness for purchase of fashion apparel is lesser viz a viz earlier period.

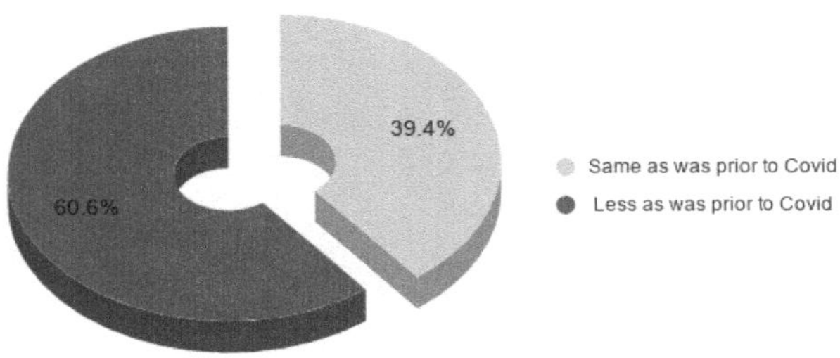

Fig. 2.9: Willingness to Spend for Purchase of Fashion Items

2.10 Occasion for Purchase of Fashion Apparel

Day to day use (40.4%) followed by family function social gathering (36.4%) are key reason for purchase of fashion apparel now.

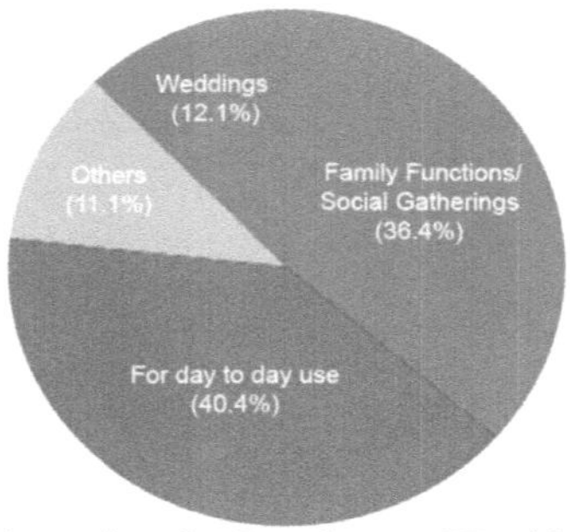

Fig. 2.10: Occasion for Purchase of Fashion Apparel

2.11 Influence of Social Media when Purchasing Fashion Apparel

Instagram (54.5% respondent) is major influencer for purchase of fashion apparel followed by other social media.

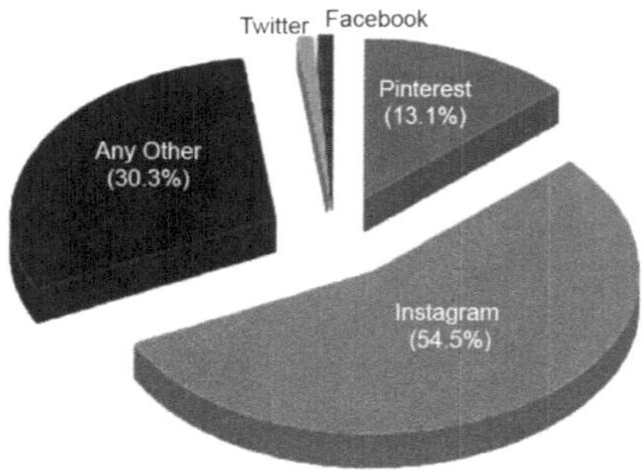

Fig. 2.11: Influence of Social Media when Purchasing Fashion Apparel

2.12 Factors Influencing Purchase of Fashion Apparel

Influence of brand is most important while purchasing fashion apparel.

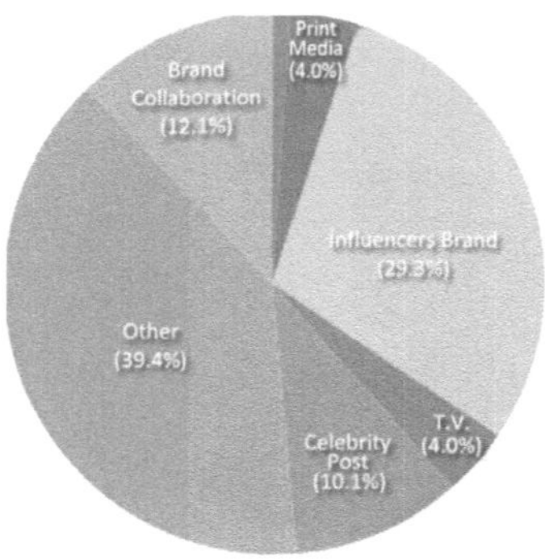

Fig. 2.12: Factors Influencing Purchase of Fashion Apparel

2.13 Purchase Pattern of Apparel Mostly Prior to Covid?

Majority (78.8%) of respondents were purchasing formal clothes prior to covid'19.

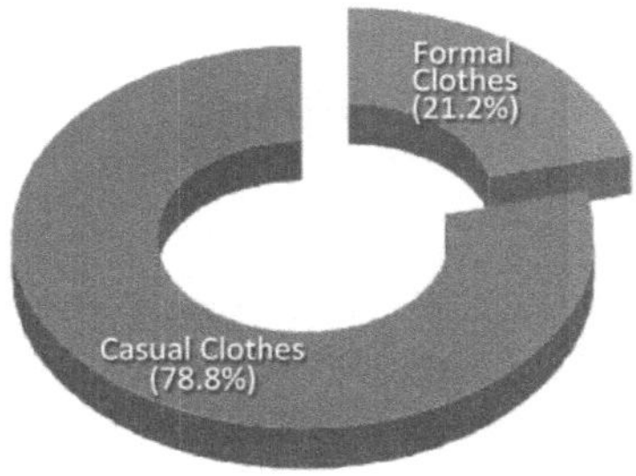

Fig. 2.13: Purchase Pattern of Apparel Mostly Prior to Covid

2.14 Purchase Pattern for Fashion Apparel Now

Most (89.7%) of respondents are purchasing casual fashion apparel now. Only a few are purchasing formal clothes now.

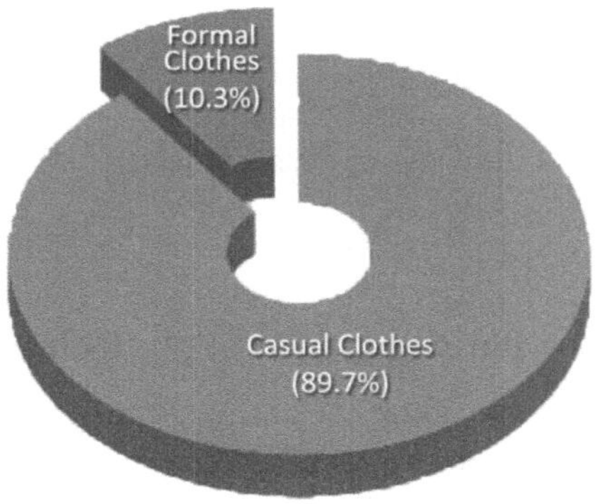

Fig. 2.14: Purchase Pattern for Fashion Apparel Now

2.15 Next Purchase of Fashion Apparel (Menswear)

Jackets/sweaters (33.3%) followed by T-Shirts (14%) & Shirts (11.8%) are preferred next item of purchase of respondents.

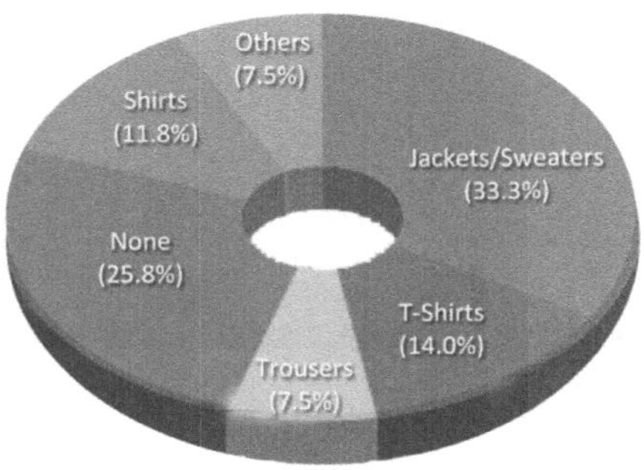

Fig. 2.15: Next Purchase of Fashion Apparel (Menswear)

2.16 Next Purchase of Fashion Apparel (Womenswear)

Jackets, Sweaters (43.6%) is most preferred item of purchase of womens wear for next purchase of respondents followed by dresses (22.3%).

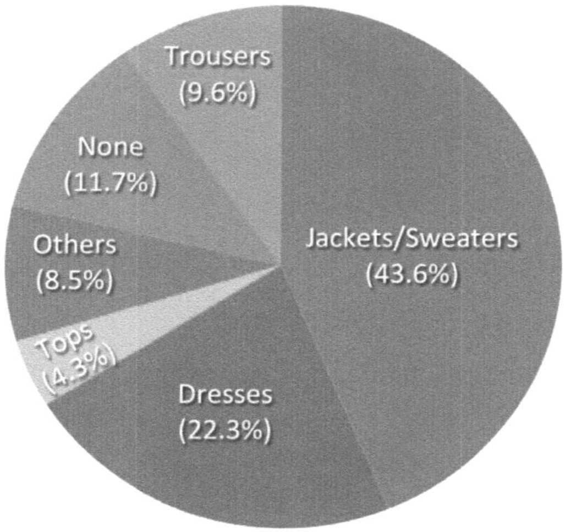

Fig. 2.16: Next Purchase of Fashion Apparel (Womenswear)

3. FINDINGS OF INDUSTRY SURVEY

3.1 Requirements for Competitiveness of Fashion Industry

Lower or competitive price, consistency in quality followed by minimum lead time and Availability of required raw material are perceived to be requirements for being competitive by fashion industry.

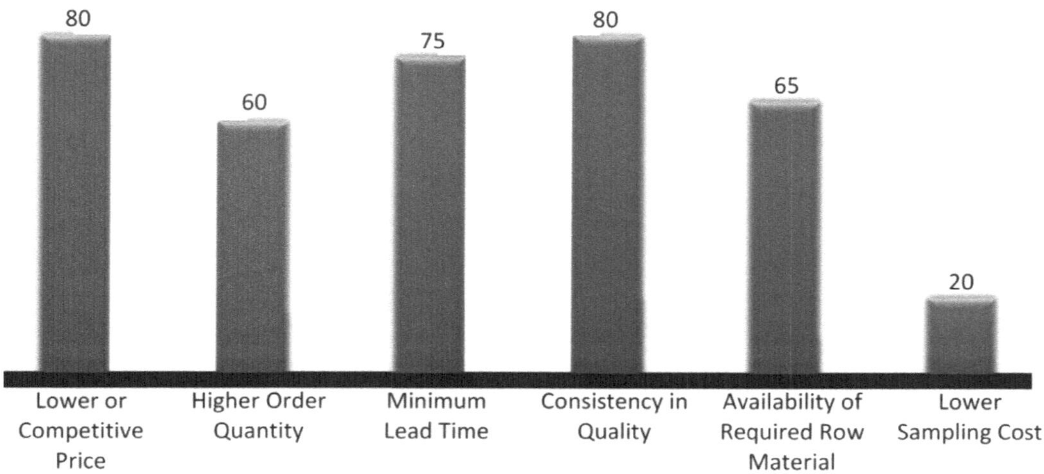

Fig. 2.17: Requirements for Competitiveness of Fashion Industry

3.2 Initiatives to be Taken by Fashion Industry for Competitiveness

Aggressive marketing requirement, innovative design followed by investment in technology (AI) is key parameters for achieving competitiveness of industry in future.

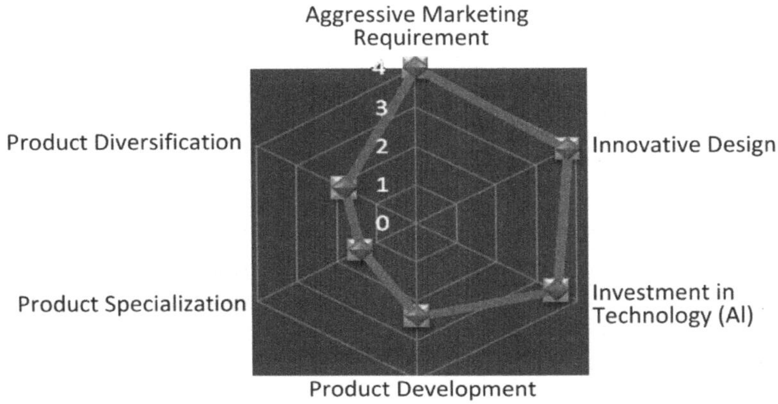

Fig. 2.18: Initiatives to be taken by Fashion Industry for Competitiveness

4. RESULTS AND DISCUSSION

Percentages of respondents are from 18–25 years age group followed by 27.3% respondents from 25–35 year age group. Hence this study primarily covers 18–35 years age group with approximately 89% representation in study. Study is primarily an attempt to understand how purchase pattern has changed post COVID-19 for young target customers who constitute large chunk of flourishing fashion apparel market.

More than 85% respondents have responded that their purchase pattern has been affected by COVID. 39.4% respondents purchased 6–12 pieces of fashion apparel last year followed by 23.2% respondents purchasing 12–24 pieces 44.9% respondents purchased less than 6 pieces of fashion apparel. Whereas, this year 33.7% respondents have purchased 6–12 pieces of fashion apparel. Reduction in consumption is quite steep which indicates as of now fashion apparel are non-priority items for purchase for target customers.

Last decade has witnessed the increasing usage of online platform by brand's retailers. Online as well as offline was equally preferred as mode of purchase prior to COVID-19. Majority of respondents 52.5% preferred both modes as per convenience. Online is preferred as mode of purchase of fashion apparel by 72.7% respondents. Only 8% respondents preferred to purchase offline. It is certainly the impact of restricted movement of targeted customers besides market closure due to pandemic. It will further affect manpower requirement for retails of fashion apparels besides having an impact on rentals and property rates for retail which will cast its shadow on economy.

Majority (more than 60%) of respondent's willingness for purchase of fashion apparel is lesser in comparison to earlier period. Day to day use (40.4%) followed by family function social gathering (36.4%) are key reason for purchase of fashion apparel now. Pandemic has changed our lifestyle with families being confined to themselves and lesser social gatherings due to government restrictions and fear of pandemic too. Requirement and purchase of fashion apparel for young customers was driven with occasions and not need based, which is visible strongly as customers are not much willing to spend on fashion apparel in current time. It will be required from industry to attract customers back to stores for revival of industry. Instagram (54.5% respondent) is major influencer for purchase of fashion apparel followed by other social media. Influence of brand is most important while purchasing fashion apparel 78.8% Majority of respondents were purchasing formal clothes prior to COVID-19. Most (89.7%) of respondents are purchasing casual fashion apparel now. Only a few are purchasing formal clothes now. Jackets/ sweaters (33.3%) followed by T-Shirts (14%) and Shirts (11.8%) are preferred next item Jackets, Sweaters (43.6%) are most preferred item of purchase of women's wear for next purchase of respondents followed by dresses (22.3%). Purchase pattern is influenced by lifestyle of customers. Brands, retailers are required to plan their offerings to cater taste, preferences of target customers for fashion apparel in current time.

Lower or competitive price, Consistency in quality followed by minimum lead time and availability of required raw material is perceived to be requirements for being competitive by fashion industry. Aggressive marketing requirement, Innovative design followed by Investment in technology (AI) is key parameters for achieving competitiveness of industry in future.

5. CONCLUSION

Study reflects that there is large impact of COVID-19 on purchase of Fashion Apparel. Quantity (number of pcs) being purchased have reduced considerably in the Covid period. There is more purchase of basic and necessary items and not much of Fashion and Apparel. Online purchase has become preferred mode of purchase as either target customers are avoiding visit to marketplace or some of the markets are still closed. Due to COVID-19 target customers are not much willing to spend higher amount as there are lesser outings social gatherings, family functions etc. Fashion market has considerably grown due to increasing high disposable income, willingness to spend on fashion apparel and buying latest range of clothes, which is missing now. As target customers are taking due care of their finances based upon experiences during COVID period. There is more of purchase of casual clothes rather formal one due to prevalent practice and culture of online meetings and online offices becoming stronger. It indicates that in the times to come requirement or purchase of formal fashion apparel will be lesser and customers will spend lesser amount also for purchase. Brands/retailers are to plan their inventory as per changing customers taste preferences for purchase of fashion and apparel items.

The textile and apparel industry are one of the largest industrial sectors in India and a leading foreign exchange earner. Industrial liberalization in the domestic economy since 1991 has been followed by changes in the global environment. Abolition of licensing controls on the industry was followed by the Uruguay round of negotiations resulting in a ten year phase out of textile export quotas under the 'Agreement on Textile and Clothing'. The roll down of textile quotas was accompanied by roll down of custom duties and quantitative restrictions on import of wide range of textile and apparel products into the

Indian market. These sequential steps are leading the convergence of the domestic and international markets of textiles and apparel. The domestic industry is no longer insulated from global competition. The phased removal of textile quotas since 1995 has catapulted Indian export firms into a new competitive environment.

Aggressive marketing requirement, Innovative design followed by Investment in technology Artificial Intelligence (AI) is key parameters for competitive future industry. Lower or competitive price, consistency in quality followed by minimum lead time and availability of required raw material are perceived to be requirements for being competitive by industry. Better supply chain for manufacturing and retail, reduction in the manpower cost, in depth knowledge and understanding of taste and preferences of target customer are found to be the main areas with likely impact of AI. Product design is found to have least likely impact of AI. Fashion Industry catering to basic product category or to mass segment will get benefitted by reduction in the manpower cost and increase in the production (volume) due to artificial intelligence. However, industry segment catering to designer/luxury items will provide customer delight by virtue of utilizing information of taste and preference of target customer and delivering it in time. Fashion Industry catering to basic product category or to mass segment will get benefitted by reduction in the manpower cost and increase in the production (volume) due to artificial intelligence. However, industry segment catering to designer/luxury items will provide customer delight by virtue of utilizing information of taste and preference of target customer and delivering it in time.

REFERENCES

[1] Arpana, D. (2020). A Study on Changing Buying Habits of the Customers Using E-Commerce Platform, *Academy of Marketing Studies Journal,* 24(3), 1–10.

[2] Barr, S. (2020). Should Customers be Stockpiling a Mid Corona Virus Outbreak?, *Independent Digital News.*

[3] Garcia, D. (2020). Employment in the Great Recession: How Important Were Household Credit Supply, *Journal of Money, Credit and Banking,* 52(1), 165–203.

[4] Halan, D. (2020). Impact of COVID-19 on *Online Shopping in India.*

[5] India Brand Equity Foundation (2017). *About Indian Retail Industry, Market Size, Export & Investment.*

[6] James, F. (2018). Cyber Physical Systems, *Birckbeck University, D. Mourtzis,* 23, 129–134.

[7] Joshi, P. (2006). Apparel and Textile Exports: Strategies for WTO Era, *CBS Publishers and Distributers, India.*

[8] Rajendran, M. (2020). "It's Corona Time" With No End to Covid-19 Lockdown in Sight, *Education Reboots to 4.0 With E-learning*, Frankfurt, Germany.

Changing Paradigms in Apparel Manufacturing: Industry 4.0

Manoj Tiwari
Department of Fashion Technology, National Institute of Fashion Technology, Jodhpur, India
E-mail: manoj.tiwari@nift.ac.in

ABSTRACT: *With the ever-growing competition at all levels in the supply chains, the industries are devising newer ways to operate and get a competitive edge over others. The newer technological solutions are redefining the ways of working, and traditional approaches are becoming obsolete or not relevant in the current scenario. The apparel industry is also not an exception to this cut-throat competition. The fashion world has always been dynamic, and there is always pressured to work better, faster, and cheaper. To survive and thrive, technology is playing a decisive role and its applications are helping industries in achieving the newer height of success in terms of value generation, improved efficiency, process transparency, and quick response with effectiveness. Industry 4.0 is powered by some key AI (Artificial intelligence), ML (Machine learning), IoT (Internet of Things), BC (Block Chain), etc. This paper presents an overview of changing paradigms of apparel manufacturing in the context of Industry 4.0. It also discusses, some key advancements such as real-time data monitoring, Intelligent Quality Management Systems (IQMS), and Smart Machine Maintenance Systems (SMMS) that are impactful and bringing a new wave of significant improvements in apparel manufacturing.*

Keywords: Apparel Manufacturing, Global Competition, Industry 4.0, IQMS, SMMS.

1. INTRODUCTION

Apparel manufacturing is primarily a labor-intensive sector, where the role of human involvement is decisive in creating garments. Mass production of apparel also presents a complex manufacturing environment, not because of the manufacturing process but due to several other factors causing uncertainties and negatively impacting the performance and efficiencies. Typical apparel manufacturing setups commonly witness scenarios, to name a few include labor absenteeism, late arrival of raw materials, delayed approvals, overloading and under-loading of resources, skill mismatch, quality issues, poor productivity and efficiency, unplanned machine breakdowns, shipment delays, etc. These are day-to-day challenges which one has to encounter, and most of the valuable time is consumed in struggling to solve these problems. Such fire-fighting is purely a non-value-added activity and brings in efficiencies in the system. This study discusses the competitive apparel manufacturing environment and day-to-day challenges of the shop-floor in its initial sections. In the subsequent sections, technological transformations (that industry is witnessing) in the context of Industry 4.0, and the next technologies (that may be impactful in the future) have been discussed.

2. COMPETITION AND SURVIVAL

In the era of globalization and digital technologies, consumers are exposed to better products and services at competitive prices with wider options. The dynamic nature of the fashion business (with reduced order quantities, increased product diversity, and lesser lead-times), cut-throat competition is another key challenge from the buyers' end. At the same time increased labor cost and expectations (due to increased living expenses and life-style of the workers), cost of raw material, cost of infrastructure (machinery, spare parts, water, electricity, fuel, space, etc.) are the pressures felt by the apparel manufacturers.

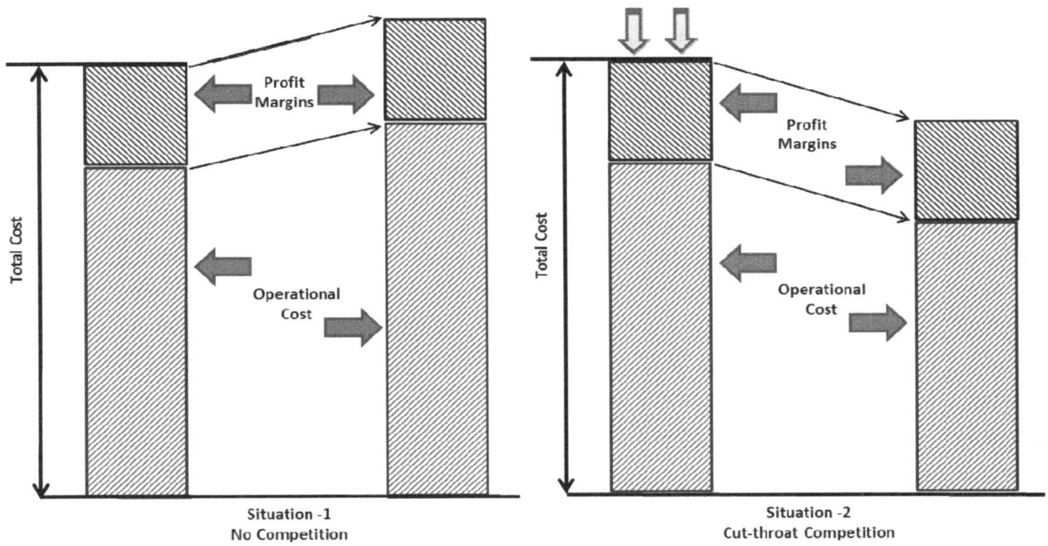

Fig. 1: Competition and Survival (Jana & Tiwari, 2018)

In the context of competition and survival in the apparel manufacturing business (refer to Figure 1, Situation-1), in the past (around the year 1990 to 2000) when there was lesser competition in the global and domestic market. Assuming an instance of increased operational costs, a manufacturer was able to hike the price of the products manufactured to earn its profit margins. However, in another instance of the current high competition (refer to Figure 1, Situation-2) among the apparel manufacturing vendors, there is an immense pressure to remain competitive by offering faster, better, and cheaper products (Jana & Tiwari, 2018). This has resulted in putting real efforts towards efficiency enhancement, increasing productivity, and improving resource utilization to reduce or control the operational costs. In present times, improving operational efficiency by adopting innovative and sustainable practices has become the prime objective. This is necessary not only to survive but thrive and achieve manufacturing excellence.

3. INFORMATION FLOW IN APPAREL MANUFACTURING

The day-to-day apparel manufacturing operations involve a significant transmission of information from one point to another. The information about the product, process, quality, approvals, material, management, etc. is essential to run the show. However, the traditional way of data capturing and its communication to all concerned involves a lot of loopholes and are sources of errors as well. Another important concern is delay of information received, which many a time is of no use and eventually, we end up doing post-mortem which is a futile exercise. In the absence of the right data at right time, the concerned authorities are not aware of ground realities or the current status of the operations. This becomes the biggest barrier in timely decision-making and paralyzes the entire system resulting in fatal failures and huge losses in different aspects.

On the other side, timely and accurate information equips the management to take an informed decision without wasting any time and helps immensely in damage control. Most significantly, capturing the data right from the source in a real-time manner makes the system quick to respond, live, transparent, accurate, and effective.

4. CHANGING MANUFACTURING PATTERNS

As the business environment is becoming dynamic and competitive, the apparel manufacturers are adapting to the newer ways of working. The ever-increasing customer expectations are the key thrust behind the transformation of manufacturing practices through various evolutions and paradigm shifts. Such transformation can be witnessed clearly as traditional craftsmanship based manufacturing systems are turning into the state of the art technology-driven intelligent systems. The application of Industry 4.0 based technological solutions in apparel manufacturing is a milestone in this never-ending journey of advancement.

5. INDUSTRY 4.0

Industry 4.0 is one of the hot topics in modern manufacturing. The term Industry revolution 4.0 was coined in 2011 and refers to the fourth industrial revolution. Industry 4.0's has its roots in manufacturing, where around 2011 German research union for economy and science presented a proposal (regarding a research program for computers in the industry to maintain technological advantage) to the German government. However, the core idea is not at all new. According to Roser (2015) there have been some terms with the same concept which have been used in the past. Some of such terms include Digital manufacturing in the 1970s, Computer Integrated Manufacturing in the 1990s, Digital factory in the 2000s, Factory 2.0 in 2005 by the European Union, and Smart factory in 2007. As Industry revolution 4.0 was not considered as an industrial revolution, the name is generally referred to as Industry 4.0. The first industry revolutions have their origins from mechanization, electricity, and Information and Computer Technology (ICT) (Roser, 2015).

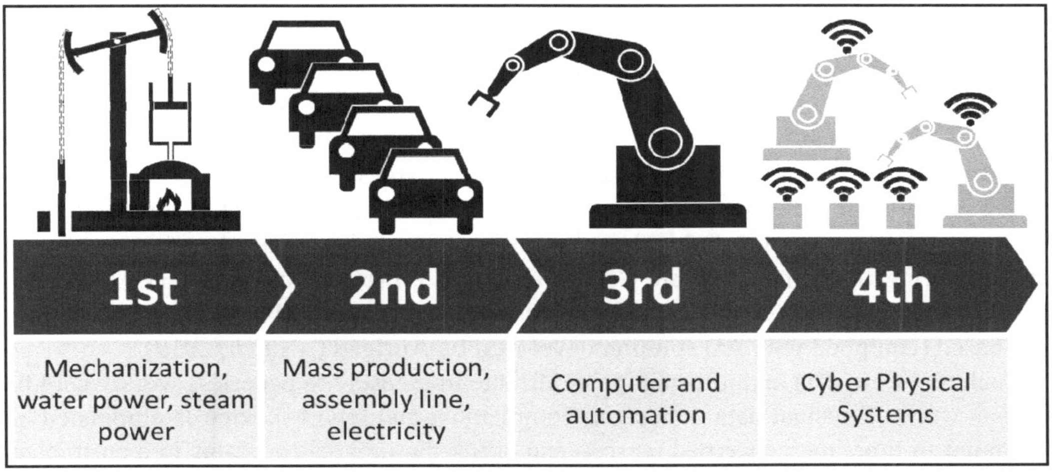

Fig. 1: Industry Revolutions (Roser, 2015)
Image courtesy: By Christoph Roser at AllAboutLean.com under the free CC-BY-SA 4.0 license.

According to Gilchrist (2016), it was General Electric that coined the term "Industrial Internet" for the Industrial Internet of Things (IIoT), while Cisco termed it as the "Internet of Everything" and others called it Internet 4.0 (Gilchrist, 2016). Industrial Internet of Things (IIoT) may be considered as a synonym to Industry 4.0, as it is referred to widely in the United States and other parts of the world as well. The Industry 4.0 applications have been widely adopted by several nations worldwide including the European Union, China, India, and other Asian countries.

6. INDUSTRY 4.0 AND APPAREL MANUFACTURING

Industry 4.0 applications are gaining popularity in the apparel industry worldwide. There have been impactful transformations in different domains of the global apparel business supply chains right from the consumers, buyers, vendors, and other stakeholders. The global networks of industrial businesses are getting connected seamlessly as Cyber Physical Systems (CPS). The machinery, factories, warehouses are getting connected and controlled intelligently by sharing information enabling effective and efficient decision making. In recent years, these cyber-physical systems are transforming traditional set-ups into smart organizations with smart factories, smart machines, smart storage facilities, and smart supply chains. Gökalp *et al.,* (2018) conceptualize smart apparel factory as Apparel 4.0 that is in line with Industry 4.0 and smart factory visions. Tsai (2018) proposes green production planning and control for the textile industry using mathematical programming and Industry 4.0 technique. The scope of this paper is limited to Industry 4.0 applications in apparel manufacturing only. Some key applications have been discussed in the coming paragraphs.

7. REAL-TIME DATA MONITORING

Real-time capturing of the progress is achieved by solutions such as RFID or QR placed on each product/bundle. The information about the stage of the process or product can be tracked and collected in real-time. Real-time employee performance, machine runtime or downtime, maintenance time can also be captured. Tracking the location and status of orders and allowing the customer to monitor the product status instantly. There are many solutions available in this field, though there are limited service providers in apparel manufacturing. To name a few are CGS (BlueCherry), ProCon, and STITCH. BlueCherry by CGS Inc. and ProCon by Crystal Consultancy is based on RFID, while STITCH is an IoT based (equipped with AI) solution developed by Auriga IT (Stitch, 2019).

Such real-time data and information contribute immensely to paperless work where the tedious work of manual data collection, compilation, and report making is eliminated. At any point in time, the concerned persons can check the progress or status in a customized manner. This greatly helps in timely and informed decision making and results in huge benefits.

8. SMART MAINTENANCE MANAGEMENT SYSTEMS

Maintenance of the resources has never been given due importance in apparel manufacturing. It has always been treated as secondary to the production. However, unplanned machine breakdowns and subsequent corrective actions result in huge financial losses, which we ignore most of the time. According to Online Clothing Study (OCS) for an average apparel manufacturing unit, such maintenance-related downtimes are responsible for 5% to 20% productive capacity loss and more than 75% of the factories are unable to predict maintenance-related downtime costs and related reasons (OCS, 2020). In most of the cases, data related to machine and equipment failures, machine waiting time due to break down and maintenance, skill sets of the mechanics, spare parts availability, etc. are not recorded.

There are smart machine maintenance solutions based on integration of Manufacturing, AI, and IoT. Such solutions are bringing in revolutionary changes with effective control of the unplanned maintenance. Most of such intelligent maintenance systems use past data (such as failure patterns, machine breakdown data, machine repair data, etc.) to forecast issues and follow predictive maintenance management. Tiwari & Chaudhary (2019) discuss Reliability-Centered Maintenance (RCM) using artificial intelligence. The research was based on the deep learning model to learn the defects and its analysis as per the specific requirements of an apparel manufacturing environment. The nature of machine breakdown and frequency of the breakdowns were recorded from sewing floors. For system development, the month-wise data before and after system implementation (for machine breakdown occurrence, machine breakdown, time, and major defects) was collected and categorized according to machine types and number of machines (Tiwari & Chaudhary, 2019). Trend forecasting of breakdowns, determining the reasons for sudden breakdowns, calculation of various matrices such as FR%, MTBF, MTTR, OEE, etc., real-time monitoring of maintenance activities and their status, real-time machines status visibility, real-time skill matrix creation of the mechanics, smart spare parts management, automatically assigning a specific task to the most appropriate mechanic are some notable features of intelligent maintenance systems available for apparel manufacturing.

9. INTELLIGENT QUALITY MANAGEMENT SYSTEMS

Quality management is another area that is witnessing revolutionary transformations using Industry 4.0 solutions. The traditional practice of collecting quality data (while garment checking) using paper-based printed formats is not an efficient way. An advanced paperless quality data (refer to Figure 3) collection using mobile phone or tablet results in saving of time as well as human efforts.

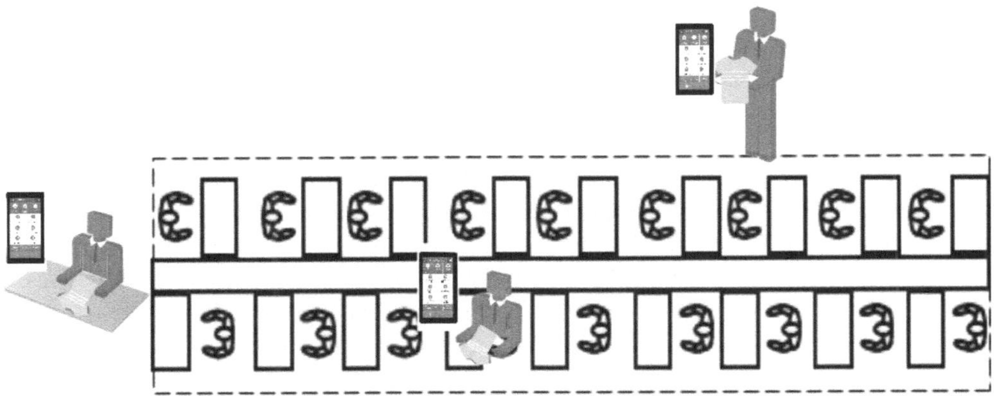

Fig. 3: Intelligent Quality Management System

Image Courtesy: CGS Inc (BlueCherry® Shop Floor Control)

Quality assessment using a mobile application also ensures almost error-free data collection storage and quick response whenever there is an issue. As a standard practice, user logins are created for quality checkers, and by entering the credentials (user ID, password, order or PO, style, bundle number, etc.) quality checkers can perform their responsibility. These systems may also be integrated with shop-floor control systems where one can check the garment being produced by the sewing operator. The checker is required to enter the sewing operator number or scan the RFID or QR code. This enables automatically pick the bundle number (with operation code and several pieces in the bundle) on which the operator is currently working. This results in live quality status along with real-time process details.

10. AUTOMATIC LINE BALANCING

An industrial engineer and production in charge consume a significant amount of time in line balancing. This is a task performed frequently, where work allocation has to be done according to the requirements and available skill-sets. An optimum line balance (and process balance) is a challenge in the complex apparel manufacturing environment which is heavily human dependent. Real-time data capturing enables automatic recording of operators' performance and efficiencies and creating a skill database (also called a skill matrix). The available solutions provide the most appropriate operator to perform a specific task. Apart from operators' performance and efficiency records, such outcomes are generally based on algorithms that consider the complexity of work, the skill level required, past performance, the pattern of defects generated by the operator, ability to rectify the errors or defects, punctuality, etc. These solutions ensure the most optimum line balance and can follow any approach (as decided by the management) i.e. resource-based or target-based balancing.

11. OPERATOR ATTENDANCE MANAGEMENT AND WORK-ALLOCATION

In a typical apparel manufacturing environment, a supervisor or production in-charge consumes the very crucial first hour in worker searching and allocation. It is a daily exercise where a supervisor checks whether all the operators of his lines are present or not. In case any regular operator is absent, then the supervisor has to seek some appropriate alternatives. This is a purely non-value-added exercise and results in production loss. It is observed that the first-hour output is generally 30% (40% in best-case scenarios) of the hourly output at peak (Sadh & Tiwari, 2018). According to Surender Jain, Business Director, Wazir Advisors, "Every minute lost in the first hour would mean the operators have to stay back beyond their shift timings to meet targets pending from the first hour, which, apart from being unethical, promotes lethargy and callousness in the manufacturing environment". The first-hour productivity sets the pace of the day, but unfortunately not given due attention (despite an hour contributes to 12.5% of output in an eight-hour shift) (Deshpande, 2016).

Automatic operator attendance management and work-allocation can be well-taken care of by the Industry 4.0 based solutions, where the information about the operator attendance is captured well in advance (may be based on the intimation as received from the operators going to be absent on that day or an early attendance monitoring application installed in the mobile or at the vehicles carrying the operators). Such information is processed further, and the skill requirement for the day is determined. Using the algorithms, best-fit skill replacements are suggested by the system itself to the supervisor. As this entire process is done in an advanced manner, it results in a huge saving of valuable time and efforts and controls productivity loss. It is important to note, that such solutions are still in the trial and experimentation phase and yet to be adopted in a full-fledged manner.

12. DIGITAL TWINS

Digital Twins can be considered as one of the most advanced Industry 4.0 solutions. It can be defined as the convergence of the virtual world with the physical world. To the best, of knowledge, as of now no apparel manufacturing organization in India is using this technology, but surely it has promising applications in near future. The core idea of Digital Twins is creating a complete virtual factory (covering every aspect right from simulating products and processes to controlling operations and its working). The changes (or deviations) in the real factory are judged in advance, and the issues are taken care of before it even occurs.

The entire concept of Digital Twins is based on three functions as 1. Conceptualize, 2. Compare and 3. Collaborate. Digital twins can be used for physical products in real space, virtual products in virtual space, and for the connections of data and information that ties the virtual and real products together. Digital Twins incorporate Big Data, Artificial Intelligence (AI), Machine Learning (ML,) and the Internet of Things (IoT) and create virtual copies of processes. These virtual copies provide control over the products or processes right from the development to deployment stages (Apparel Resources, 2018).

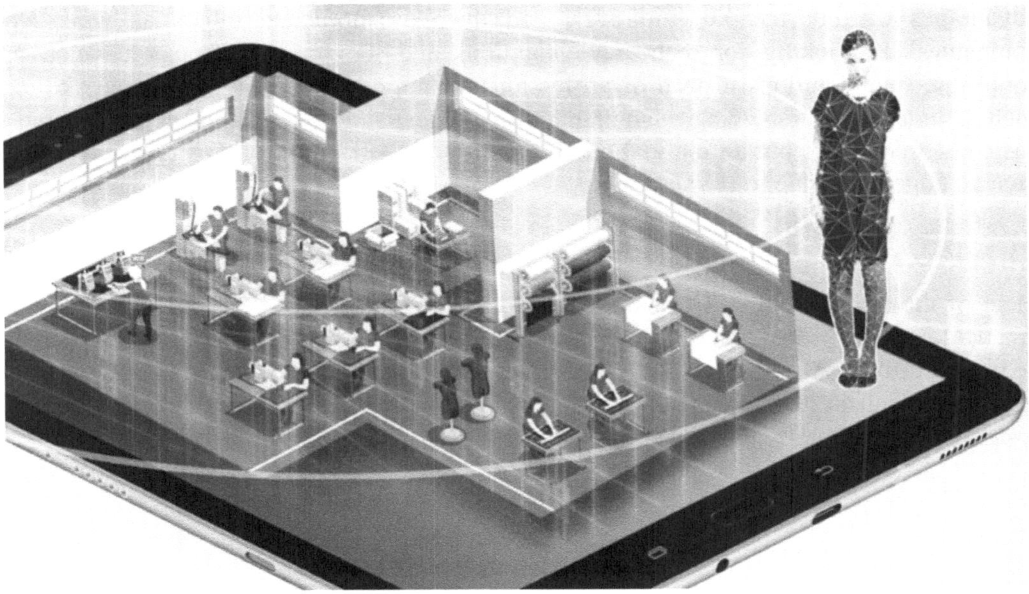

Fig. 2: Digital Twins

Image Courtesy: Apparelresources.com (URL: apparelresources.com/technology-news/retail-tech/digital-twin-overlapping-real-world-with-virtual)

13. FUTURE OF INDUSTRY 4.0 IN CONTEXTS OF INDIAN APPAREL MANUFACTURING

13.1 Scope

The core manufacturing of apparel is losing its shine in India and manufacturers do not find it as lucrative as it was a few years back. The key reasons for this trend may be tough competition and reduced profit margins. This can be confirmed with shrinking apparel export in the last few years. In such difficult times, adopting a knowledge-based solution using advanced technology is not a choice but a need. Industry 4.0 applications have been instrumental in bringing the new wave of transformations and restoring the faith in apparel manufacturing through its promising benefits. Some of the notable advantages include improved agility and customer-centricity in manufacturing with quick identification of customer requirements, improved supply-chain resilience to gain competitive advantage, increased speed and productivity, and achieving sustainability and eco-efficiency in the system (Betti *et al.*, 2020). Ahmad *et al.,* (2020) explored the Role of Business Intelligence Systems in the Era of Industry 4.0 for the textile and apparel industry. The qualitative research was based on a survey from 12 of the world's high-end T&A companies. As a finding of the research, it was revealed that Industry 4.0 improvements are perceived in the apparel retail business because apparel companies are more prone to adopt the Industry 4.0 technologies with advanced business intelligence (BI) solutions (Ahmad *et al.*, 2020). There have been several apparel manufacturers in India, who have witnessed the advantages of Industry 4.0 initiatives, and this has motivated others as well. Now companies are looking for tech-savvy solutions and the brains that can develop and contribute such solutions instead of professionals equipped with only traditional skill sets such as industrial engineering, planning, and merchandising. Such skills are considered as basic essential skills but an inclination and ability to contribute to Industry 4.0 initiatives give an edge over others.

13.2 Precautions

Most of the industry 4.0 solutions are relatively new and yet to be fully explored in the practical aspects. This makes these solutions a little costly and requires careful and dedicated efforts. Any Industry 4.0 intervention should be handled with care and caution, as it may result in huge financial losses, wastage of time and efforts, damage of existing systems, and eventually loss of reputation.

Fig. 3: P-D-C-A Cycle

It is suggested to follow the P-D-C-A (Plan-Do-Check-Act) approach while going for Industry 4.0 transformations. The improvement should be done in a phased manner rather than going to complete transformation at one go. Here, it is imperative to note that human handling and mental barriers to change are some of the biggest challenges for any improvement, as after all these transformations will be driven by people only. Given this, extra efforts should be put into overcoming the "inertia to change" to make the improvement sustainable.

14. CONCLUSION

Industry 4.0 applications are revolutionary and that is bringing significant positive transformations in the apparel manufacturing environment. However, in the context of apparel manufacturing in India, adopting Industry 4.0 seems an advanced level practice and only a few established manufacturers have taken initiatives in this regard. Most of the apparel manufacturers are still struggling to ensure smooth day-to-day functioning to meet the targets. In the Indian apparel manufacturing context, some of the key pre-requisites to adopt Industry 4.0 should be:

- Fixing the basics right,
- Establishing good practices and standard procedures, and
- Adopting a lean way of working (efficient and effective).

Adherence to the above three points helps in preparing better to adopt advanced Industry 4.0 practices and may guide in achieving operational excellence.

REFERENCES

[1] Ahmad, S., Miskon, S., Alabdan, R. and Tlili, I. (2020). Towards Sustainable Textile and Apparel Industry: Exploring the Role of Business Intelligence Systems in the Era of Industry 4.0, *Sustainability*, 12(7), 2632.

[2] Apparel Resources (2018). Digital Twin-Overlapping Real World with Virtual. (Apparel Resources Pvt. Ltd.) Retrieved September 2018, from Apparel Resources: *http://apparel resources.com/technology-news/retail-tech/digital-twin-overlapping-real-world-with-virtual/.*

[3] Betti, F., Boer, E.D. and Giraud, Y. (2020). The Fourth Industrial Revolution and Manufacturing's Great Reset, *McKinsey Insights,* 1–5.

[4] Deshpande, A. (2016). Combatting First-hour Productivity Woes. (Apparel Resources Pvt. Ltd.) Retrieved January 2017, from Apparel Resources: *https://apparelresources.com/ business-news/manufacturing/combatting-first-hour-productivity-woes.*

[5] Gilchrist, A. (2016). Industry 4.0: The Industrial Internet of Things.*United States: A press.*

[6] Gökalp, E., Gökalp, M.O., Eren, P.E., Gülseçen, S., Reis, Z.A., Gezer, M. and Erol, Ç. (2018). *Industry 4.0 Revolution in Clothing and Apparel Factories: Apparel 4.0. in Industry 4.0 from the MIS Perspective,* 169–184.

[7] Jana, P. and Tiwari, M. (2018). Value Engineering, *Industrial Engineering in Apparel Manufacturing—Practitioner's Handbook,* 232–233.

[8] OCS (2020). Increase Factory Capacity Utilization with Predictive Machine Maintenance Tools (O.C. Study, Producer, & Online Clothing Study) Retrieved Septemeber 2020, from Online Clothing Study: *https://www.onlineclothingstudy.com/2020/07/increase-factorycapacityutili zation.html*

[9] Roser, C. (2015). A Critical Look at Industry 4.0. Retrieved 2020 October, from All About Lean: *https://www.allaboutlean.com/industry-4-0.*

[10] Sadh, S. and Tiwari, M. (2018). Line Balancing through Attendance Detection System, Department of Fashion Technology, *National Institute of Fashion Technology,* Jodhpur.
[11] Stitch (2019). Retrieved December 2019, from Stitch ERP: *https://www.stitcherp.com/about-us*
[12] Tiwari, M. and Chaudhary, M. (2019). Development and Implementation of Reliability Centered Maintenance (RCM) System Using Artificial Intelligence, *Oorja*, 17(1), 22–32.
[13] Tsai, W.H. (2018). Green Production Planning and Control for the Textile Industry by Using Mathematical Programming and Industry 4.0 Techniques, *Energies,* 11(8).

Impact of Sweat and Water on Different Seam Patterns on Vapour Moisture Transmission of Multilayered Ensembles

Agya Preet[1], Arunangshu Mukhopadhyay[2] and Vinay Midha[3]

Department of Textile Technology, Dr. B.R. Ambedkar National Institute of Technology, Jalandhar, Punjab, India

E-mail: [1]agyapreet13@gmail.com; [2]arunangshu@nitj.ac.in; [3]midhav@nitj.ac.in

ABSTRACT: *Presences of seam in multilayered ensembles lead to significant changes in moisture vapour transmission behaviour through the layered ensembles. Performance of multilayered ensembles in seamed and unseamed conditions depend on the behaviour of sweat and are found different from test water. It has been observed that seamed fabrics when tested in presence of sweat and water showed greater water vapour transmission rate as compared to un-seamed fabric. Unidirectional seam shows higher water vapour transmission rate. In all the cases of un-seamed and seamed layered ensembles, layered ensemble polyester spacer fabric as middle layer exhibits better water vapour permeability with water and sweat both in comparison to layered ensemble with fleece fabric as middle layer. In case of water vapour transmission behaviour, impact of layered ensemble is found predominant.*

Keywords: Comfort, Layered, Liquid, Moisture, Seam.

1. INTRODUCTION

Heat and moisture transfer properties of the outdoor apparels clothing is of considerable importance as they play a major role in determining the performance of garment in cold and moist environmental conditions. Not only proper selection of textile material but it is also important to ensure the integrity of the constituent layers, in order to ensure good characteristics of the layered clothing system. Parameters like structural properties of seams influences the integrity of layers (Grinevičiūtė *et al.*, 2014). Hence, overall designing which includes stitching of garment play an important role. Fabrics are joined to form garments using seams (Jones & Stylios, 2013). Depending upon the type of garment prepared and utility of the clothing, different types of seams are used. For making cold weather jackets, generally, unidirectional, and bidirectional seams are prevalent. Earlier studies show (Rogina-car *et al.*, 2018; Beaudette & Park, 2017), that the overall product performance of protective clothing, is determined by the characteristics of seam. Insertion of seams may leave holes in the fabric. It also leads to formation of thin spots, called 'cold spots', which are in turn responsible for heat loss from the human body to its outer climate (Saeed *et al.*, 2017). Many researchers suggested that the functional performance of clothing is affected especially in areas around the seams as they may lead to changes in resistance of heat and moisture transport from garment to environment (Beaudette & Park, 2017). Till date, the impact of the seam insertion on performance behaviour of fabrics in terms of the moisture transmission behaviour of the multilayer ensembles has not been investigated. Studies on moisture transmission behaviour done so far are restricted to pure water. In addition, impact of sweat composition in seamed multilayer structures are yet unknown. Hence, present study is intended to explore the aforementioned aspects with simulated sweat solution and distilled water in order to investigate moisture transmission behaviour of layered structures in seamed condition. In the previous study (Mukhopadhyay *et al.*, 2018), it is observed that composition of sweat has influential impact on water vapour transmission behaviour of multilayered ensemble. Moisture transmission behaviour of the fabrics varies highly when tested with sweat and water in unseamed conditions. Vapour permeability both in form of pure water and sweat solution is maximum in the case of spacer fabric followed by the knitted fabric and then fleece fabric. PU coated nylon exhibits lowest level of permeability among other fabrics. It is concluded that multi-layered ensemble with spacer fabric as the middle layer shows higher vapour permeability (using pure water) as compared to the multi-layered fabric with fleece. Hence in, this study the impact of sweat through seamed multilayered ensembles is investigated.

2. MATERIALS AND METHOD

2.1 Material

Multi-layer ensembles were formed using Polyester knitted PK as inner layer, either of Polyester spacer PS or Polyester fleece PF is used as middle layer and Polyurethane coated nylon PUN is taken as outer layer (where testing is done on inner side of polyurethane nylon). The specification of constituent layers is shown in Table 1.

Table 1: Specifications of Fabrics Used

Fabric Type	*Polyester Knit*	*Polyester Fleece*	*Polyester Spacer*	*PU Coated Nylon*
Construction	Knitted (Double jersey/interlock)	Knitted with fleece on one side (fibres raised on the inner side)	Knitted 3D structure (Single jersey on both sides with spacer yarn in between)	Woven Fabric (Nylon filament) with PU coating
GSM (g/m^2)	80	320	450	75
Filament/spun type	Multifilament (150D/144F)	Spun Polyester	(Outer) Multifilament (Inner) Monofilament	Multifilament (70D/24 F)
Porosity (%)	79.4	74	82.5	45.6

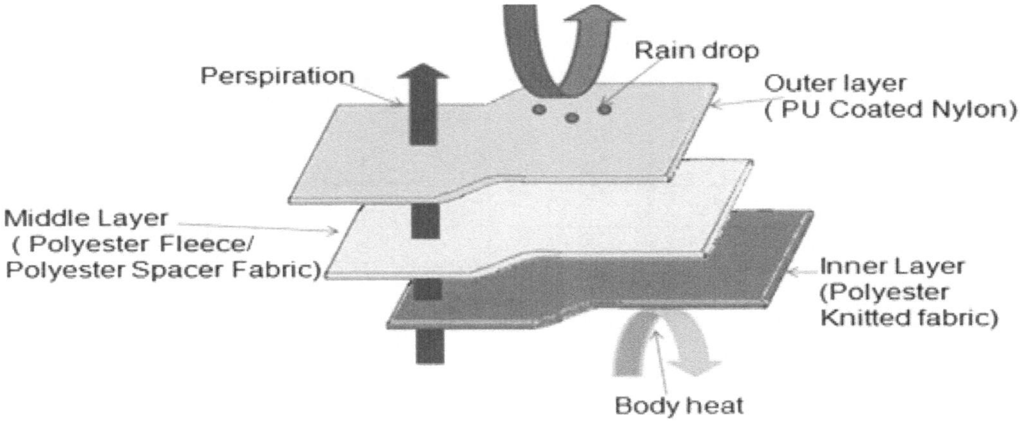

Fig. 1: Schematic View of Layered Ensemble

2.2 Stitching of Samples

To form seamed layered ensembles, stitching was done using the lockstitch machine (JUKI ddl-8100) at 3000 rpm. Specification of yarn used for stitching were polyester spun, 2 ply with twist direction Z/S (ply/single) with overall tpm of 578 and 28 tex. The selection of the stitch density, distance between seams and sewing yarn is made on the basis of stitch density and thread used in existing commercial jackets. Directions and pattern of stitching as unidirectional and bi-directional are shown in Figure 2(a) and (b), respectively. The distance between the adjacent seams is set 2 cm wide.

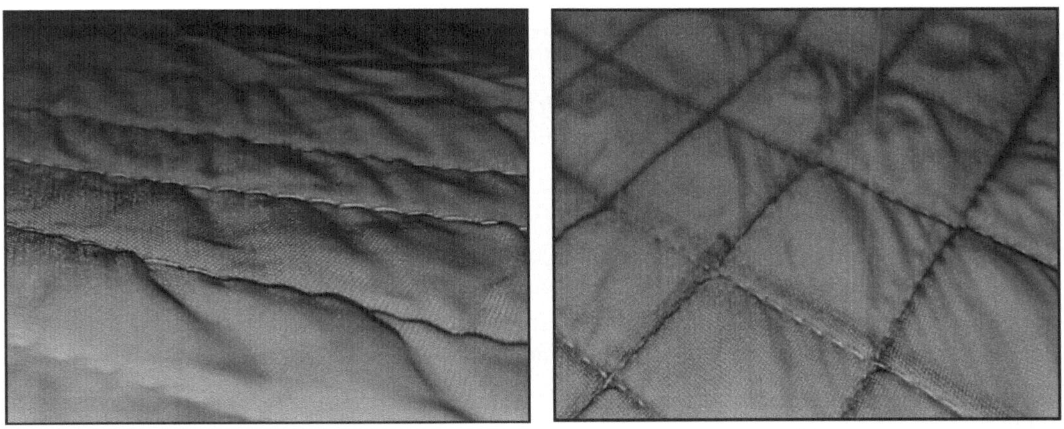

(a) (b)

Fig. 2: Seam Patterns (a) Uni-Directional Seams (b) Bi-directional Seams

For detailed analysis of behaviour of moisture inside the structure, thickness of the seamed (unidirectional and bidirectional seam) and unseamed multi-layered ensembles was measured using thickness measurement tester ASTM D 16777 shown in Table 2.

Table 2: Thickness (mm) of Multilayered Ensembles

Fabric Type	*Unseamed*	*Unidirectional Stitched Sample*		*Bidirectional Stitched Sample*	
		At Stitch Junction	*In between Junction Points*	*At Stitch Junction*	*In between Junction Points*
Multi-layered spacer	2.91	2.50	3.10	2.24	3.50
Multi-layered fleece	2.25	2.15	2.89	2.08	4.02

Before performing the test, all the fabric samples were first conditioned at temperature 27°C ± 2°C and relative humidity level of 65% ± 2% in order to minimise the variation between the samples and to reduce the error percentage. For each test, fifteen samples were tested to minimise the error below 2%.

2.3 Evaluation of Moisture Transmission

Fabric samples were tested for their water vapour permeability as per BS 7209 using the evaporative dish method.

2.4 Statistical Method

Further, all results were statistically analysed and ANOVA analysis was done to see the effect of individual factors on the thermo-physiological comfort performance. The percentage contribution of the different factors (as derived through ANOVA) was analysed. The evaluation was based on following expression.

$$\%\ \text{Contribution} = \frac{(Ssf - df_f \,.\, V_e)}{SS_T} \times 100$$

Where, SS_f= Sum of square of the factor, df_f= Degree of freedom of the factor, Ve = Mean square of pooled error, and SS_T= Total sum of squares.

3. RESULTS AND DISCUSSION

In order to assess the thermo-physiological comfort of the multi-layered system design in garment form, fabric structures were tested with different seam pattern and without seam, for their water vapour permeability behaviour (WVP) and moisture management properties. The effect of the constituent layers of multilayer ensembles, seam and solution type is quite significant on moisture transmission behaviour of the material. The detailed has been mentioned below.

3.1 Water Vapour Transmission Behaviour

Results of water vapour permeability shows that the role of junction between the stitch points is predominant over type of constituent layer. It is observed that the water vapour permeability behaviour of the multilayered ensemble is influenced when seams are inserted in the layered structures. Role of seam is found significant nearly 69% (Table 3) followed by type of layer ensemble (18.9%) whereas solution type had the least impact (6.7%) on water vapour permeability of seamed multi-layer ensembles.

Table 3: ANOVA Table for Water Vapour Permeability

Parameters	*F-Calculated*	*(%) Contribution*
Type of layered ensemble	1031.64	18.9
Seams Type	1887.02	69.20
Solution Type	368.47	6.7
Type of layered ensemble X seams	89.77	3.2
Type of layered ensemble X solution type	64.62	1.1
Seams X solution type	9.93	0.03

Further, the results (Figure 3 a, b) show that seamed specimen have greater water vapour permeability as compared to un-seamed samples. This is due to the fact that the water vapour permeation becomes easier at the stitch line due to reduced thickness and presence of stitch hole (Table 2). On the other hand, stitching also leads to bulging of fabric between stitch lines causing increased air gap within layer (Figure 2). This leads to higher air draft in dynamic situation which in turn results in easier movement of moisture vapour across the fabric. The combined effects of seam holes and fabric deformation between the seams are thought to be responsible for the aforementioned behaviour.

On comparing the behaviour of seamed structures it is found that maximum water vapour permeability rate is found in the unidirectional seam as compared to the bi-directional seams. This can be explained on the basis that in case of unidirectional seams, bulging of the fabric material in between the seam junction is less as compared to that of bi-directional seams (Table 2) which is more prominent over the effect of increased stitch line.

In addition to that it is found that rate of water vapour permeability is significantly affected by the presence of outer layer. WVP is reduced drastically with insertion of the PU coated nylon layer. It may be added that polyurethane coated nylon is stiffer component and therefore responsible for greater bulging. It appears that outer layer (poly urethane coated nylon) plays a major role in influencing the water vapour permeability.

WVP of the layered ensemble with spacer as the middle layer is higher as compared to the multi-layer structure with fleece as the middle layer in cases of both unidirectional and bi-directional seams. This is ascribed to greater porosity of the spacer fabrics which assist in better water vapour transmission through the fabric (Table 1). In addition, it is found that difference in WVP of sweat and water is lower in case of unseamed fabric as compared to the unidirectional and bi-directional stitched layered ensembles.

It has been also observed that sweat is showing less WVP as compared to water in all the cases (Figure 3a & 3b). This is explained on the basis of difference in the pressure gradient build up by both water and sweat across the fabric because they posses different surface tension and density (Water: 72 dyne/cm and 0.99 g/cc; sweat: 62 dyne/cm and 1.04 g/cc).

Additionally, presence of the salt in sweat lowers the vapour pressure and resulted in less pressure gradient to be built across the fabric as compared to water (Boguslawska-Baczek & Hes, 2014).

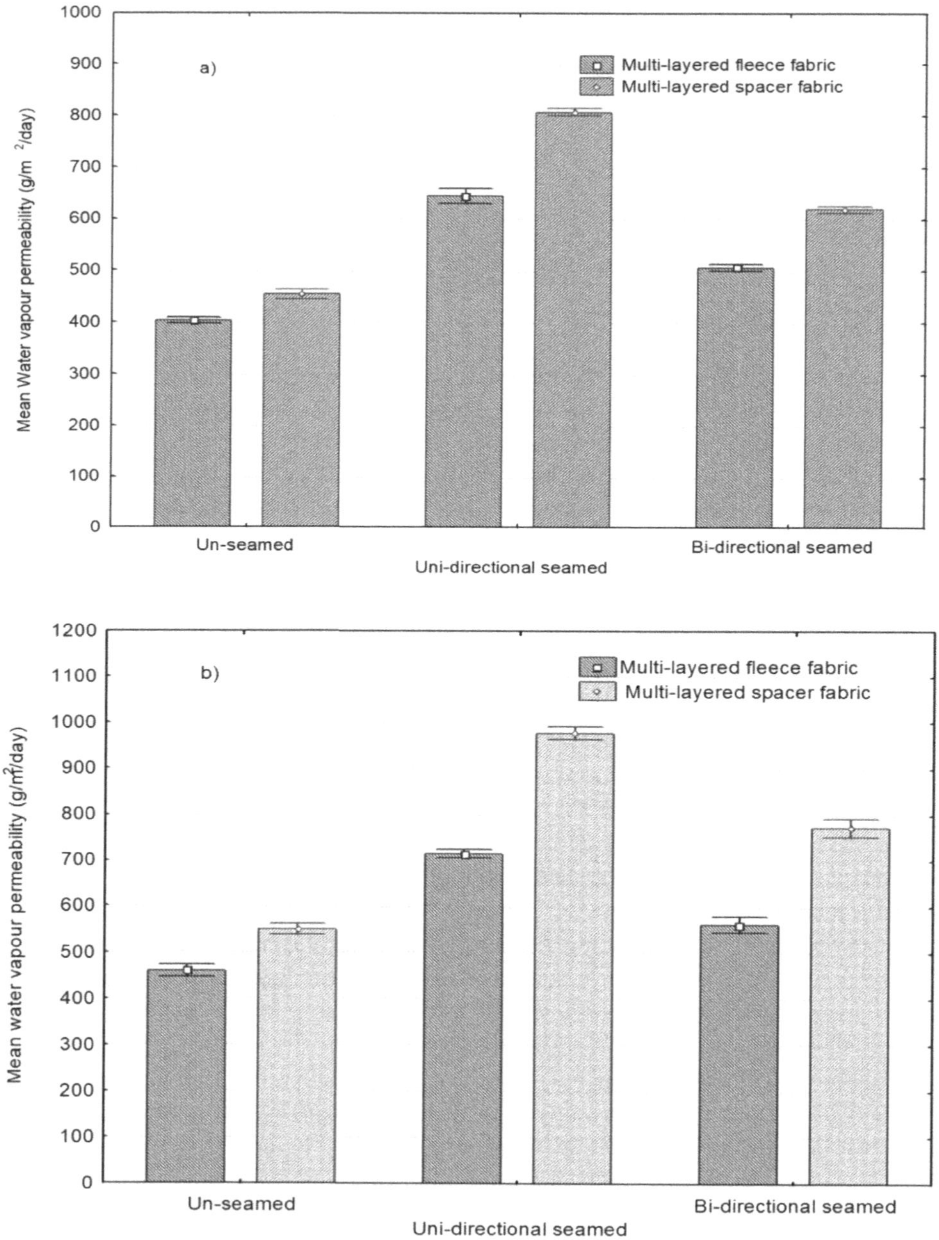

Fig. 3: Water Vapour Permeability of Seamed and Un-Seamed Sample; (a) with Sweat (b) with Pure Water

4. CONCLUSION

- Based on the present study of the vapour moisture transmission behaviour through the seamed and unseamed fabric ensemble, following conclusions can be drawn which are mentioned below.
- Seamed samples showed higher water vapour permeation rate as compared to the un-seamed samples. On insertion of the seam in the layered ensemble, permeability of vapours through the fabric changes. This is due to formation of stitch hole and change in air gap between the layers which in turn affected the passage rate of vapour moisture through the seamed layered structures.
- Unidirectional seams showed higher transmission rate as compared to un-seamed and bi-directional seamed samples with both sweat and water. This is due to greater impact of bulging of layers over the stitch density.
- Investigation showed that WVP of the multi-layered spacer fabric both in case of un-seamed and seamed condition with both sweat and water is higher as compared to the multi-layered fleece fabric. Role of the layered component is detrimental in affecting the permeation rate of vapours through the fabric.
- From statistical analysis, it can be inferred that in case of water vapour transmission behaviour, the role of layered ensemble is predominant whereas impact of solution type was found marginal.
- From the study, it can be concluded that unidirectional stitched multi-layer structures performed better in terms of water vapour permeation rate with distilled water.

REFERENCES

[1] Beaudette, E. and Park, H. (2017). Impact of Seam Types on Thermal Properties of Athletic Bodywear, *Textile Research Journal*, 87(9), 1052–1059.

[2] Boguslawska-Baczek, M. and Hes, L. (2014). Thermal Conductivity and Resistance of Nomex Fabric Exposed to Salty Water, *Tekstilve Konfeksiyon,* 24(2), 180–185.

[3] Grinevičiūtė, D.; Valasevičiūtė, L.; Narvilienė, V.; Dubinskaitė, K. and Abelkienė, R. (2014). Investigation of Sealed Seams Properties of Moisture Barrier Layer in Firefighters Clothing, *Materials Science*, 20(2), 198–204.

[4] Jones, I. and Stylios, G. K. (2013). *Joining Textiles: Principles and Applications,* Elsevier.

[5] Mukhopadhyay, A.; Preet, A. and Midha, V. (2017). Moisture Transmission Behaviour of Individual Component and Multi-Layered Fabric with Sweat and Pure Water, *The Journal of the Textile Institute,* 109(3), 383–392.

[6] Rogina-Car, B.; Schwarz, I. and Kovačević, S. (2018). Analysis of Woven Fabric at the Place of the Sewn Seam, *Autex Research Journal,* 18(3), 216–220.

[7] Saeed, H.; Rödel, H.; Krzywinski, S. and Hes, L. (2017). Spacer Stitching: An Innovative Material Feeding Technology For Improved Thermal Resistanc, *In IOP Conference Series: Materials Science and Engineering* (Vol. 254, No. 13, 132004), *IOP Publishing.*

Different Types of Face Masks and their Effectiveness for Prevention of COVID-19

Ritu Sharma

The Technological Institute of Textile and Sciences, Bhiwani, Haryana, India

E-mail: ritu.kaushik33@gmail.com

ABSTRACT: *Mask is an object worn on the face, typically for protection, disguise, performance or entertainment. Mask has been in use since the pre-ceramic Neolithic period and dates to 7000 BC for various ceremonial and practical purposes. In modern times with increase in pollution level in the air, masks have become an essential component of protective gear especially in the large cities. It has become evident that masks are very effective in preventing the spread of COVID-19 disease. With the outbreak of novel corona virus, masks have become even a legal necessity in most parts of the world. With the availability of many types of masks in the market it is necessary to compare and analyse which type is most suitable and provide best protection from COVID-19 disease. This paper focuses on a comparative study on different types of masks available for prevention of COVID-19.*

Keywords: Corona Virus, Masks, N-95 Respirators, Nonwoven, Surgical Masks.

1. INTRODUCTION

The use of masks for various rituals as a decorative piece or for protection is a very ancient practice across the globe. The oldest masks that have been discovered are 9000 years old although it is believed to be in use since 30000–40000 years. Mask has been in use for either functional purposes or for ritual purposes. For the ritual purposes the masks have been used in a variety of shapes and depicting various characters depending upon the society in which they are used across the world whereas functional masks have been used for the protective properties or sometimes (Leung *et al.*, 2020) for disguise also. There has been a drastic increase in the use of functional masks in recent times but there is a long history of protective masks as armour and even medical masks. The medical masks have different purposes like to filter the outside air as respirator and dust masks or to minimise the risk of infection or to protect medical professionals from bodily fluids.

It has been established that the use of face masks can drastically reduce the chances of spreading COVID-19 (Offeddu *et al.*, 2017) disease which can spread by direct contact or fomite, droplets infection through coughing or sneezing or through transmission of airborne particles. Hence, it was advised by WHO and government agencies to use the face mask especially in public places along with other measures to curb the spread of COVID-19, but with the availability of variety of types of face masks in the market the general public is confused as which type of face mask provides best protection against the disease and which is ineffective. This paper is an effort in the direction to clarify the types of masks available in the market and their uses, and what type of the mask is most suitable to provide the protection (Offeddu *et al.*, 2017) against the COVID-19.

2. TYPES OF MASKS

2.1 Dust Mask

This mask is used for protection against non-toxic dust generated by grass gardening, sweeping, fibre glass and sandstorms. It is very flexible, mouldable, disposable mask made of paper pad. It does not provide any protection (Offeddu *et al.*, 2017) against Covid-19.

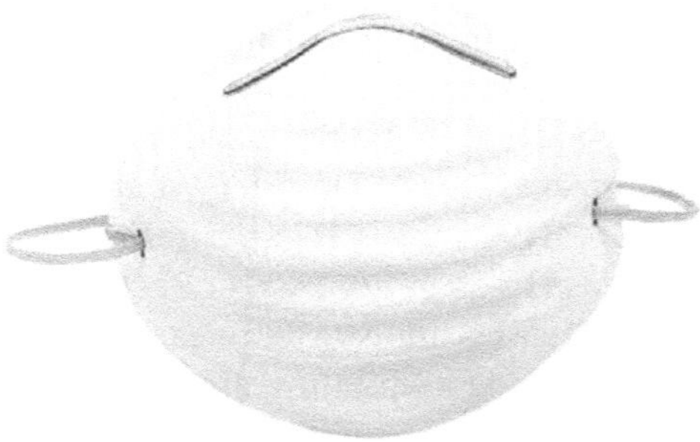

Fig. 1: Dust Mask

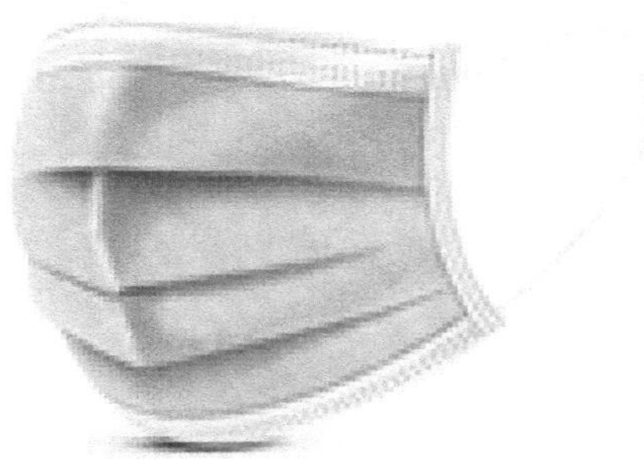

Fig. 2: Single Layer Mask

2.2 Single Layer Mask

The single layer masks are made up of wood pulp tissue paper and non-woven fabric which is used in food processing industry. It also does not provide any protection against COVID-19. It can be used only once and should not be washed (Long *et al.*, 2020) or reused.

2.3 Surgical Mask or Medical Mask

Surgical masks are used by surgeons during operation or any medical procedure to protect themselves from catching the patient's infection. It also protects the patient in case the surgeon coughed or sneezed during surgery. This type of mask is made up of three layers. The innermost layer is made of an absorbent material (W.H.O., 2020) that absorbs moisture from the wearer's breath. Middle layer is made up of melt-blown material which acts as a filter and the outer layer repels liquid. Bacterial filtration efficiency of surgical mask is greater than 98% and particulate filtration efficiency is less than 99% for 0.1 micron particle size. After used once these masks are disposed of.

Medical masks provide better protection from COVID-19 hence it is suggested by the WHO to use medical masks specifically for health workers, anyone having any symptoms (Loeb *et al.*, 2009) of COVID-19 and the elderly people who are above 60 years of age.

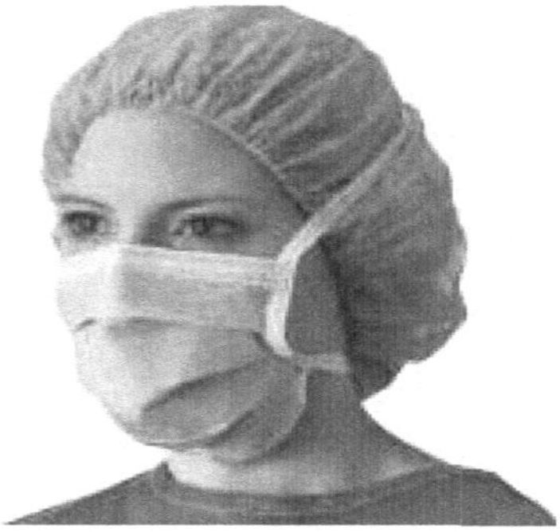

Fig. 3: Surgical Mask

Fig. 4: Cotton Mask

2.4 Respirator Mask

Respirator mask filters the toxious particles, gases, oils or microorganism, bacteria, viruses and fungi by filtration system and does not allow microorganism to go inside the respiratory tract and does not allow the air from the sides into the nose and mouth. These respiratory masks are called N95, N99, N100. N95 means a mask can filter off at least 95% of particles which are greater than 0.3 microns while N99 means it can filter off 99% of these particles. These masks can filter the corona virus (Loeb *et al.*, 2009) effectively. This type of mask can create difficulty in breathing and often get moist after wearing for more than 30 minutes. So these types of masks are recommended while travelling outside and for health care providers who come in close contact with the Covid-19 patients. N95 masks maybe valve type or non-valve type. The valve type N95 masks are not suitable for Covid (Yan *et al.*, 2019) patients as during exhalation they will allow the virus to pass out easily and can infect neighbouring people.

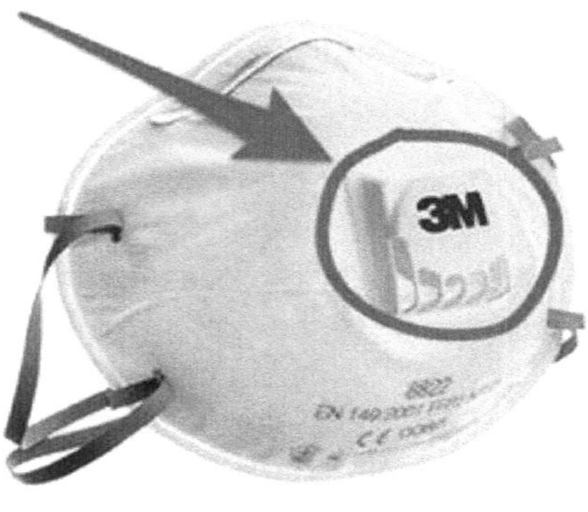

Fig. 5: N95 Respirator

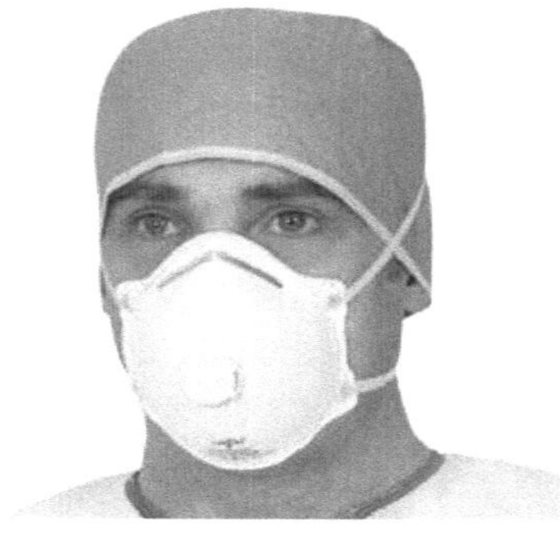

Fig. 6: N95 Respirator with Valve

2.5 Cloth Mask

Surgical masks need to be replaced every day. N95 masks are expensive and recommended for frontline healthcare providers. So, there is a need of cheap, easily available, washable and reusable mask. Cloth mask seem to fulfil this criteria. It can be made from an old T-shirt or cotton (W.H.O., 2020) vest by using two or three layers. WHO has recommended that cloth masks can be used by anyone under the age of 60 and who do not have an underlying health (Radonovich *et al.*, 2019) condition.

3. CONCLUSION

In this pandemic situation of COVID-19 everybody should wear a mask because it reduces the spreading and catching infection. After going through the various types of masks, 3- layered cotton mask is a reasonable mask although it is not much effective for corona virus but it is comfortable, washable, reusable and available at reasonable price and can be made at home also.

It is also suggested that the mask should be worn properly and should cover both the nose and mouth. Apart from that transparent face shields, medical goggles, and other types of personal protective equipment should also be used especially when you are in crowded places.

REFERENCES

[1] Leung, C., Lam, T. and Cheng, K. (2020). Mass Masking in the COVID-19 Epidemic: People Need Guidance, *Lancet*, 395(10228), P945.

[2] Loeb, M., Dafoe, N., Mahony, J. *et al.* (2009). Surgical Mask v/s N95 Respirator for Preventing Influenza among Health Care Workers: A Randomized Trial, *JAMA*, 302(17), 1865–71.

[3] Long, Y., Hu, T., Liu, L. *et al.* (2020). Effectiveness of N95 Respirators Versus Surgical Masks against Influenza: A Systematic Review and Meta-analysis, *Journal of Evidence-Based Medicine.*

[4] Offeddu, V., Yung, C.F., Low, M.S.F. *et al.* (2017). Effectiveness of Masks and Respirators against Respiratory Infections in Healthcare Workers: A Systematic Review and Meta-Analysis. *Clinical Infectious Diseases*, 65(11), 1934–42.

[5] Radonovich, L.J., Simberkoff, M.S., Bessesen, M.T. *et al.* (2019). N95 Respirators v/s Medical Masks for Preventing Influenza among Health Care Personnel: A Randomized Clinical Trial. *JAMA*, 322(9), 824–33.

[6] W.H.O. (2020). Advice on the Use of Masks in the Community, During Home Care, and in Health Care Settings in the Context of COVID-19, Geneva: WHO 2020.

[7] W.H.O. (2020). Advice on the Use of Masks in the Community, During Home Care and in Health Care Settings in the Context of the Novel Corona-Virus (2019–Nov) Outbreak: Interim Guidance, 29 Jan, 2020. *https://apps.who,int/iris/handle/10665/330987*(Accessed Mar, 2020).

[8] Yan, J., Guha, S., Hariharan, P. and Myers, M. (2019). Modelling the Effectiveness of Respiratory Protective Devices in Reducing Influenza Outbreak, *Risk Analysis*, 39, 647–661.

A Review of Impulse Buying Behavior for Apparel Consumer

Vishal Trivedi[1], **Pradeep Joshi**[2] **and K.N. Chatterrjee**[3]
[1,2]Amity School of Fashion Technology, Amity University, Noida, Uttar Pradesh, India
[3]The Technological Institute of Textile & Sciences, Bhiwani, Haryana, India
E-mail: [1]vtrivedi@amity.edu

ABSTRACT: *The purpose of this study is to provide detailed understanding of the impulse buying behavior of apparel consumers by analyzing numerous researchers work in the field of retail (both online and offline mode). The investigation gives a broad overview of impulse buying behavior and other aspects of purchase behavior related to apparel. Apparel shoppers browse the retail store to get an idea about latest fashion trend and style, rather than relying on shopping list which increases the greater the opportunity of impulse purchasing. Moreover, apparel consumers are likely to be more divergent on what they want than what they need. This suggests apparel as a product category as a stimulus effecting impulse purchases. This review paper focuses on numerous characteristics which influence impulse buying like consumer, retail store, E-commerce website, situational and product related characteristics. The multiple aspects of the subject are categorized for future research works in the area of impulse buying with the suggestions.*

Keywords: Apparel Consumer, Buying Behavior, E-commerce Website, Impulse Buying, Retail.

1. INTRODUCTION

Indian textile and apparel industry have ancient history of a fine craftsmanship and global presence. The textile and apparel industry is one of the largest industrial sectors in India and a leading foreign exchange earner. India is the second largest textile exporter in the world. In 2017, India's share in global trade of textiles and apparels was approximately 5 percent. In FY19, total textile and clothing exports stood at US$ 36.62 billion while Apparel (Ready Made Garment) exports reached at US$ 16.27. In the future, India's apparel exports are expected to increase considerably. Total exports of textiles and apparel are expected to touch US$ 82 billion by 2021 with CAGR of 12.06 per cent. The major market for Indian textile and apparel exports are: USA, Europe, Asia and Middle East (Anon, 2019). The Indian retail industry plays a vital role to increase the sale of apparel as an end consumer product. The Indian e-commerce and retail industry is one of the fastest growing industries in the world and played a critical role for overall development of the economy. The apparel consumer buying behaviors are changing rapidly, so the growth of apparel retail sectors is not only limited to urban areas but it is also growing in rural areas. Many global brands are entering in Tier-I and II cities, while domestic brands are strengthening their position. Inflow of international brands is creating deep changes in Indian apparel market. The apparel market consists of three categories: Men's wear, Women's wear and Children wear. The marketing strategies of the apparel segment are developed for multiple purpose viz. brand building and creating awareness about fashion and lifestyle-oriented image among the targeted customers. The young Indian consumers deliberate the appearance and image as important. While the youth consorts fashionable apparels as the basic factor for the lifestyle image, their buying consideration depends on affordability and its relevance with the latest fashion trends. The inspirational Young Indians are focused upon designed apparels (Shukla *et al.*, 2018).

Apparel market is a homogeneous in terms of new styles and design, the apparel manufacturer attempts to distinguish the product through various method and visual practice is one such step that creates appeal of products thus enhancing the impulse buying behavior of consumer. This present study focuses on impulse buying behavior of consumer and its impact on apparel industry. Various external and internal factors influence the impulse buying behavior of apparel consumer.

2. REVIEW OF LITERATURE

Many research scholars and marketing practitioners have taken much interested in impulse buying for the past many decades (Hawkins, 1962). Impulse buying plays a significant role in case of apparel shopping. Apparel shopper can get more benefits during impulse purchase, which is more important for apparel retail industry. Abratt & Stephen (1990) investigated that impulse buying in a supermarket could be of much interest to the manufacturer as well as retailer worldwide. Piron (1991) defining the impulse purchasing as unplanned purchasing, which is widely known as "Sudden or Unplanned buying" a term used for any purchase made by a consumer without further planning. The young researchers and modem theorists are observant of impulse buying behavior of consumers and consider it as a unique and a continuous phenomenon in the lifestyle of the consumers living in this modem urban areas (Youn & Ronald, 2000).

2.1 Definitions of Impulse Buying

Many researchers have given the definition of impulse buying but no integrated definition can be approved in both academic and commercial literature (Rook & Robert, 1995). According to the various retailers outlook, impulse buying is a sudden or unplanned buying. In an academic literature the definition of impulse buying focused on consumer standpoint and emphasis is on explaining the experience of consumer impulse purchase (Chang *et al.*, 2014). Various definitions of impulse buying are given by many researchers, which are mentioned in Table 1.

Table 1: Definition of Impulse Buying

Author	*Definition*
(Hawkins, 1962)	Any Purchase which a shopper makes but has not planned in advance.
(Weinberg & Wolfgang, 1982)	Impulse buying is categorized as buying with great emotional instigation, with little cognitive influence and mainly responsive behavior.
(Rook & Hoch, 1985)	Sudden and Spontaneous desire to act.
(Cobb & Hoyer, 1986)	Impulse buying is when an individual makes the decision to buy items while in the store.
(Rook, 1987)	The impulse buying is unintended, non-reflective reaction which occurs after being exposed to stimuli inside the retail store.
(Gardner & Dennis, 1988)	Impulse buying is quick decision making and relatively biased in favour of instant acquisition.
(Piron, 1991)	Impulse buying is a purchase that is unplanned, the result of an exposure to a stimulus, and decided on-the-spot. After the purchase, the customer experiences emotional and/or cognitive reactions
(Rook & Gardner, 1993)	Impulse buying as an unplanned behaviour involving quick decision-making and tendency for immediate acquisition of the product
(Beatty & Ferrell, 1998)	Impulse buying refers to immediate purchases which are without any pre-shopping objective either to purchase the specific product category or to fulfil a specific desire.
(Block & Morwitz, 1999)	Impulse purchase as consumer buying an item with little or no deliberation after the result of a sudden, powerful urge.
(Kacen & Lee, 2002)	Impulse buying behaviour is more arousing and irresistible but less deliberative when compared to planned purchasing behaviour.

2.2 Categories of Impulse Buying for Apparel

Hawkins (1962) identified four classification of impulse buying: Pure, Reminder, Suggestion and Planned impulse buying but Han *et al.,* (1991) modified the classification of impulse mix in context of apparel products and developed a fifth type of impulse buying that is planned buying.

1. Pure Impulse buying
2. Reminder Impulse buying
3. Suggestion (Fashion oriented) Impulse buying
4. Planned Impulse buying
5. Planned buying.

Pure impulse buying occurs without any previous thought or plan to buy. It can be described as a novelty or escape buying and results from a sudden urge to buy something, not necessarily something new or fashionable. In Reminder impulse buying, the shopper remembers the previous decision or experiences which causes an on the spot impulse purchase. Suggestion or fashion oriented impulse buying occurs when the consumer sees the new product (Apparel) in terms of new style, color, design and fabric, and suddenly decides to buy it.

Planned impulse buying occurs when consumers purchase is dependent upon sale/ promotional conditions such as special less prices, coupon offers, and the other promotional offers; that is, this shopper waits to see what is available at what prices and then decides. The buying decision is not planned and is made suddenly in store.

Planned buying occurs when the purchaser has a list of specific items in mind before entering the store and then buys only those items which are available in list.

According to Belleneger *et al.,* (1978) apparel product is considered and found to be most often purchased on impulse. Prasad (1975) found that Mens and Womens wear were bought most on impulse situation. In the study of Belleneger *et al.,* (1978) sportswear of women and jewelry were two most frequently bought products impulsively. According to Han *et al.,* (1991) students of textile and clothing were more likely to be impulse purchaser than non-textile background students. In the study of Chen-Yu & Yoo (2002) about half of respondents were impulsive buyer, they spent more time and money in their shopping duration than non-impulse buyer.

3. FACTORS AFFECTING IMPULSE BUYING BEHAVIOR

Most of the research related to impulse buying behavior studied the causes of an impulse purchase. According to Sirhindi (2010) two basic types of variables that can cause impulse buying behavior are:

1. External Factors of Impulse buying
2. Internal Factors of Impulse buying.

External factors or stimuli refer to marketing stimuli or product characteristics that are controlled and placed by marketers in attempt to attract the consumer to make buying (Dawson & Minjeong, 2009). Further these factors can also be described as product related

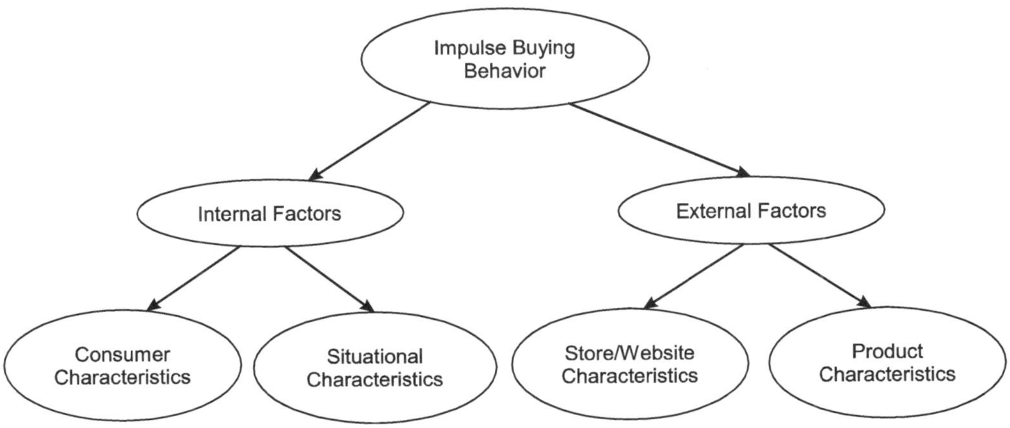

Fig. 1: Different Factors of Impulse Buying Behavior

characteristics and store related characteristics. Internal factors or stimuli emphasize on the individual, it refers to internal stimuli and characteristics of a consumer that make them engage in impulse buying. Situational stimuli and related factors also play a significant role for consumer impulse buying behavior.

3.1 Consumer Characteristics

Consumer characteristics include age, gender, mood, income level, materialism, shopping enjoyment and impulse buying tendency. Previous studies on consumer characteristics have found that young purchasers do more impulse buying then elder shoppers (Helmers *et al.*, 1995). According to Logue & Chavarro (1992) age and impulse buying have negative relationship between each other; that means young shoppers have less self-control over their shopping behavior rather than elder shoppers. Another characteristic is gender which also affects the impulsiveness. Dittmar *et al.*, (2004) found that men and women do not have similar kind of preferences, when they do purchase. Men generally pay for products with practical and instrumental application and they mostly consider personal identity when they do shopping, on the other hand women select products which are associated with emotions and relationships and they mostly consider social identity when they do shopping (Bayley & Clive, 1998). Previous studies also found that consumer's mood also influence the impulse buying behavior of shopper. According to Faber (1996) negative moods affects impulse buying behavior of consumer. So, impulse buying can help the shopper to change their mood from negative to positive. Shoppers who consider value with a materialism point of view mostly avoid impulse buying. They mainly choose to save money rather than spending money on impulse buying (Mariam, 2003). These consumers usually do not affect by impulse buying. Another important characteristics is impulse buying tendency (IBT), which is associated with the lifestyle of consumers (Rook, 1987). Consumer with high amount of impulse buying tendency (IBT) are more probable to have desire to purchase various types of products impulsively (Jones *et al.*, 2003). According to Bellenger & Korgaonkar (1980), purchasing is fun for some customer and they sometimes spend their spare time on shopping, but they do not purchase any product. Therefore, customers are shopping and enjoying their time, vendors may encourage them to stay longer at their shops and stimulate them to do impulse purchasing (Bellenger & Korgaonkar, 1980).

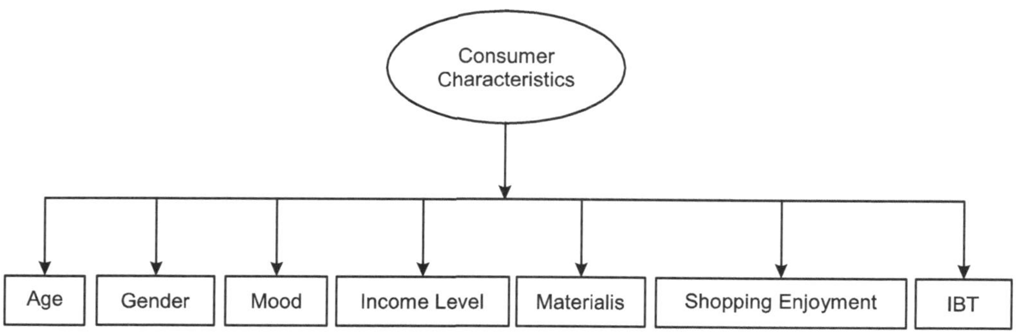

Fig. 2: Consumer Characteristics

3.2 Store Characteristics

Earlier studies related to store characteristics have suggested that the environment of the store is directly linked to individual buying behavior. Beatty & Elizabeth (1998), proposed the idea of store environment and defined it as an attempt to design the environment of the store to generate certain emotional and motivating effects in the consumer's mind that ultimately increase buying chances. According to Faber & Gary (1995), the tangible factors a store environment comprises of are equipment, color scheme, store cleanliness, store layout, product display and arresting interior decoration. On the other side the intangible factors consist of fragrance, temperature, lighting and music (Donovan *et al.*, 1994).

Design factors in a retail environment such as visual elements and physical facilities are deemed as direct sensory stimuli and general design factors include color, shelf space and store layout (Baker *et al.*, 2002). Visual merchandising display is exhibition of products that influence individuals' buying intentions and through the right product display retailers intend to offer consumers the right product at the right time (Baker *et al.*, 1992). Store layout may increase a consumer's interest in store environment and therefore attain customer attention (Bloch *et al.*, 1989).

Social factors in a store consist of consumers and sales people present in the store. Behavior of salespeople in a retail outlet has a key influence on individuals buying decisions (Baker *et al.*, 2002). Another factor is crowding which is complex as well as essential to store environment (Michon *et al.*, 2005). Crowding along with store ambience, design and social factors are more significant element that is considered to influence individuals buying behavior. Crowding in a store is the outcome of social, personal and physical factors that stimulate individual towards potential problems evolved due to limited space (Eroglu & Karen, 1990).

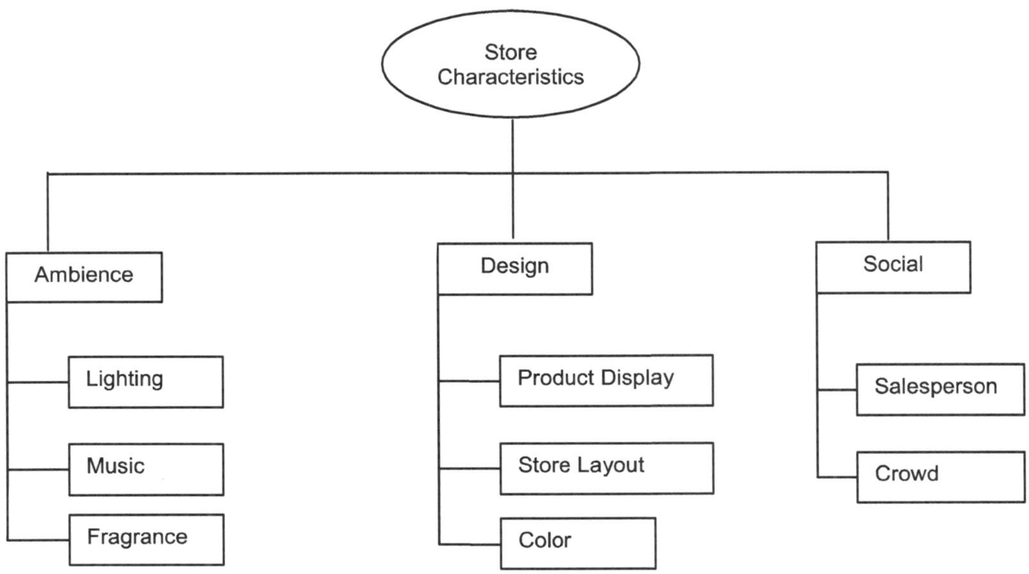

Fig. 3: Store Characteristics

3.3 Website Characteristics

Now a day's internet (E-commerce website) is alternative convenient channel for doing a shopping, which will play a decisive role to encouraging the impulse buying on internet. According to Phau & Chang (2004), internet permits customer to shop any time, offering 24 × 7 days a week shopping. Bellini *et al.,* (2017) investigate that researchers are now identifying the numerous aspects of impulse buying and developing different tactics to inspire buyer to make impulse buying online. Apparel shopping website characteristics play a vibrant role during a process of impulse buying because all online consumers have to interact with it, and it delivers features that can encourage the impulse buying.

Perceived enjoyment means that consumer develop desire in their online apparel shopping and if they enjoy the shopping atmosphere, they devote extra time for shopping there are the main causes of impulse purchasing (Odekerken-Schröder *et al.*, 2003). Impulse buying is highly provocation driven in nature, the environment of E-store is expected to play a significant role in seducing individuals to buy impulsively (Vonkeman *et al.*, 2017). Website store design can be described as the extent to which as an individual considers the E-store visually attractive (Gulas & Peter, 1995). Media format and visual appeal consist of the element that help in development the overall presentation of E-store website such as text, images, video, graphics and display fonts (Parboteeah *et al.*, 2009). The usability (Ease of use) of website can be demonstrated as the level to which a potential consumer expects an online website should be effort free (Castaneda *et al.*, 2009), such as

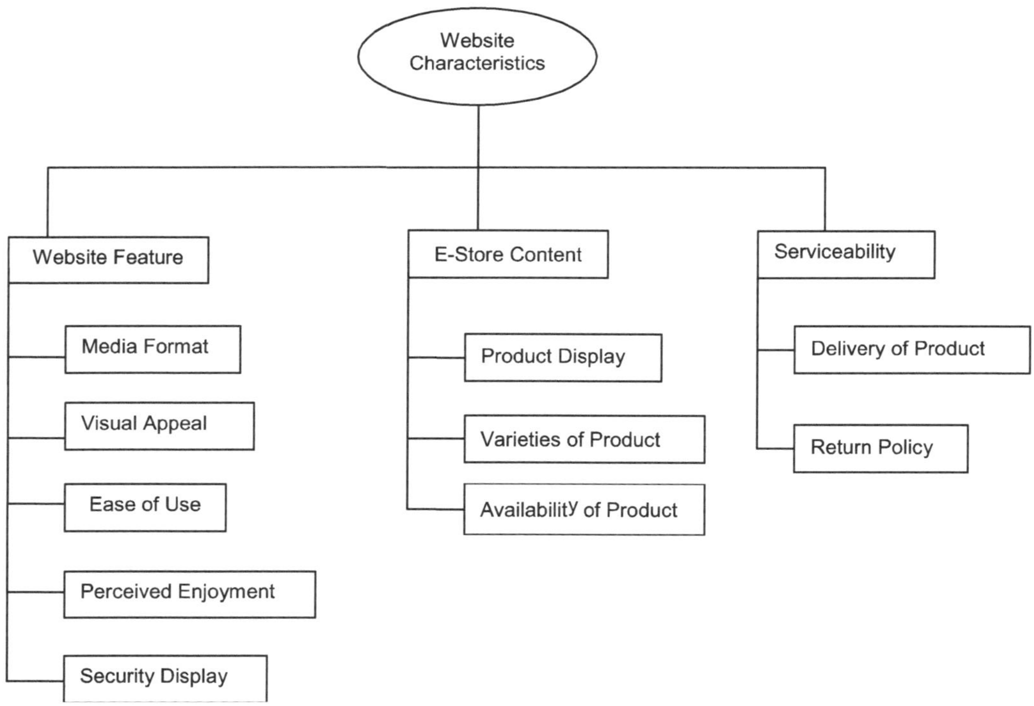

Fig. 4: Website Characteristics

easy functions and ease of moving between pages by scrolling movement and familiarizing with the internet, so individual can access easily through server (Louis *et al.*, 2016). Display of product in a visual attractive way, for example, displaying a number of images, information and videos with customized view and numerous varieties available of product in E-store content specially in case of clothing, which will help the individual get involved in purchase of the product (Liu *et al.*, 2013). The fast delivery and easy return policy come in serviceability, also encourages the impulse buying during promotional activities on a website.

3.4 Situational Characteristics

Situational stimuli or characteristics consist of an individual and environmental aspect that encourage consumer towards impulse buying. Situational factors include availability of time/money, presence of others and sales promotion (Dholakia, 2000). Duarte *et al.*, (2013) suggested that impulse buying behavior of consumer may be affected by special discounts and promotional offers, these activities fulfill the psychological requirement of consumer. Availability of time and money are another situational factors and main key element in a process of impulse buying (Beatty & Elizabeth, 1998). Availability of time during shopping determines whether an individual will get involved in impulse buying or not and limited time during shopping has negative effect on impulse buying as individuals will possibly feel frustrated because of limited time. Therefore, if individuals have extra time then they will spend additional time in a store and they will browse further (Beatty & Elizabeth, 1998).

Availability of money enhances the consumer's buying power and limited availability of money can decrease the chances of impulse purchasing (Beatty& Elizabeth, 1998).

In a shopping situation, presence of others will possibly influence the purchase decision; such effects may possibly occur irrespective of whether the others are family members or peers (Aribarg *et al.*, 2002).

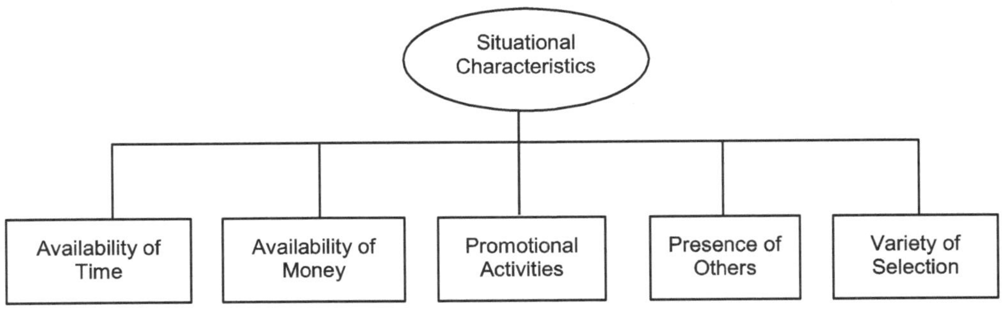

Fig. 5: Situational Factors and Characteristic

3.5 Product Characteristics

Earlier studies on impulse buying categorize products as impulse and non-impulse items. Many studies examined the frequency of impulse purchasing indifferent product class

(Applebaum, 1951). Nowadays Impulse buying behavior is still discussed regarding different types of products which are or are not impulsive (Belleneger *et al.*, 1978). According to Jones *et al.,* (2003) product characteristics can also influence the impulse buying behavior and variables such as the appearance of products in the retail environment and some of the product features have been suggested to be effective on consumer impulse buying behavior (Beatty & Elizabeth, 1998). According to Dittmar *et al.,* (1995) clothing are the most likely shopper for impulse buying and clothing are "consumer goods which appear to have potential for self-presentation, self-expression, mood adjustment, diversion and entertainment.

Product price and type has also been found to be an important aspect of impulse buying (Zhou & Wong, 2003). Youn & Ronald (2000), find that shoppers are buying impulsively, when there are sales available and product is on discount prices. Products which are purchased on impulse are generally inexpensive. Moreover, previous studies show that products are generally purchased impulsively based on their characteristics (Hausman, 2000).

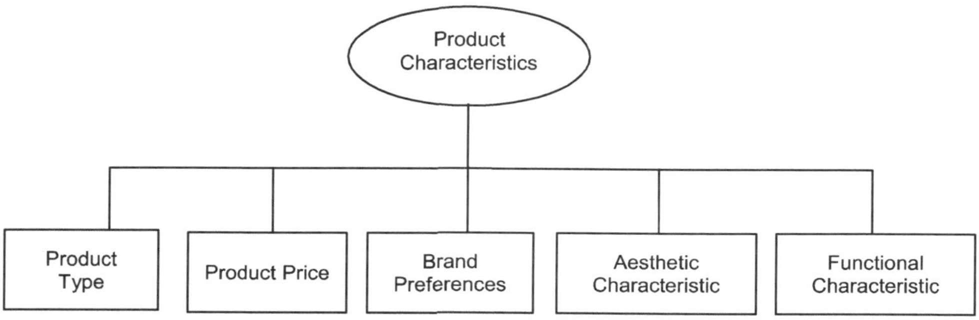

Fig. 6: Product Characteristics

4. RESEARCH FRAMEWORK AND FUTURE RESEARCH

After the detailed review of the literature and analyzation of many research papers, we have derived some predictions for the future research endeavors in the field of impulse buying. In this paper, we propose a framework to increase the understanding of impulse buying behavior. After analyzing the various factors studied in the past studies, we formulized four different groups which could accommodate the various factors. Therefore, we have categorized the various factors under the broad categories of "External Factors or Stimuli", "Internal Factors or Stimuli". In this paper four broad categories evaluated under external factors and internal factors, are Consumer, Store/Website, Situational and Product related characteristics.

Based on the literature review, we now produce a set of research propositions that could help in developing the knowledge on this area of impulse buying behavior. Our proposed model captures the different aspects of impulse buying of the consumers. In this part, we have explored four propositions that arise from the above derived conceptual framework. The future researches could be done under the following lines:

E1: Effect of various Consumer Characteristics on Impulse Buying Behavior
E2: Effect of various Situational Characteristics on Impulse Buying Behavior
E3: Effect of various Store Characteristics on Impulse Buying Behavior
E4: Effect of various Website Characteristics on Impulse Buying Behavior
E5: Effect of various Product Characteristics on Impulse Buying Behavior.

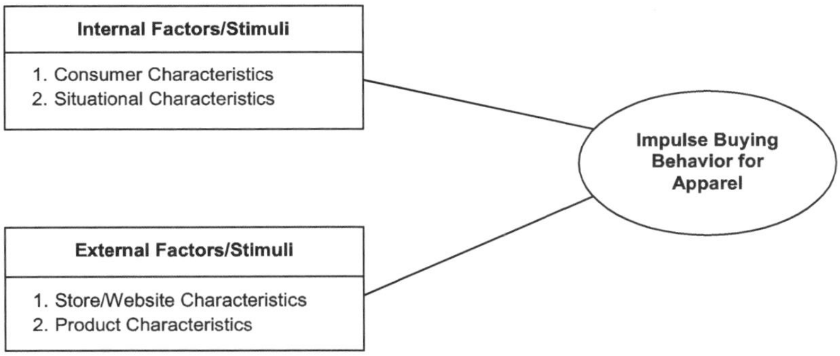

Fig. 7: Research Framework for Impulse Buying Behavior for Apparel Consumer

These above research frame works could be further hypothesized in the future research attempts and studied through exploratory studies. Among these five implications of research, the most challenging implication of our research is the effect of the store or website characteristics and situational factors during impulse purchasing. With the technological development of the retail store formats, only store related aspect is fully under the control of the retailers. External stimuli or factors could be leveraged by the retailers by framing suitable retail strategies to tap the potential consumers inside the store. Future research studies could be extended in this area by evaluating the interactive effects of the various factors and the apparel shopper's personal traits.

5. CONCLUSION

Impulse buying is a composite phenomenon which is a challenge for apparel retailers and researchers to understand the apparel shopper's buying behavior. Marketing practitioner and researchers have mainly focused on evaluating the different characteristics that encourage impulse buying. In the growing economies, there is a need to understand the shopper's impulse buying behavior due to recent technological development in retail atmosphere and huge societal traditional differences when compared to different developed countries. Increases in individual income, lifestyle and easy credibility of money have made an impulse purchasing ubiquitous phenomenon across the online and offline apparel retail formats. Attractive store ambience and visual display of product in retail environment and in-store atmosphere is important in case of offline shopping to increase the sales through unplanned impulse buying. Various website features including media format, visual appeal, e-store content, security and service ability of website play a significant role in online shopping during impulse buying. Both online and offline shopping situation, retailers needs to acknowledge the buying behavior of apparel consumer. Other socio-cultural factors namely culture, tradition and religion also play imperative role in affecting the impulse buying. The present Indian retail scenario provides with a huge opportunity for the end consumer to gravitate towards impulse buying in both contexts i.e. offline and online mode. The detailed literature review can be concluded that the phenomenon of impulse buying has various proportions such as store/website characteristics, situational factors and relationship with the consumer, can lead to strengthen the knowledge and implication in the field of consumer buying behavior. Based upon the frequent change in fashion trends and styles of apparel, it is imperative that impulse buying may turn into a growing area of future investigation and could be seen across the various retail formats.

REFERENCES

[1] Abratt, R. and Stephen, D.G. (1990). Unplanned Buying and In-Store Stimuli in Supermarkets, *Managerial and Decision Economics* 2(1), 1–12.

[2] Anon (2019). Apparel and Garment Industry and Export, Retrieved December 31, 2020 *https://www.ibef.org/exports/apparel-industry-india.aspx.*

[3] Applebaum, W. (1951). Studying Customer Behavior in Retail Stores, *Journal of Marketing,* 16(2), 172–78.

[4] Aribarg, A., Arora, N. and Onur, H.B. (2002). Understanding the Role of Preference Revision and Concession in Group Decisions, *Journal of Marketing Research,* 39(3), 336–49.

[5] Baker, J., Michael, L. and Dhruv, G. (1992). An Experimental Approach to Making Retail Store Environment Decisions. *Journal of Retailing,* 68(Winter), 445–60.

[6] Baker, J., Parasuraman, A., Dhruv, G. and Glenn, B.V. (2002). The Influence of Multiple Store Envi Ronment Cues on Pe Received Me Rchandise Value and Patron Age Intentions, *Journal of Marketing,* 66 (April), 120–41.

[7] Bayley, G. and Clive, N. (1998). Impulse Purchasing: A Qualitative Exploration of the Phenomenon, *Qualitative Market Research: An International Journal,* 1(2), 99–114.

[8] Beatty, S.E. and Elizabeth, M.F. (1998). Impulse Buying: Modeling Its Precursors, *Journal of Retailing,* 74(2), 169–91.

[9] Belleneger, D.N., Dan, H.R. and Elizabeth, C.H. (1978). Impulse Buying Varies by Product, *Journal of Advertising Research,* 18(6), 15.

[10] Bellenger, D. and Korgaonkar, P. (1980). Profiling the Recreational Shopper, *Journal of Retailing,* 56(03), 77–92.

[11] Bellini, S., Maria, G.C. and Benedetta, G. (2017). A Structural Equation Model of Impulse Buying Behaviour in Grocery Retailing, *Journal of Retailing and Consumer Services,* 36, 164–71.

[12] Bloch, P.H., Nancy, M.R. and Daniel, L.S. (1989). Extending the Concept of Shopping: An Investigation of Browsing Activity, *Journal of the Academy of Marketing Science,* 17(1), 13–21.

[13] Block, L.G. and Vicki, G.M. (1999). Shopping Lists as an External Memory Aid for Grocery Shopping: Influences on List Writing and List Fulfillment, *Journal of Consumer Psychology,* 8(4), 343–75.

[14] Castaneda, A.J., Frias, D.M. and Rodriguez, M.A. (2009). Antecedents of Internet Accepatnce and Use as an Information Source by Tourist, *Online Information Review,* 33(3), 548–67.

[15] Chang, H.J., Ruoh, N.Y. and Molly, E. (2014). Moderating Effects of Situational Characteristics on Impulse Buying, *International Journal of Retail and Distribution Management,* 42(4), 298–314.

[16] Chen-Yu, J.H. and Yoo, K.S. (2002). Adolescents Clothing Purchase Motivations, Information Sources, and Store Selection Criteria: A Comparison of Male/Female and Impulse/Nonimpulse Shoppers, *Family and Consumer Sciences Research Journal,* 31(1), 50–77.

[17] Cobb, C.J. and Hoyer, WD. (1986). Planned versus Impulse Purchase Behavior, *Journal of Retailing,* 62(4), 384–409.

[18] Dawson, S. and Minjeong, K. (2009). External and Internal Trigger Cues of Impulse Buying Online, *Direct Marketing,* 3(1), 20–34.

[19] Dholakia, U.M. (2000). Temptation and Resistance: An Integrated Model of Consumption Impulse Formation and Enactment, *Psychology and Marketing,* 17(11), 955–82.

[20] Dittmar, H., Jane, B. and Susanne, F. (1995). Gender Identity and Material Symbols: Objects and Decision Considerations in Impulse Purchases, *Journal of Economic Psychology,* 16(3), 491–511.

[21] Dittmar, H., Jane, B. and Susanne, F. (2004). Objects, Decision Considerations and Self-Image in Mens and Womens Impulse Purchases, *Leleannec. Free. Fr,* 1–13.

[22] Donovan, R.J., John, R.R., Gilian, M. and Andrew, N. (1994). Store Atmosphere and Purchasing Behavior, *Journal of Retailing,* 70(3), 283–94.

[23] Duarte, P., Mário, R. and Marlene, F. (2013). Drivers of Snack Foods Impulse Buying Behaviour among Young Consumers, *British Food Journal,* 115(9), 1233–54.

[24] Eroglu, S.A. and Karen, A.M. (1990). An Empirical Study of Retail Crowding: Antecedents and Consequences, *Journal of Retailing,* 66(2), 201–21.

[25] Faber, R.J. and Christenson, G.A. (1996). In the Mood to Buy: Differences in the Mood States Experienced by Compulsive Buyers and Other Consumers, *Psychology & Marketing,* 13(8), 803–19.

[26] Faber, R.J. and Gary, A.C. (1995). Can You Buy Happiness?: A Comparison of the Antecedent and Concurrent Moods Associated with the Shopping of Comulsive and Non-Compulsive Buyer, *AMA Winter Educators Conference Proceedings*, 378–79.

[27] Gardner, M.P. and Dennis, W.R. (1988). Eflfects of Impulse Purchases on Consumers Affective States, *Advances in Consumer Research,* 15, 127–30.

[28] Gulas, C.S. and Peter, H.B. (1995). Right under Our Noses: Ambient Scent and Consumer Responses, *Journal of Business and Psychology,* 10(1), 87–98.

[29] Han, Y.K., George, A.M., Antigone, K. and Jikyeong, K.P. (1991). Impulse Buying Behavior of Apparel Purchasers, *Clothing and Textiles Research Journal,* 9(3), 15–21.

[30] Hausman, A. (2000). A Multi-Method Investigation of Consumer Motivations in Impulse Buying Behavior, *Journal of Consumer Marketing,* 17(5), 403–19.

[31] Hawkins, S. (1962). The Significance of Impulse Buying Today, *Journal of Marketing,* 26(2), 59.

[32] Helmers, K.F., Young, S.N. and Pihl, R.O. (1995). Personality and Individual Differences and Undefined, Assessment of Measures of Impulsivity in Healthy Male Volunteers, *Personality and Individual Differences,* 19(6), 927–35.

[33] Jones, M.A., Reynolds, K.E., Weun, S. and Beatty, S.E. (2003). Journal of Business Research and Undefined, The Product-Specific Nature of Impulse Buying Tendency, *Journal of Business Research*, 56(7), 505–11.

[34] Liu, Y., Hongxiu, L. and Feng, H. (2013). Website Attributes in Urging Online Impulse Purchase, An Empirical Investigation on Consumer Perceptions, *Decision Support Systems,* 55(3), 829–37.

[35] Logue, A.W. and Adolfo, C. (1992). Self-Control and Impulsiveness in Preschool Children, *The Psychological Record,* 42(2), 189–204.

[36] Louis, L.S., Sheng, W.L. and Li, Y.H. (2016). Motivation for Online Impulse Buying: A Two-Factor Theory Perspective, *International Journal of Information Management,* 36 (5), 759–72.

[37] Mariam, T. (2003). The Art of Buying: Coming to Terms with Money and Materialism, *Journal of Happiness Studies,* 4, 405–35.

[38] Michon, R., Jean-Charles, C. and Turley, L.W. (2005). Mall Atmospherics: The Interaction Effects of the Mall Environment on Shopping Behavior, *Journal of Business Research* 58, 576–83.

[39] Odekerken-Schröder, G., Kristof, D.W. and Patrick, S. (2003). Strengthening Outcomes of Retailer-Consumer Relationships, The Dual Impact of Relationship Marketing Tactics and Consumer Personality, *Journal of Business Research,* 56(3), 177–90.

[40] Parboteeah, D.V., Joseph, S.V. and John, D.W. (2009). The Influence of Website Characteristics on a Consumer's Urge to Buy Impulsively, *Information Systems Research,* 20(1), 60–78.

[41] Phau, I. and Chang, C.L. (2004). Profiling Fashion Innovators, A Study of Self-Concept, Impulse Buying and Internet Purchase Intent, *Journal of Fashion Marketing and Management,* 8(4), 399–411.

[42] Piron, F. (1991). Defining Impulse Purchasing, *ACR North American Advances,* NA-18.

[43] Prasad, V.K. (1975). Unplanned Buying in Two Retail Settings, *Journal of Retailing,* 51(3), 3–12.

[44] Rook, D.W. and Hoch, S.J. (1985). Consuming Impulses. ACR North American Advances., *ACR North American Advances* 23–28.

[45] Rook, D.W. (1987). The Buying Impulse, *Journal of Consumer Research,* 14(2), 188–89.

[46] Rook, D.W. and Robert, J. F. (1995). Normative Influences on Impulsive Buying Behavior, *Journal of Consumer Research*, 22(3), 305–13.

[47] Rook, D.W. and Gardner, M.P. (1993). In the Mood: Impulse Buying's Affective Antecedents, *Research in Consumer Behavior*, 1, 1–28.

[48] Shukla, P.S., Vyas, P.H. and Pandya, M.N. (2018). Investigating the Relationship between Visual Merchandising and Impulse Buying of Apparel Products by Selected Retail Shoppers in Vadodara City of Gujarat, *Review of Professional Management—A Journal of New Delhi Institute of Management,* 16(1), 17.

[49] Sirhindi, A. (2010). A Critical Review of In-Store and Online Impulse Purchase Behavior, *ProQuest Dissertations and Theses,* 48.

[50] Vonkeman, C., Tibert, V. and Willemijn, V.D. (2017). Role of Local Presence in Online Impulse Buying, *Information and Management,* 54(8), 1038–48.

[51] Weinberg, P. and Wolfgang, G. (1982). Impulsive Consumer Buying as a Result of Emotions, *Journal of Business Research,* 10(1), 43–57.

[52] West, J.C. (1951). Results of Two Years of Study into Impulse Buying, *Journal of Marketing,* 15(3), 362–63.

[53] Youn, S. and Ronald, J.F. (2000). Impulse Buying: Its Relation to Personality Traits and Cues, *ACR North American Advances,* NA-27.

[54] Zhou, L. and Wong, A. (2003). Consumer Impulse Buying and In-Store Stimuli in Chinese Supermarkets, *Journal of International Consumer Marketing,* 16(2), 37–41.

SECTION-IV

Technical Textiles and Composite Materials

Bacterial Cellulose Production by *Acetobacter xylinus* 2526 for Medical Textile Applications

Mukesh Kumar Singh[1], **Ashutosh Pandey**[2] **and Annika Singh**[3]

[1,2]Uttar Pradesh Textile Technology Institute, Kanpur, Uttar Pradesh, India

[3]Institute of Biosciences and Biotechnology, CSJMU, Kanpur, India

E-mail: [1]mukesh70ster@gmail.com

ABSTRACT: *The bacterial cellulose possesses unique properties over plant cellulose like high purity level, high water retention power, high crystallinity, and tensile strength, making it a suitable compound to produce various biocomposites of medicinal uses. The production cost of bacterial cellulose is much higher, which become a major hurdle to be used BC as a component in biocomposites.*

The present study is focused on the production of low-cost BC by using different media composition and waste material. The quality of BC is also monitored by using different analytical techniques like Fourier transform infrared spectroscopy (FTIR), X-ray diffraction (XRD), scanning electron microscopy (SEM), differential scanning calorimetry (DSC), thermo gravimetric analysis (TGA), tensile testing, and a digital moisture meter. BC's maximum yield was found 6.2 g/l under shaking condition for 96 h under Hestrin Schramm medium. Bacterial cellulose bio films are characterized as a transparent porous mesh of micro fibrils used as a drug carrier for various wound treatment. The porous structure becomes appropriate material to absorb fluids, and fibrous morphology becomes suitable to work as a successful preform for biocomposites. This study can establish a new protocol to synthesize bacterial Cellulose from Acetobacter xylinus 2526 NCIM to achieve maximum production yield. The revival of bacteria is performed in standard broth HS medium. Produced small pellicles are selected further to produce bacterial cellulose in both dynamic and static conditions to engineer the morphology of bacterial cellulose to achieve desired physical properties and morphology.

Keywords: *Acetobacter xylinum*, Bacterial Cellulose, Differential Scanning Calorimeter, Scanning Electron Microscopy, Thermal Gravimetric Analyzer.

1. INTRODUCTION

The bacterial cellulose was first reported and characterized by Brown in 1988. The Bacterial Cellulose is quite identical to plant Cellulose in chemistry, but the purity level of BC is higher than plant cellulose (Andriani *et al.*, 2020). Both plant and bacterial Cellulose is a linear polymer in which D-glucose units are linked by β (1→4)-glycoside bonds (Revin *et al.*, 2018). It is established from different studies that the chemical structure of Bacterial Cellulose is similar to plant cellulose; both are linear polymer of D-glucose unit linked by β (1→4)-glycoside bonds. BC structure exhibits highly pure unique 3-D nanostructure, transparent, enough mechanical strength, high crystallinity, high water holding capacity, biodegradable and biocompatible. Various bacteria produce bacterial Cellulose with different yield and morphology.

Bacterial Cellulose is produced in the form of a pellicle of the desired size and shapes on a culture medium's surface in a dynamic condition or gel-like pulp in static condition (Jagannath *et al.*, 2008). The cost of BC-based biomaterial is high due to low yield production. The reported yield 2.6 g/L of BC was produced in 1954 after 67 hours, and the cost of developed medical products 50 times higher than plant cellulose (Shi *et al.*, 2014, Liu *et al.*, 2018, Gullo *et al.*, 2018, Revin *et al.*, 2018). Therefore, researchers have tried to increase bacterial cellulose yield from Acetobacter species by using various biochemical and natural sources (Son *et al.*, 2001). The previous study shows that bacterial cellulose yield depends on several factors such as Carbon source, Nitrogen source, Incubation temperature, Incubation time, type of bacterial strain, condition for revival of bacteria, time of shaking, rpm of shaking, pH, the concentration of carbon sources and nitrogen source (Zhang *et al.*, 2016).

Therefore, in this study, bacterial Cellulose was produced from *Acetobacter xylinum* NCIM 2526 under both statistic and dynamic conditions. Large pellicles were produced in dynamic condition, and 1.5 mm thick membrane was produced in dynamic condition. Bacterial cellulose pellicles and membrane have ribbon-like micro fibril structure connected with physical force.

2. MATERIALS AND METHOD

2.1 Revive the *Acetobacter xylinus* NCIM 2526

The bacterial strain was purchased in the form of agar tube. The bacterial cellulose was required to revive first with maintenance medium containing D-sorbitol 5 g/l, Yeast extract 0.5 g/l, Agar 2 g/l. at pH 6.2. One-third of sterilized maintenance medium was poured in 90 mm Petri plates followed by UV radiation to avoid possible contaminations and solidification of agar media as similarly opted by Ullah *et al.,* (2019). Solidification, subsequently A. sylinum inoculated on agar plates with the help of sterilized inoculum loop. Streaked agar plates were stored in an incubator at 30°C for 72 h. The colonies of A. xylinum appeared as shown in Figure 1.

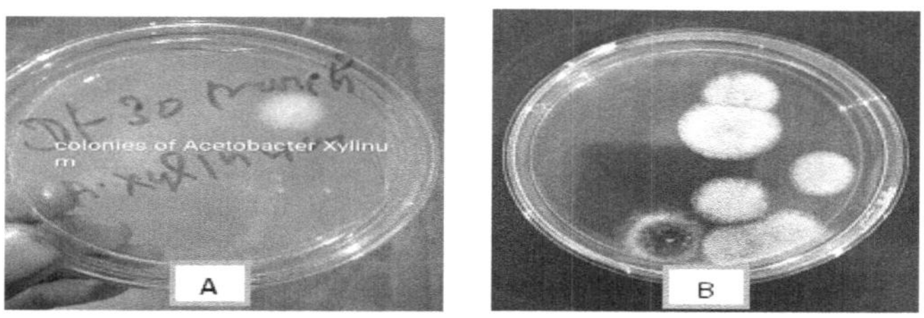

Fig. 1: *Acetobacter xylinum* NCIM 2526 Colony Formation

2.2 Bacterial Cellulose Production

Hestrin and Schramm (HS) medium containing 20 g/l glucose, 5 g/l yeast extract, 5 g/l peptone, 2.7 g/l K_2HPO_4 and 1.15% g/l citric acid was prepared and divided into two parts, first part is carbon source which contains Glucose only and second parts is a nutrient source with pH between 5.5 to 6.2. Both carbon source and nitrogen source were sterilized at 121°C under 15 psi pressures for 22 min. After cool down, both carbon and nitrogen sources were mixed by stirring.

The culture was performed by opting 150 ml HS in a conical flask, followed by 3 mm revive colonies by sterilized inoculum loop. The inoculation followed by shaking at 30°C and 110 rpm. Small pellicles of BC appeared after 72 h as shown in Figure 2.

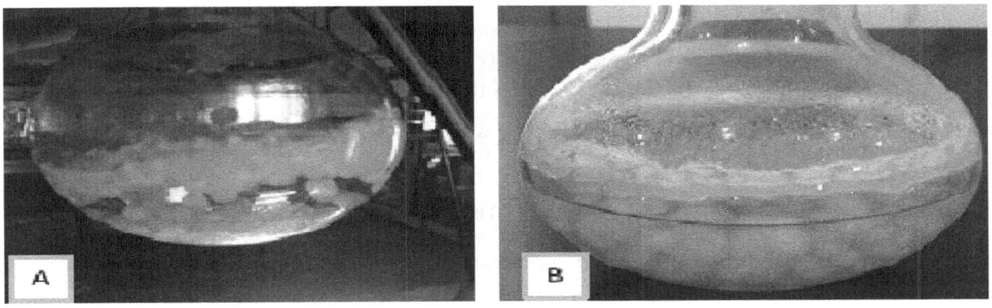

Fig. 2: BC Pellicles Synthesized by Acetobacter Xylinum Bacteria

The small BC pellicles were inoculated in fresh HS media and kept at 110 rpm, and 26°C for 72 h big size pellicles appeared which further converted to the thick membrane after 96 h static storage Figure 3.

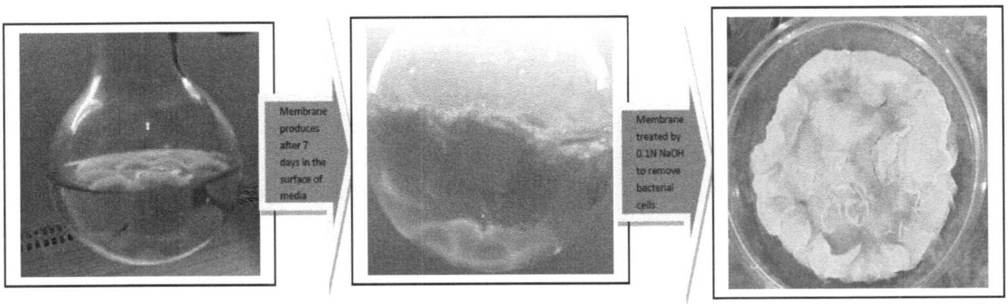

(A) (B) (C)

Fig. 3: A is Showing Large Pellicles of BC after 4 Days and B is Showing Membrane of BC Developed on the Surface and C is Showing Dried BC Membrane

After the incubation of 168 h large white pellicles and BC membrane appeared on the media surface, it was removed with 0.1 N NaOH for 30 min, followed by two hot and two cold wash with distilled a neutral pH was attained in the drained water. The cleaned BC membranes were dried and conditioned for characterization. The complete process of BC production is shown in Figure 4.

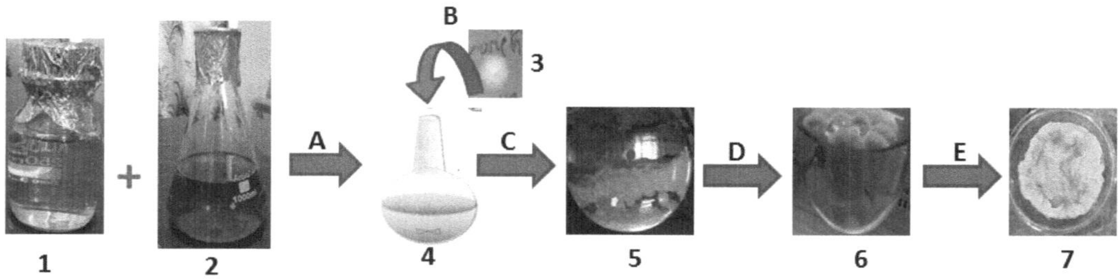

1. Sterilized Carbon Source
2. Sterilized Nitrogen Source
3. Revive Acetobacter Xylinus Bacteria
4. HS media (Carbon + nitrogen source)
5. Small pellicles of Bacterial Cellulose
6. Large Pellicles of Bacterial Cellulose
7. Membrane of Bacterial Cellulose

A. Cool and mix both Carbon and Nitrogen sources.
B. Inoculum Revive Acetobacter Xylinus Colony.
C. Small pellicles produce after 2 day incubate at 30°C.
D. Again Inoculum small pellicles in fress HS media and Large Pellicles Produce after 3 days incubate at 30°C.
E. After 3 day bacterial cellulose membrane produce on the surface of HS media.

Fig. 4: Flow Chart of Bacterial Cellulose Production in HS Media

2.3 Characterization of Bacterial Cellulose

The chemical purity of bacterial cellulose was characterized in transmission mode using attenuated total reflection (ATR) facility in an Alpha-model FTIR spectrometer of Bruker. BC's FTIR spectra were recorded in the range of 100 cm^{-1} to 4000 cm^{-1} wave number.

Scanning Electron Microscope JEOL JCM-7000 model observed the surface morphology of synthesized BC. BC Samples were coated with a gold layer for 1 min by coater. Zero mag smooth imaging, live analysis of constituents and three-dimensional image creations became possible with Neo Scope software's assistance in Jeol JCM-7000 SEM.

The crystallinity index and crystallite size of the bacterial cellulose were characterized by PANalytical Multipurpose X-ray diffraction machine. XRD patterns of the BC powder were recorded at the CuKα radiation wavelength (λ = 1.54 Å), generated at a voltage of 45 kV and a filament emission of 40 mA. Samples were scanned in 2θ range from 0° to 70°. The scan speed was $0.5°\cdot min^{-1}$ with a step size of 0.02°. High score software was used to analyze XRD data. The Scherer equation estimated the crystallite size (L) of bacterial cellulose.

$$L = K\lambda/\beta \cos\theta$$

K is constant, which depends upon lattice direction and crystallite morphology, and β is the full width at half maximum (FWHM).

The thermal transitions of BC samples were recorded by differential scanning calorimeter model DSC 8000 of Perkin Elmer under nitrogen atmosphere. The BC sample containing crucibles first cooled to –90°C, held isothermally for 1 min and then ramped to 30°C and finally to 400°C at a heating rate 10°C/min. Pyris DSC 8000 software was used to record and calculate Tg, Tm and crystallinity. The weight loss and derivative curves were recorded and analyzed Pyris TGA 4000 software.

The tensile behavior of BC membrane was examined by Instron universal tensile testing machine with 1 kN static load cell, and 1.5 mm thick sheets of BC were cut into 30 × 5 mm strips. The tensile test was performed at a gauge length of 25 mm with a 2 mm/min extension rate. The tensile data were analyzed by using Instron blue hill lite software.

3. RESULTS AND DISCUSSION

The yield of Bacterial Cellulose (BC) is recorded after weighing in dried condition, and it was found 6.56 ± 0.57 g/L under static conditions and 7.31 ± 0.79 g/L under shaking conditions.

3.1 FTIR Analysis

The FTIR spectra of bacterial cellulose prepared from *Acetobacter Xylinum* is shown in Figure 5. The transmittance band at 3278 cm^{-1} is assigned to hydroxyl groups stretching, and transmittance band at 2925 cm^{-1} and 1373 cm^{-1} to 1319 cm^{-1} establishes the stretching and deformation vibrations of C-H group in glucose unit. The absorbance bands at 1640 cm^{-1} and 1543 cm^{-1} are assigned to stretching and high IR wave absorption by C=O group in glucose unit. The absorption band at 551 cm^{-1} is characteristic of β-glycosidic linkage between glucose units. The signal at 1028 cm^{-1} is assigned to the C-O group of secondary alcohols and ethers functions existing in the backbone's bacterial cellulose chain. The analysis of bacterial cellulose produced from *Acetobacter xylinus* 2526 NCIM showed peaks at 3278 cm^{-1}, 2925 cm^{-1}, 1640 cm^{-1}, 1543 cm^{-1}, 1373 cm^{-1}, 1319 cm^{-1} and 1028 cm^{-1} which confirming the purity of the bacterial cellulose (Revin *et al.*, 2018).

Slight variation in fingerprint absorption peaks is possible depending on the origin of cellulose bacterial species. The produced BC is pure cellulose as confirmed by the above FTIR results.

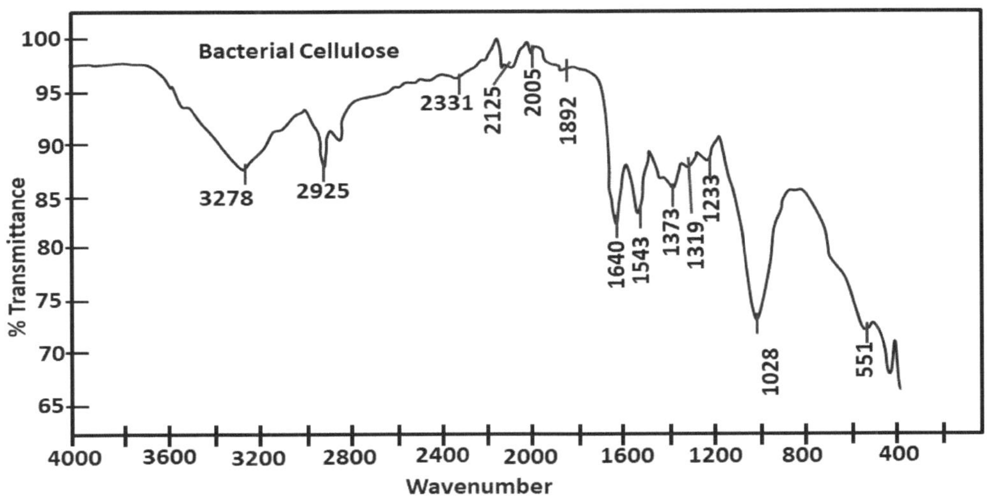

Fig. 5: FTIR Graph of Bacterial Cellulose

3.2 X-ray Diffraction Analysis

The X-ray diffraction pattern of BC is illustrated in Figure 6. The diffraction diagrams of the BC produced in HS media revealed three characteristic diffraction peaks, which indicates the presence of both Iα and Iβ cellulose crystals.

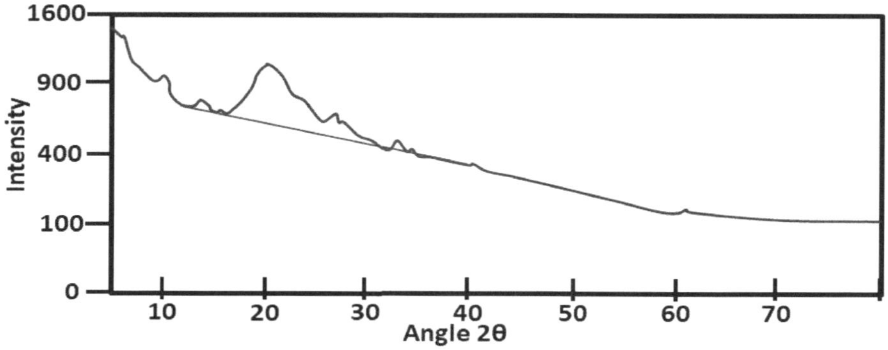

Fig. 6: X-ray Diffracting Graph of Bacterial Cellulose

The crystallinity and crystal size data are summarized in Table 1. The percent crystallinity as obtained by X-ray diffraction is defined as the ratio of intensity from the crystalline peaks to the sum of the crystalline and amorphous intensities:

Percent Crystallinity = Int. Crystalline/(Int. Crystalline + Int. Amorphous)

The highest crystalline content in BC produced by Acetobacter Xylinum in HS Media was 68%, and the average crystallite size was 2.328 nm.

Table 1: X-ray Diffraction Peak

Pos.[°2θ]	*Height [cts]*	*FWHM Left.[°2θ]*	*d-spacing[Å]*	*Rel. Int.[%]*
10.1123	129.60	0.8817	8.74757	82.18
27.4244	157.69	0.7557	3.25228	100.00
32.9059	90.42	0.5038	2.72197	57.34

3.3 Scanning Electron Microscopy

In Figure 7A is an SEM image of bacterial cellulose pellicles shows ribbon-like Micro fibril structure connected with physical force at 15 KV and 11000 magnificence. The average diameter of fibril in micron is 4.7. Ribbon like micro fibril structure arranged in 3-D order.

In Figure 7B is an SEM image of bacterial cellulose membrane shows highly connected micro fibril pore structure. During observation irregular pores and a continuous, inter-connected porous network shown in BC membrane. The porous structure is not uniformly distributed throughout the entire cross-sectional area of samples and imaged their structure using SEM. As displayed in Figure 7 at 1-micron range, BC Membrane has pore architecture, in which the pore size was not uniform throughout the substrate.

Bacterial cellulose morphology is affected when this is synthesized in both dynamic and static condition. Bacterial cellulose pellicle shows ribbon-like fibril structure with large diameter arranged in irregular 3-D structure. While Bacterial cellulose membrane shows, ribbon likes micro fibrils structure developed irregular pours geometry. It reduces pore size as well as pores density.

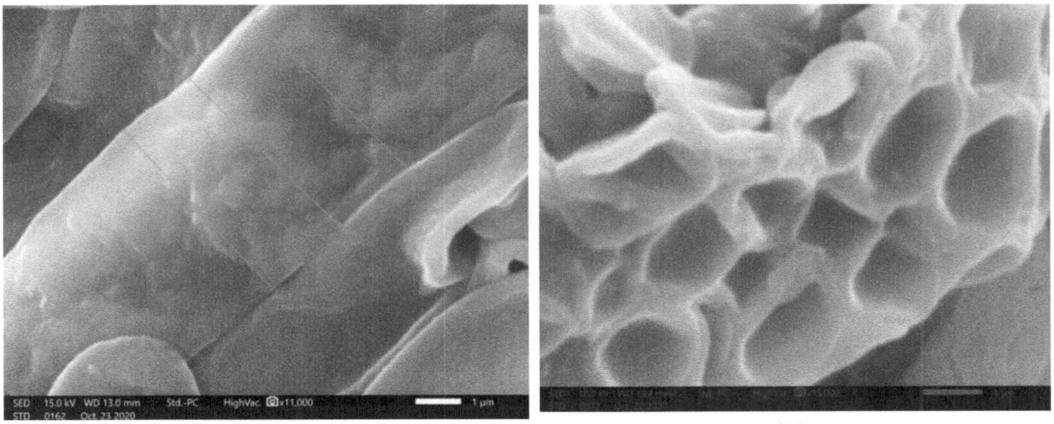

(A) (B)

Fig. 7: (A) is SEM Image of Bacterial Cellulose Pellicle, (B) SEM Image of Bacterial Cellulose Membrane

3.4 Tensile and Water Holding Potential

1 g of Bacterial cellulose pellicles and membrane was considered sample and put into the sample pan to measure the moisture-holding ability. After 24 hour conditioning of bacterial cellulose pellicles and membrane at 65% RH and 27°C, the water holding potential was found 69% and 63% respectively for pellicles and membrane.

3.5 Tensile Strength

Twenty samples were considered for tensile testing. The average tensile strength of membrane BC was found 21.6 MPa with 55% elongation.

4. CONCLUSION

This effort of increasing the yield of bacterial synthesis has been improved up to 6.8 g/l from 3.4 g/l. The bacterial cellulose is an open structure for enormous medical science applications from tissue engineering to drug carrier for controlled drug delivery. Bacterial cellulose has found quite suitable material for biosensor because it has a very low transition at 18°C. Higher water retention potential extends new aspects of green smart materials for medical textiles.

REFERENCES

[1] Andriani, D., Apriyana, A.Y. and Karina, M. (2020). The Optimization of Bacterial Cellulose Production and its Applications: A Review, *Cellulose*, 27, 6747–6766.

[2] Barud, H.S., Ribeiro, C.A., Crespi, M.S., Martines, M.A.U., Dexpert-Ghys, J., Marques, R.F.C., Y. Messaddeq, S.J. and Ribeiro, L. (2007). *Journal of Therm. Anal. & Calo.*, 87(3), 815–818.

[3] Chen, S.Q., Lopez-Sanchez, P., Wang, D., Mikkelsen, D. and Gidley, M.J. (2018). *Food Hydrocolloids,* 81, 87–95.

[4] George, J., Ramana K.V., Sabapathy, S.N., Jagannath, J.H. and Bawa, A.S. (2005). *Int. Journal of Bio. Macro.* 37, 189–194.

[5] Gullo, M., La-China, S., Falcone, P.M. and Giudici, P. (2018). *Appl. Micro. and Biotech*, 102(16), 6885–6898.

[6] Klemm, D., Schumann, D., Udhardt, U. and Marsch, S. (2001). *Progress in Polymer Science,* 26(9), 1561–1603.

[7] Liu, M., Li, S., Xie, Y., Jia, S., Hou, Y., Zou, Y. and Zhong, C. (2018). *Applied Micro-bio. and Biotech*. 102(3), 1155–116.

[8] Revin, V., Liyaskina, E., Nazarkina, M., Bogatyreva, A. and Shchankin, M. (2018). *Brazilian Journal of Microbio,* 49, 151–159.

[9] Son, H.J., Heo, M.S., Kim, Y.G. and Lee, S.J. (2001). *Biotech. and Applied Biochem.*, 33(1), 1–5.

[10] Stumpf, T.R., Yang, X., Zhang, J. and Cao, X. (2018) *Material Science and Engineering,* C, 82, 372–383.

[11] Ullah, M.W., Manan, S., Kiprono, S.J., Ul-Islam, M. and Yang, G. (2019). *Fundamentals to Advanced Materials,* First Edition, Wiley-VCH Verlag GmbH & Co. KGaA, 81–110.

[12] Zhang, H., Chen, C., Zhu, C.H.U.N.L.I.N. and Sun, D. (2016). *Cellu. Chem. and Tech., 50*(9–10), 997–1003.

Development of Needle Punched Nonwovens from Textile Waste for Thermal Insulation Application

N. Muthu Kumar, G. Thilagavathi and V. Vinoth
Department of Textile Technology, PSG College of Technology, Coimbatore, Tamil Nadu, India
E-mail: *nmk.textile@psgtech.ac.in

ABSTRACT: *The growing environmental awareness throughout the world has triggered a shift towards developing environmentally friendly materials from renewable resources. The aim of this work is to develop needle-punched non-woven fabrics from textile waste for thermal insulation application and also to analyze the influence of density and porosity of the nonwovens on the thermal insulation performance. The needle punched nonwovens were manufactured using comber noil, silk cocoon waste, and P/C flat waste. Also, a nonwoven has been developed by blending comber noil with silk cocoon waste (50/50%). It was found that the nonwoven developed from 100% silk cocoon waste has better thermal insulation value (0.201 m^2K/W) compared to 100% comber noil and P/C flat waste. It was also found that blending of silk cocoon with comber noil improves thermal insulation of the nonwovens. 50/50 silk cocoon/comber noil nonwoven has good thermal insulation value (0.212 m^2K/W) among the developed nonwovens. Decrease in thermal insulation was observed with increase in porosity of nonwovens. The test results showed that all the developed nonwovens have an excellent thermal insulation performance and are comparable with that of conventional insulation materials.*

Keywords: Air Permeability, Comber Noil, Needle Punched Nonwoven, Silk Cocoon, Thermal Insulation.

1. INTRODUCTION

In recent years, recycling has been an attractive issue for researchers considering the sustainability of the wastes arising from different engineering applications. Huge amount of waste occurs in textile industries, such as fibres, yarns and fabrics while converting one form of raw material into other form. Despite that, a small amount is recycled or incinerated by charities or companies while the remaining waste is thus wasted. Recycling textile waste is not yet a widespread activity, but researchers suggest that it will be developed rapidly (Ghosh, 2016; Jabbar, 2020). Already, more large companies are integrating the situation and many are calling on specialized providers for the treatment of textile waste as thermal insulator. The use of textile waste for thermal insulation would add value to this local resource as an interesting alternative to the conventional insulators (e.g. glass wool, stone wool, etc.).

Silk, 'The Queen of Textiles' is a fibre extruded by the mature silkworm, *Bombyx mori*. The silkworm spins the cocoon around itself as a protective covering while it transforms into pupa and then moth. The cocoon is hard and strong yet elastic and soft to allow movement inside the cocoon. The silk in the cocoon consists of filamentous fibroin bonded by gummy sericin. Silk thread is made by unwinding silk from the cocoons. However, about 25% to 35% of silk in a cocoon can not be reeled and is removed as silk waste. In general, the silkworm cocoon is expensive. However, most of the pierced cocoons or very dirty cocoons of the silkworm are discarded as industrial wastes and can be obtained as a low cost material (Zhang, 2013). Cotton is one of the most important commodity fibres and is widely employed in apparels. At present, the share of natural fibres in production of nonwoven fabrics is low and are used in opt applications. The cotton fibre is conventionally converted into woven and knitted fabrics by short staple spinning methods. The comber noil is short fibre waste produced when cotton yarns are combed.

The thermal insulation properties of nonwovens made from textile waste fibres have been investigated by some researchers. The thermal insulating mats were manufactured from waste wool and recycled PET fibres for building industry applications by Patnaik *et al.,* (2015) and characterized in terms of thermal insulation, acoustic, moisture absorption, fire properties. The results showed that developed mats had good thermal insulation, acoustic absorption, moisture absorption and good fire properties. Wazna *et al.,* (2019) have developed needle punched nonwovens from acrylic and wool wastes and showed that all developed nonwovens had an excellent insulation performance. Rajkumar *et al.,* (2018) have manufactured needle punched nonwoven fabrics by blending silk waste fibres with wool fibres and characterized the comfort properties such as air permeability, wickability and thermal conductivity of the developed nonwovens. The results showed that the blend proportion has a significant effect on the comfort properties of the nonwoven fabrics. In our previous study, needle punched nonwovens have been developed from 100% recycled PET fibres and also by blending with comber noil for thermal insulation applications. The results showed that the addition of comber noil with r-PET fibres decreases the thermal insulation performance of the nonwovens (Muthukumar *et al.*, 2020).

The objective of this work is to develop needle punched nonwovens from textile industry wastes for thermal insulation application. Three types of wastes i.e. comber noil, polyester/cotton flat waste and silk cocoon waste were used. Nonwovens were developed from 100% comber noil, 100% silk cocoon and P/C flat waste. Also a nonwoven was developed by blending silk cocoon waste fibres and comber noil in 50/50 proportion. In total, four needle punched nonwoven samples were developed and thermal insulation performance of the developed nonwovens have been studied.

2. MATERIALS USED

The silk cocoon waste fibre was sourced from Central Silk Board, Bangalore, India. The comber noil and 70/30 P/C flat waste were sourced from in and around Coimbatore, India. The properties of the fibres used in this study are given in Tables 1 and 2. The images of the silk cocoon and silk cocoon waste fibres are shown in Figure 1.

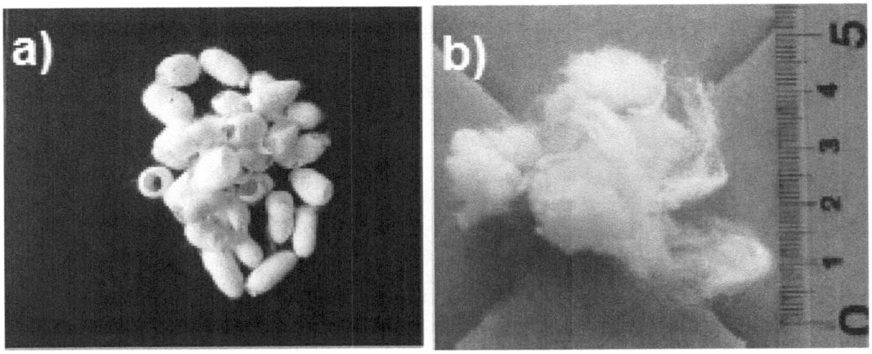

Fig. 1: Images of Silk Cocoon (a) and Silk Cocoon Waste (b)

Table 1: Properties of Silk Cocoon Waste Fibres and Comber Noil

Silk Cocoon Waste Fibre		*Comber Noil*	
Fibre length (mm)	38.00	2.5% span length (mm)	19.38
Fibre fineness (denier)	1.03	50% span length (mm)	8.42
Tenacity (g/denier)	4.68	Fineness (micrograms/inch)	2.54
Elongation (%)	19.90	Strength (g/tex)	15.50
		Elongation (%)	7.40

Table 2: Properties of 70/30 P/C flat Waste

Polyester Fibre		*Cotton Fibre*	
Fibre length (mm)	38.00	2.5% span length (mm)	29.50
Fibre fineness (denier)	0.80	Fineness (micrograms/inch)	3.90
Tenacity (g/denier)	6.70	Strength (g/tex)	24.70
Elongation (%)	20.00	Elongation (%)	8.00

3. NONWOVEN MANUFACTURING

Nonwovens were developed from 100% silk cocoon waste, 100% comber noil and P/C flat waste. Also, a nonwoven was developed by blending silk cocoon waste with comber noil at 50/50 blend ratio. Totally four nonwovens were developed in this study. The laboratory scale carding machine used in this study was able to produce the web in a size of 50 cm × 70 cm by using the known web area (0.35 m^2). The total required weight of fibres was calculated to produce web layers with a basis weight of 100 g/m^2. Considering the fibre loss during the carding stage and drafting of the web during the needle punching stage, approximately 10% extra fibre weight was added. The fibres were weighed and prepared manually under the standard atmospheric condition of 60–70% relative humidity and 20–25°C temperature, then fed into the laboratory scale carding machine. Several carded webs of 100 g/m^2 were produced and combined in the needle punching process to obtain the basis weight of 400 g/m^2. Needle punching was performed with needle Loom -DI-Loom OUG-II 6 with a total punch density of 200 punches/cm^2 and needle penetration depth of 10 mm. The developed nonwovens are shown in Figure 2.

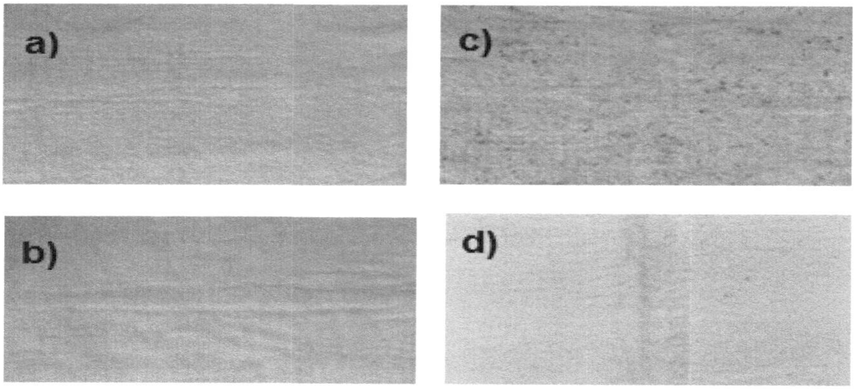

Fig. 2: Images of the Developed Nonwovens (a) 100% Comber Noil (b) 100% Silk Cocoon Waste (c) 70/30 P/C Flat Wastes, and (d) 50/50 Silk Cocoon/Comber Noil

4. NONWOVEN CHARACTERIZATION

The developed nonwoven fabrics were conditioned in standard testing temperature (20°C ± 2°C) and humidity (65% ± 2%) for 24 hours according to ASTM standard before testing. An electronic weighing balance with a capacity of 0.001 g was used to measure the areal density of the developed nonwoven fabrics according to ASTM D-1910 standard. The thickness of the developed nonwovens has been measured under 2 KN load using fabric thickness gauge with a capacity of 0.01 mm according to ASTM D-1777 standard. The Textest FX 3300 air permeability tester was used to characterize the air permeability of the developed nonwoven fabrics as per ASTM D 737 standard.

4.1 Thermal Insulation Characterization

The thermal resistance of the developed nonwovens was evaluated by using a heat flow meter apparatus HFM 436 Lambda supplied by NETZSCH (NETZSCH-Gera¨tebau GmbH, Selb, Germany) according to the guidelines of ISO 8301:1991 standards to determine the thermal insulation performance of the developed nonwovens. In this method, samples size 30 × 30 cm^2 are placed between a hot and a cold plate, and the heat flow created by the well-defined temperature difference is measured with a heat flux sensor, thus allowing the calculation of thermal resistance.

5. RESULTS AND DISCUSSION

The physical properties of the developed nonwovens are shown in Table 3. It was observed that the developed nonwovens had the thickness in the range of 1.74 to 2.10 mm. The variation in thickness of nonwovens may be due to the random distribution of fibres during nonwoven formation. The developed nonwovens had a density in the range of 193.19–263.55 (Kg/m^3). The developed nonwovens had the porosity value in the range of 79– 86%.

Table 3: Properties of the Developed Nonwovens

Sample	*Thickness (mm)*	*Density (Kg/m^3)*	*Porosity (%)*	*Air Permeability (cc^3/cm^2/sec)*	*Thermal Resistance (m^2 K/W)*
100% Comber noil	1.74	263.55	83	32.80	0.147
P/C Flat waste	2.10	193.19	86	55.00	0.118
100% Silk cocoon	2.10	220.72	79	53.10	0.201
50/50 (Silk cocoon/comber noil)	1.89	242.25	81	41.60	0.212

5.1 Air Permeability

The Figure 3 shows the air permeability of the developed nonwoven fabrics. It was observed that among the developed nonwoven fabrics, the nonwoven fabric made from P/C flat waste has highest air permeability value compared to other nonwoven fabrics (Saleh, 2011). It may be due to the presence of higher inter-pores in the P/C flat waste nonwovens due to the trash content. Following that nonwoven made from 100% silk cocoon has higher air permeability value. The nonwoven made from 100% comber noil has lowest air permeability value among the developed nonwovens. The nonwoven made from 50/50 silk cocoon waste/comber noil has higher air permeability than 100% comber noil and lower air permeability than 100% silk cocoon waste.

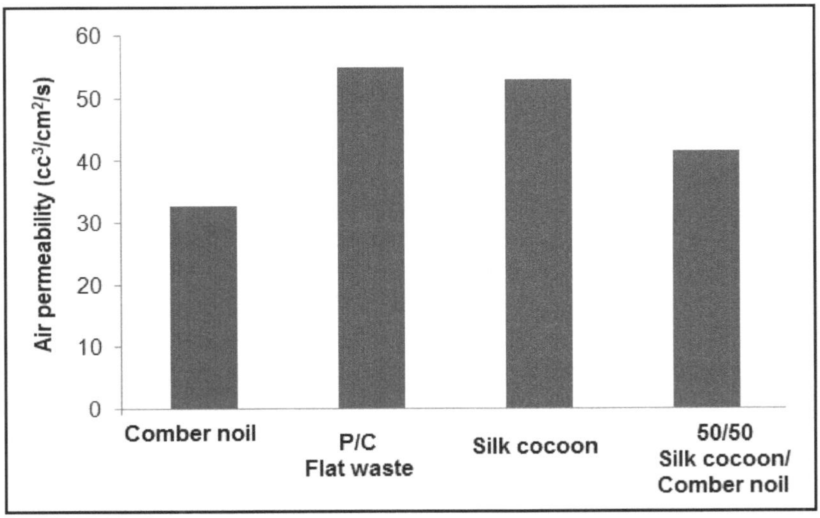

Fig. 3: Air Permeability of the Developed Nonwovens

5.2 Thermal Insulation Properties of Nonwovens

The thermal resistance value of the developed nonwovens is given in Figure 4. It was observed that the nonwoven made from P/C flat waste had lowest thermal resistance value compared to others. It may be due to the lower density and higher porosity value of the sample (Yachmenev *et al.*, 2006). Though the nonwoven made from 100% comber noil had higher density and lower air permeability, it had lower thermal resistance compared to nonwoven made from 100% silk cocoon waste. It may be due to lower thermal conductivity value of silk compared to cotton. The nonwoven made from 100% silk cocoon had better thermal insulation value compared to the nonwovens made from 100% comber noil and P/C flat waste. It may be due to lower porosity value of the nonwoven and lower thermal conductivity value of silk fibres (Yamada, 1997).

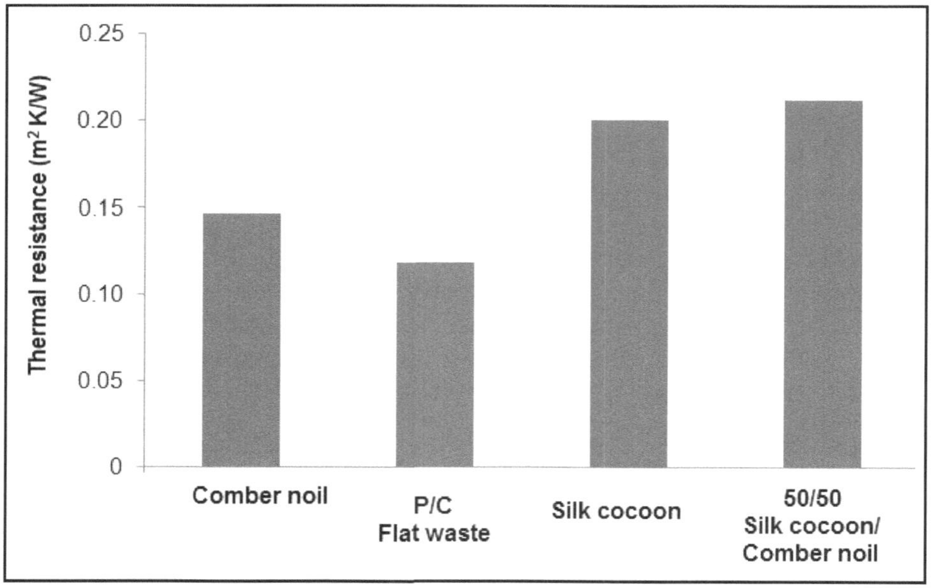

Fig. 4: Thermal Resistance Value of the Developed Nonwovens

It was also observed that the nonwoven made from 50/50 silk cocoon waste/comber noil had better thermal insulation value compared to nonwoven made from 100% silk cocoon waste and 100% comber noil. The blending of comber noil with silk cocoon further improves the thermal insulation behavior of the silk cocoon waste.

The commercial thermal insulation product named Autex Composition made of 100% polyester fibres manufactured by Autex Industries Ltd, New Zealand has been selected to compare the thermal insulation performance of the developed 50/50 silk cocoon/comber noil nonwoven. The Table 4 shows the comparison of the developed nonwoven with commercial product. It was observed that the commercial product has 5–6 times higher in thickness value compared to 50/50 silk cocoon/comber noil nonwoven. Thermal resistance converted to unified thickness of 1 mm, it turns out to be 0.022 m^2 K/W and

0.112 m^2 K/W for the commercial and developed sample respectively in which case the developed 50/50 silk cocoon/comber noil nonwoven appears to stand superior to the commercial product.

Table 4: Comparison of the Developed Nonwoven with Commercial Product

Specifications	*Commercial Product*	*Nonwoven Made from Textile Waste*
Raw material	100% polyester	50/50 silk cocoon/comber noil
Thickness (mm)	10–12 mm	1.89
GSM	1600	242.25
Thermal resistance (m^2K/W)	0.220	0.212

6. CONCLUSION

Growing amounts of textiles waste in landfills have become a concern for many. In this study, needle punched non-woven fabrics were made from comber noil, silk cocoon waste, and P/C flat waste for thermal insulation application. The nonwoven developed from 100% silk cocoon waste has higher thermal insulation value compared to 100% comber noil and P/C flat waste. The blending of comber noil with silk cocoon improves the thermal insulation behavior of the silk cocoon waste. Also blending of comber noil with silk cocoon waste will lower the product cost. The developed products have superior thermal insulation value compared to commercial product. The developed nonwoven structures are good candidates to be used as low cost and environmental friendly insulation materials for automotive, furniture and clothing industries. Thus the textile wastes are an alternative for synthetic fibres for technical textiles application. It will help to minimize the carbon foot print, during disposing-off the samples after their service life.

ACKNOWLEDGMENT

The authors thank M/s PSG TECHS COE INDUTECH, Coimbatore for providing facilities to carry out this research.

REFERENCES

[1] ASTM D-1777 (1996). Standard Test Method for Thickness of Textile Materials, *West Conshohocken,* PA, and U.S.A: ASTM Internationals.

[2] ASTM D-1910 (1996). Standard Test Method for Weight of Textile Fabrics, *West Conshohocken,* PA, and U.S.A: ASTM Internationals.

[3] ASTM D-737 (2018). Standard Test Method for Air Permeability of Textile Fabrics, *West Conshohocken,* PA, and U.S.A.: ASTM Internationals.

[4] Ghosh, S.K., Bairagi, S., Dutta, S. and Bhattacharyya, R. (2016). A Comparative Study of The Thermal Insulation Properties of Jute and Jute-Polyester Fibre Blended Nonwoven Fabrics, *American Journal of Engineering Research*, 5(8), 43–49.

[5] ISO 8301-1991 (2014). Thermal Insulation-Determination of Steady-State Thermal Resistance and Related Properties-Heat Flow Meter Apparatus, Geneva, Switzerland: *International Organization for Standardization.*

[6] Jabbar, A., Ali, M.A., Shahzad, A., Naeem, M.S., Javed, Z., Qadir, M.B., Rehman, K., Irfan, M. and Ahmad, Z. (2020). Development of kapok/recycled-PET Blended Needle-Punched Thermal Waddings, *Journal of Natural Fibres*, DOI: 10.1080/15440478.2020.1784816.

[7] Muthu, K.N., Thilagavathi, G. and Karthikka, M. (2020). Development of Recycled PET/Comber Noil Nonwovens for Thermal Insulation Application, *Journal of Natural Fibres,* DOI: 10.1080/15440478.2020.1841064.

[8] Patnaik, A., Mvubu, M., Muniyasamy, S., Botha, A. and Anandjiwala, R.D. (2015). Thermal and Sound Insulation Materials from Waste Wool and Recycled Polyester Fibres and Their Biodegradation Studies, *Energy and Buildings,* 92, 161–169.

[9] Rajkumar, G. and Srinivasan, J. (2018). Certain Properties of Needle Punched Nonwoven Fabrics Made from Silk and Wool Fibres, *Trends in Textile Engineering and Fashion Technology,* 1(2), 28–31.

[10] Saleh, S.S. (2011). Performance of Needle-Punching Lining Nonwoven Fabrics and Their Thermal Insulation Properties, *Journal of Basic and Applied Scientific Research*, 12, 3513–3524.

[11] Wazna, M. EL., Gounni, A., Bouari, A. EL., Alami, M. EL. and Cherkaou, O. (2019). Development, Characterization and Thermal Performance of Insulating Nonwoven Fabrics Made from Textile Waste, *Journal of Industrial Textiles,* 48(7), 1167–1183.

[12] Yachmenev, V.G., Negulescu, I. and Yan, C. (2006). Thermal Insulation Properties of Cellulosic-Based Nonwoven Composites, *Journal of Industrial Textiles,* 36, 73–86.

[13] Yamada, A. (1997). Thermal Conductivities of Fabrics Made of Silk or another Textile Materials *The Journal of Sericultural Science of Japan,* 66(4), 266–271.

[14] Zhang, J., Rajkhowa, R., Li, J.L., Liu, X.Y. and Wang, X.G. (2013). Silkworm Cocoon as Natural Material and Structure for Thermal Insulation. *Materials and Design,* 49, 842–9.

Jute Gypsum Composites

Pooja Katkar*, Avinash Chougle, Sammed Patil, Rushikesh Shingare and Mujammil Mulla

DKTES's Textile and Engineering Institute, Ichalkaranji, Maharashtra, India

E-mail: *pmkatkar@gmail.com

ABSTRACT: *In the adverse climatic conditions, a significant amount of energy is required to keep the indoor atmosphere comfortable. The thermal insulation of the buildings is an effective way to achieve the same. There are several ways to keep the room thermally insulated. Use of false ceilings, below the structural ceiling, is one of the alternatives to keep the room insulated, with no consumption of energy. While thinking of energy consumption, a thought about the environment is also essential in today's era. In the similar direction, the natural fibres containing lignin and cellulose such as jute, coir etc., with suitable matrix is being used in construction industry from several decades. These natural ligno-cellulosic fibres are abundantly available, at affordable price. The gypsum composites reinforced with such ligno-cellulose fibres, can be the solution towards this problem, which can be used as false ceiling, being an insulator for buildings. In the present study, jute fibre has been used as reinforcement in the form of fibre and fabric. The fibre fraction was used in three different proportions viz. 0%, 1% and 2%. Layers of the jute fabric were varied as 1, 2 and 3. The experiment was designed using Taguchi technique. The composites were tested for thermal insulation, load bearing capacity and moisture absorption. It was found that jute fibre, in both the forms, helps in improving thermal insulation of the composites.*

Keywords: Gypsum Composites, Thermal Insulation, Energy, Ligno-Cellulosic, Taguchi Technique.

1. INTRODUCTION

Specific characteristics of natural fibres such as biodegradability, low cost, high specific strength have attracted the scientists and technologists to rethink in a new direction to develop sustainable solutions in various fields, such as infrastructure, automotive and furniture. Natural fibre, especially ligno-cellulose fibre composites are the best substitute for wood-based material in many applications. As a reinforcing material, the jute fibre is an abundantly available, strong ligno-cellulosic bast fibre, comprising of bundled ultimate cells. As a matrix material, sound and thermal insulators with good strength and work-ability are the advantages associated with the products made up of gypsum. So, this project aims to study practicability of jute in different forms viz. fibres and fabric in gypsum to fabricate jute-gypsum composites for false ceiling application.

2. EXPERIMENTS

Jute fibres were dispersed in water by continuous stirring and gypsum was poured in water and continuously agitated to form slurry. This slurry was poured into the moulds and kept for drying. The proportion of fibre in composites was varied from 0%, 1%, 2% on weight of gypsum. The fibres were weighed accurately on the dry weight of gypsum. The layers of 300 GSM woven jute fabrics were cut as per the size of the samples and sandwiched between two layers of gypsum. The gypsum slurry was poured half in the mould and then fabric layer was placed, and remaining slurry was poured and kept for drying. The jute reinforced gypsum composites were tested for compressive strength, thermal insulation, moisture content, and density.

Table 1: Design of Experiment

Sample No.	*No. of Layers*	*Jute Fibre Proportion %*
1	1	0
2	1	1
3	1	2
4	2	0
5	2	1
6	2	2
7	3	0
8	3	1
9	3	2

2.1 Density

Density was calculated by direct method (weight/volume).

2.2 Compressive Strength

The specimens were prepared for compressive strength test in size of 50 mm × 50 mm × 50 mm in order to have a cuboids size according to the procedure specified in ASTM C472. The specimen was tested for compressive strength parallel to the plane of the board by applying increasing compressive load until failure occurs.

2.3 Thermal Insulation

Flat sheet of 300 mm × 300 mm × 10 mm were prepared for thermal insulations test. The size was selected based on criteria set BS5609. Sample house with brick work of the same height and width are made; the composite sheet was used as roof to evaluate the temperature difference, between atmospheric temperature and the temperature inside the houses. Thermometers are kept in the houses and temperature was recorded hourly from 10 am to 5 pm.

2.4 Moisture Absorption

The size of the specimen used for moisture content was 175 mm × 75 mm × 40 mm as shown in figure. The size was selected based on the criteria set in IS 2542.

3. RESULTS AND DISCUSSION

3.1 Density of Gypsum Composite

For estimating density of the composite, samples with size of 300 mm × 300 mm × 10 mm were prepared. Density was determined after curing the specimen, for 14 days. There is significant effect of fibre fraction as well as number of fabrics on the density of gypsum composite.

It was observed that density of gypsum composite goes on decreasing as fibre fraction and fabric layers increase, since density of gypsum powder (2.96 g/cc) is more than that of the jute fibre (1.5 g/cc). Specimen with 0% fibres and 1 layer of fabric shows highest density value (1.55 g/cc) whereas specimen with 2% fibres and 3 layers of fabrics shows lowest density value (1.43 g/cc). The effect of number of layers on the density of gypsum composites is more than that of fibre fraction.

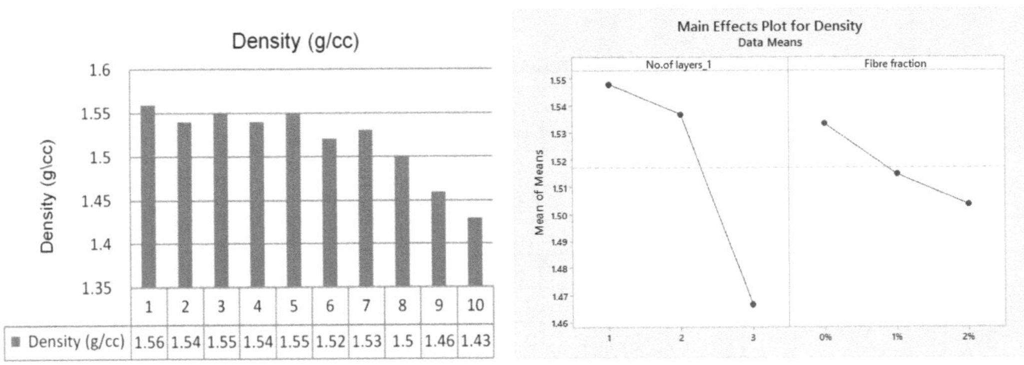

Fig. 1: Effect of Reinforcement on Density

3.2 Compressive Strength of Gypsum Composite

There is significant effect of fibre fraction as well as fabric layers on compressive strength of gypsum composite compressive strength of composite decreases initially, because of lack of bonding due to introduction of jute fibres and fabrics. As we goe on increasing fibre fraction and fabric layers, increase in compressive strength was observed. This may be because of better bonding between the reinforced material and matrix, in appropriate proportion. Specimen with 2% fibres and 2 layer of fabric shows highest strength (5.6 Mpa) and specimen with 0% fibre fraction and 1 layer of fabric shows lowest strength (2.5 Mpa) which is even lower than sample with 100% gypsum. The effect of number of layers on the compressive strength of gypsum composites is more than that of fibre fraction.

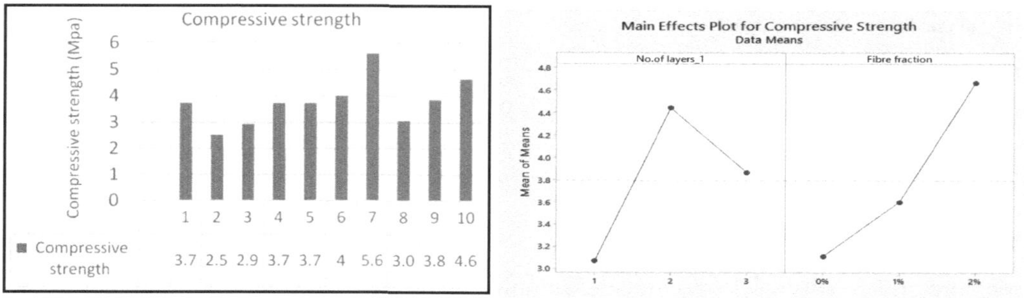

Fig. 2: Effect of Reinforcement on Compressive Strength

3.3 Thermal Insulation of Gypsum Composite

There is significant effect of different proportions of the fibres as well as the fabric layers on the thermal insulation. As we go on increasing fibre fraction and fabric layers, rise in thermal insulation was seen. Specimen with 2% fibres and 3 layer of fabric shows highest thermal insulation value (2.21) and specimen with 0% fibre fraction and 1 layer of fabric shows lowest thermal insulation value (1.57) which is even higher than sample with 100% gypsum. The effect of number of layers on the thermal insulation of gypsum composites is more than that of fibre fraction.

Table 2: Temperature Difference—Atmosphere and Sample House

Fibre %	*No. of Layers*	*10.00 am*	*11.00 am*	*12.0 am*	*1.00 pm*	*2.00 pm*	*3.00 pm*	*4.00 pm*	*Avg.*
0	1	1	1	1.5	1	2	2.5	2	1.57
1	1	2	1.5	1.5	1	2	3	2.5	1.93
2	1	2	1.5	2	1	2.5	3	2	2
0	2	2.5	1.5	1.5	1	2	3.5	2.5	2.07
1	2	1	1	2	2	3	3	2	2
2	2	1	2	2	2.5	2	2	2	1.93
0	3	2	2	2	1.5	2.5	2	2	2
1	3	2	2	2	2	2	2	2	2
2	3	2	2	2.5	2	2.5	2.5	2	2.21

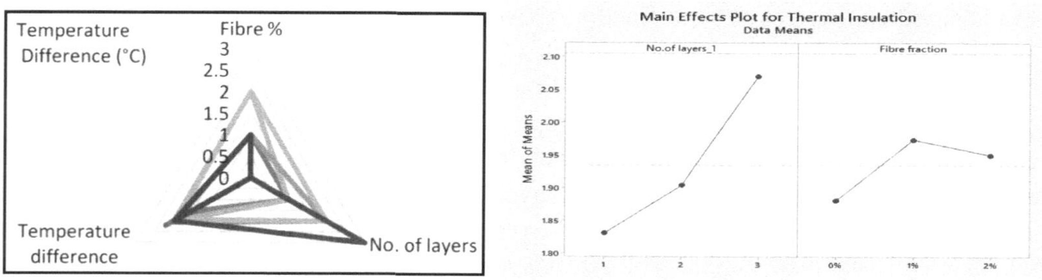

Fig. 3: Effect of Reinforcement on Temperature

3.4 Moisture Content of Gypsum Composite

There is significant effect of different proportions of the fibres as well as the fabric layers on moisture content. As we go on increasing fibre fraction and fabric layers, rise in moisture content was observed, because of hygroscopic nature of the fibre. Specimen with 3 layer of fabric shows highest moisture content (1.95%) and specimen with 0% fibre fraction and 1 layer of fabric shows lowest moisture content (0.73%). The effect of number of layers on the moisture content of gypsum composites is more than that of fibre fraction.

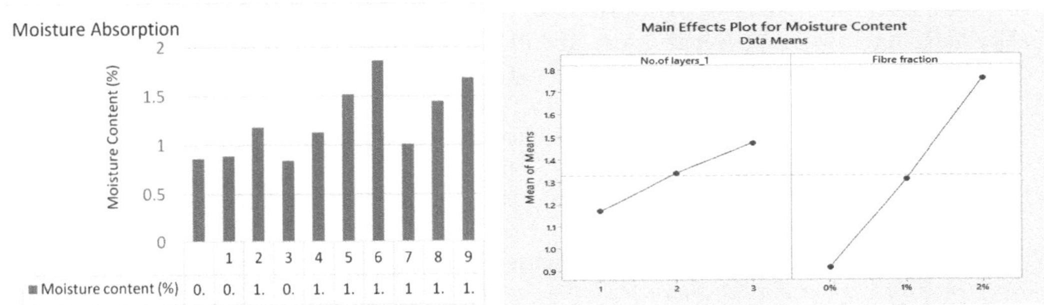

Fig. 4: Effect of Reinforcement on Moisture Absorption

4. CONCLUSION

The study concludes that, jute in the form of fibre and woven fabric can be used as reinforcement in gypsum matrix. Thermal insulation of jute reinforced composites goes on increasing with increase in fibre fraction and number of layers of jute fabric. Optimum level of fibre fraction and number of jute fabric layers results into a composite with comparatively higher compressive strength. The moisture contents of composite increase with introduction of jute fibres because of hygroscopic nature of the fibre.

REFERENCES

[1] Andrian, P.P. and Georghe, B.M. (2010). Manufacturing Process and Applications of Composites, *Annals of the Oradea University of Management and Technological Engineering,* IX(XIX), NR2.

[2] Carosena, M., Simone, B. and Giovanni, M. (2015). Carlomagno: Characterization of PLA/jute Composites with Infrared Thermography, *Published in Frontiers in Polymer Science.*

[3] Cristiano, A., Arlindo, S. and Luis, R., Pouloferrao Manuelfreitas, Sustainable Design of Automotive Components through Jute Fibre Composites, *Institute Superior Técnico, Universidade Técnica de, Lisbon, Portugal.*

[4] Das, S., Saha, A.K., Choudhary, P.K., Basak, R.K., Mitra, B.C., Todd, T., Lang, S. and Rowell, R.M. (2000). Effect of Steam Pre-treatment of Jute Fibre on Dimensional Stability of Jute Composite, *Journal of Applied Polymer Science* (76), 1652–1661, John Wiley & Sons, Inc.

[5] Denis, M.P., Michaela, D.I. *et al.* (2010). Composite Materials of Polypropylene and Waste Jute Fibres, *Published in Material Plastic,* 47.

[6] Dutta, S., Karak, N. and Baruah, S. (2009). Jute-Fibre-Reinforced Polyurethane Green Composites Based on Mesuaferrea L. Seed Oil, *Wiley Inter. Science (www.interscience.wiley.com).*

[7] Gon, D., Das, K., Paul. P. and Maity, S. (2012). Jute Composites as Wood Substitute, *International Journal of Textile Science*, DOI: 10.5923/j.textile.20120106.05 Publications.

[8] Gupta, M.K., Srivastava, R.K. and Bisariaa, H. *Potential of Jute Fibre Reinforced Polymer Composites.*

[9] Karthick, R. and Aruna, M. *Comparison of Sisal, Jute and Banana Fibre Composites: A Review*, 1, 2, Department of Mechanical Engineering, Velammal College of Engineering and Technology.

[10] Khabir, U., Khan, M.A. and Idriss, A. *Degradable Jute Plastic Composites Radiation and Polymer Chemistry Laboratory,* Institute of Nuclear Science and Technology, Bangladesh.

[11] Khalil, A., Tawfik, A., Hegazy, A.A. and El-Shahat, M.F. *Effect of Different Forms of Silica on the Physical and Mechanical Properties of Gypsum Plaster Composites*.

[12] Mir, A., Aribi, C. and Bezzazi, B. (2014). Study of the Green Composite Jute/Epoxy, *International Journal of Chemical, Molecular, Nuclear, Materials and Metallurgical Engineering*, 8(2).

[13] Mistria, E., Routhb, S., Ray, D., Saswata, S. and Misra, M. *Green Composites from Maleated Castor Oil and Jute Fibres.*

[14] Mohanty, A.K., Khan, M.A. and Hinrichsen, G. *Surface Modification of Jute and it's in Uence on Performance of Biodegradable Jute-Fabric/Biopol Composites*.

[15] Myvizhirajeshwari, G. *Manufacturing and Applications of Jute Fibre Composite.*

[16] Nerminn, F. and Magdi-El, M. Modification of the Jute Fabrics for Light Weight Composite, Textile Department, Faculty of Engineering, *Alexandria University*, Egypt.

[17] Osman, G. and Fuatkoksal, M. *Properties of Gypsum Containing Polypropylene Fibre.*

[18] Singh, M. and Garg, M. (1993). Gypsum-based Fibre-Reinforced Composites: An Alternative to Timber, *Central Building Research Institute,* Roorkee, India.

[19] Vilaseca, F., Mendez, J.A., Pelach, A., Llop, M., Canigural, N., Girones, J., Turon, X. and Mutje, P. (2006). Composite Materials Derived from Biodegradable Starch Polymer and Jute Strands, *Published in Process Biochemistry on 4th September 2006.*

[20] Vilaseca, F., Mendez, J.A., Pelach, A., Llop, M., Canigural, N., Girones, J., Turon, X. and Mutje, P. (2006). Study is the Chemical Modification of Jute Fibres Using Oleoyle Chloride to give Hydro Pohobicity and Resistance to Bio Fibre, P*ublished in Process Biochemistry on 4th Sept. 2006.*

[21] Wolfgang, L., Feng, G., Siegfried, S. and Rüdiger, F. Simulation of the Cyclic Loading and Damage Behavior of Gypsum Composites, Siemon Aicher 21, Institute for Materials Testing, *Materials Science and Strength of Materials (IMWF), University of Stuttgart,* Pfaffenwaldring 32, 70569 Stuttgart, Germany.

Ballistic Resistant Armor

Aadhar A. Mandot[1] and Sanjay B. Bambhaniya[2]

The Maharaja Sayajirao University of Baroda, Gujarat, India

E-mail: [1]aadharmandot@msubaroda.ac.in; [2]sanjubambhaniya06@gmail.com

ABSTRACT: *A ballistic resistant vest or bullet proof vest is an item of personal armor that can be worn on the torso. It helps to absorbs the impact from firearms i.e. fired projectiles and shrapnel from explosions. Generally soft armor is made from multiple layers of ballistic resistant material (woven or laminated fibres) to protect the wearer from small caliber handgun and shotgun projectiles. It is intended to be used for extended daily wear and it can be worn by police forces, private citizens, etc. Hard armor consists of metal or ceramic plates used along with soft vest to protect the wearer from rifle rounds, metallic components and stab and slash attacks from knife. It can be worn by combat soldiers, police tactical units, and hostage rescue teams. We have reviewed the science of body armor material with emphasis upon current understanding of relevant energy absorbing mechanism in fibres, fabrics, polymer laminates and ceramics. The trends in ongoing developments in ballistic fibres are then reviewed, analyzed and future projection offered. The provision to improve robustness and multiple strike performance of hard armor plates has been made. We have also reviewed possible failure modes in armor materials which is essential in designing improved armor materials and armor systems. Study on the arrangement of fabric layers for soft body armor with the drive for lighter and more protection vest has been reported.*

Keywords: Armor, Ballistic, Blunt Trauma, Monolithic Structure, Polymer Laminates.

1. INTRODUCTION

A ballistic resistant vest has been a feature of soldiers' personal equipment since the earliest periods of organized warfare. It is necessary to protect the personnel who are in the front line of threat for the security of the country. The main purpose of ballistic resistant vest is to provide good ballistic and edged weapons protection for vital organs such as heart, lungs, liver, kidneys, spleen, and backbone (Abdulgaffar, 2015). Basically, a ballistic vest uses a multiple layers of very strong materials to catch and deform a projectile, mush rooming it into a dish shape, and spreading its force over a large portion of the vest. The vest absorbs the impact from the high velocity deforming projectile, bringing it to a steady state before it can completely penetrate the object. Some layers of a vest may be penetrated but as a projectile penetrates it also deform as well, and as a result the energy is being absorbed by, a larger and larger area of the vest (Indi & Patel, 2015).

Many attempts have been made that can prevent the penetration of projectile, but with the modern technology of the weapons, the threats have been increased and therefore, the vest and the wearer can still absorb the bullet's energy. Modern projectile contains enough energy to cause blunt trauma under the impact point. These demands more protection and therefore the vest should be meant for the requirement of protection against resistance to the penetration and limits on the amount of blunt trauma to the wearer (Abdulgaffar, 2015).

The soft vest which are made of multiple layers of the ballistic resistant material (woven or laminated fibres) mainly focus on protecting the body from fragments of explosives such as hand grenades and short distance firing using a pistol, small caliber handgun and shotgun (Abdulgaffar, 2015). This type of vest is normally lighter in weight and better in comfort, therefore it can be used as daily wear (Fernando *et al.*, 2015). This can be worn by police forces, private security guard, private citizens, who are not involved in heavy threat level. Hard armor plates may be constructed from metallic plates, ceramics, compressed laminate sheets or composites that can incorporate with soft vest to provide extra protection to vital organs in the torso. It can withstand the rifle rounds, metallic components and stab and slash attacks from a knife. It should be worn by the personnel who are regularly subjected to heavy threats like combat soldiers, police tactical units and hostage rescue teams (Abdulgaffar, 2015). This type of vest is generally heavy in weight and not comfortable for user. This can affect the performance during their normal duties, resulting in discomfort which increases irritability and leads to exhaustion (Fernando *et al.*, 2015).

With improving technology of weapons, the level of threat is increasing continuously. Therefore, the armor should be made with the aim to withstand all the challenges in terms of protection and comfort.

2. MECHANISM OF BODY ARMOR

The principle on which the ballistic protective materials work can be broadly divided into two categories: (i) Absorption of impact energy and (ii) Redistribution of impact energy (Wang *et al.*, 2014). A protective material should absorb the energy of a projectile before it completely penetrates the material. Energy absorption is achieved by stretching, compressing or destroying the material. In other words, the principle on which body armor operates is based on the rapid conversion and dispersion of the kinetic energy from a striking bullet into strain energy within the ballistic body armor (Wang *et al.*, 2014). Modern body armors provide protection by three different methods:

(i) The armor completely bounces the projectile, which is very rare;
(ii) The armor decelerates and stops the projectile by dissipating its kinetic energy along the plane of the impacted material; and
(iii) A possible combination of the above (i) and (ii).

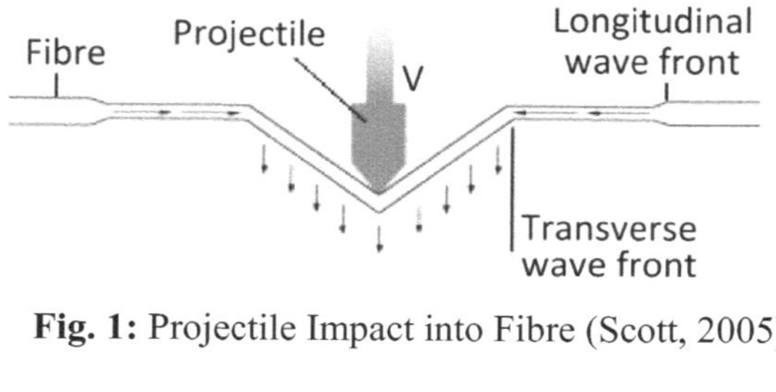

Fig. 1: Projectile Impact into Fibre (Scott, 2005)

When a projectile strikes a fabric, or a number of layers of fabric, two waves, namely longitudinal and transverse waves, are propagated from the point of impact. The longitudinal wave travels in the plane of the fabric and the transverse wave propagates perpendicular to the fabric, as shown in Figure 1. The material behind the wave front moves towards the point of impact and deflected in the direction of projectile. The velocity at which these waves propagate in a ballistic panel is dependent on the energy absorption capability of the fabric layer (Scott, 2005). The major parameters that affect the way in which impact energy is dissipated are dependent on the tensile strength of the fabrics and yarns, the fabric structures and the number of layers of fabric (Wang *et al.*, 2014). When a bullet strikes on body armor, it is caught in a 'web' of very strong fibres. These fibres in the fabric absorb and disperse the impact energy that is transmitted to the vest from the projectile, causing the bullet to deform or to 'mushroom'. Each successive layer in the vest absorbs energy, until the bullet is stopped. Because a large area of the garment becomes involved in preventing the bullet from penetrating, this helps in dissipating the forces that can cause non-penetrating injuries (commonly referred as 'blunt trauma') to internal organs as shown in Figure 2 (Scott, 2005).

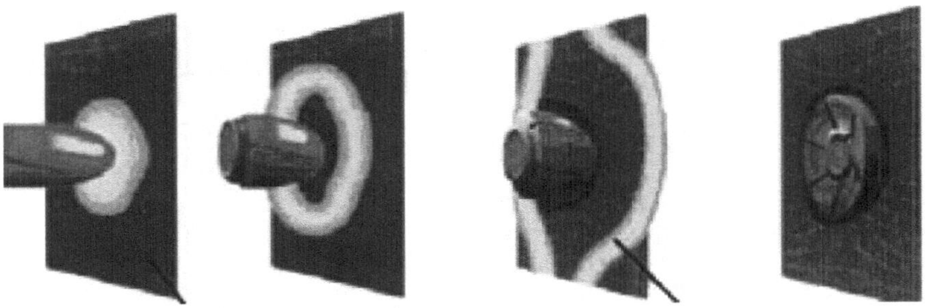

Fig. 2: Bullet Penetration Process (Scott, 2005)

In the case of hard body armor, metal and reinforced plastic provide protection by partially bouncing the bullet as well as absorbing the impact energy. Upon impact, the impact pulse on the surface of the armor is reflected as a tensile wave. When the amount of tensile stress exceeds the tensile strength of the armor material, fractures occur leading to mechanical failure of the hard body armor and the creation of a hole. If the magnitude of stress pulse still exceeds the tensile strength of the material, multiple fractures can occur. If the material is driven beyond its elastic limit as the applied load increases, the material becomes plastic (Wang *et al.*, 2014).

3. TEXTILE FIBRES USED FOR BALLISTIC PROTECTION

The fibres used for ballistic protection should have low density, high strength and high energy absorption capability. The ballistic performance of a material depends on its ability to absorb energy locally and to distribute it across a whole structure. Table 1 indicates the type of fibres and their properties currently in use for the production of soft body armor.

Table 1: Textile Fibres and their Properties Used for Ballistic Protection

Fibre Name	*Type*	*Tensile Strength (MPa)*	*Break ext. (%)*	*Density (g/cc)*	*Chemical Structure*
Kevlar®	p-aramid	3400	3.5	1.44	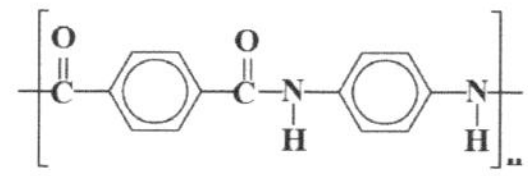
Twaron®	p-aramid	3000	3.3	1.44	
Technora®	p-aramid	3400	4.5	1.39	
Spectra®	HPPE	3000	3.0	0.97	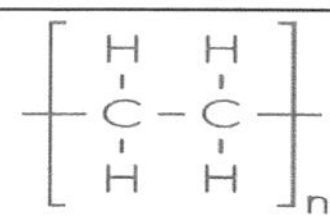
Dyneema®	HPPE	2700	3.5	0.97	
Zylon®	PBO	5200	3.1	1.56	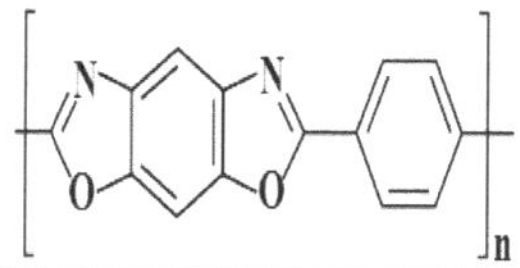
M5®	PIPD	3960	1.4	1.70	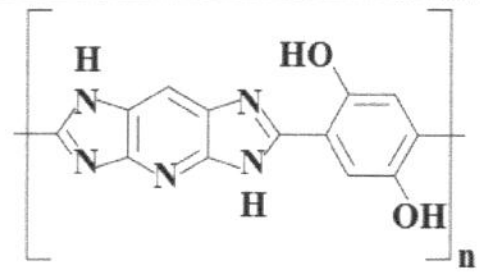

A brief outline of different high performance fibres used for ballistic soft armor is given below:

3.1 Kevlar

It is consists of long molecular chains produced from poly-paraphenylene teraphtal amide. The high degree of alignment of the molecular chains results in unique properties, such as high tensile strength, low elongation at break, high modulus, high chemical resistance, low thermal shrinkage, high toughness, dimensional stability and flame resistance. With unique high tensile strength-to weight ratio, this fibre is 5 times stronger than steel. However, the ballistic properties of this fibre are affected by moisture.

3.2 Twaron

This fibre is composed of large number of fine micro filaments, as a result vest made of Twaron® is 23% lighter than standard aramid yarn and a structure enhance to absorb bullet's impact and dissipates energy quickly. It is believed to be lighter, softer and more flexible (Scott, 2005).

3.3 Technora

It has good tensile strength and modulus, chemical and hydrolytic resistance, excellent fatigue resistance, good dimensional stability and good heat resistance. Stiff and highly oriented molecular structure enables low creep and stress relaxation (Scott, 2005).

3.4 Spectra

This fibre has good strength-to-weight ratio, are ten times stronger than steel, durable and has a specific strength that is 40% greater than that of aramid fibres (Scott, 2005). It has extremely low moisture absorption, low density, low coefficient of friction and very high abrasion resistance (Wang *et al.*, 2014).

3.5 Dyneema

This fibre has an extremely high strength-to-weight ratio and low density. It has high-energy absorption capability, which dissipates the shock waves faster than other ballistic materials (Scott, 2005).

3.6 Zylon (PBO)

It consists of rigid-rod chain molecules of poly p- phenylene-2, 6-benzobisoxazole. It has high tensile strength, which is 1.6 times higher than Kevlar and modulus which is almost double combined with good thermal stability compared to p-aramid fibres. Body armor made with this fibre is lighter, more comfortable, and stronger than products made from aramid fibres. However, the tensile strength degrades under certain extreme environmental conditions, such as high ultraviolet radiation (Wang *et al.*, 2014).

3.7 M5 (PIPD)

It can enable the weight of soft body armor to be reduced by 40–60% compared with Kevlar at same level of protection. All the earlier fibres have poor transverse compression and shear properties, while PIPD have greatly improved transverse properties. However, M5® fibres suffer some degradation in the mechanical properties required for ballistic protection after exposure to ultraviolet light and elevated moisture. In future, M5® is a strong competitor to be used as a reinforcing fibre in hard armor backing composites (Cunniff & Auerbach).

4. FABRIC STRUCTURES USED FOR BODY ARMOR

Ballistic-resistant fabric is made from high-strength and high-modulus fibres that we have seen above. The most commonly used ballistic resistant fabrics are woven and non-woven. The brief introduction about woven and nonwoven fabric structure is given below.

4.1 Woven Structures

The ballistic performance of a woven fabric is dependent on the dynamic mechanical properties of constituent fibres as well as the fabric geometry (i.e. type of weave, fibres per yarn, weave density, etc.). Usually, the ballistic fabric is densely woven with square plain weave and basket weave. It is recognized that fabrics should possess cover factors from 0.6 to 0.95 to be effective when utilized in ballistic applications. Cover factors greater than 0.95, degraded by the weaving process and when cover factors fall below 0.6, the fabric may be too 'loose' to be protective (Scott, 2005).

4.2 Non-woven Structures

In this fabric all the fibres are laid in parallel, in the same plane, rather than being woven together. This is made of several layers of fibres, with the direction of fibres in each layer at 90° to the direction of the fibres in the adjacent layers as shown in Figure 3. The unidirectional configuration of the fibres allows the energy transferred from the impact of a bullet or other threat to be distributed along the fibres much faster and more efficiently than in conventional woven fabrics. This is because the absorption power of the yarn in woven fabrics is lost at the crossover points, as these points reflect rather than absorb the shockwaves of the impact. In addition, it has been reported that nonwoven ballistic fabric has strong weight advantages over woven fabrics. However, this results in significant increase in the bulk of the system, and also unidirectional materials are less flexible than their woven counterparts (Scott, 2005).

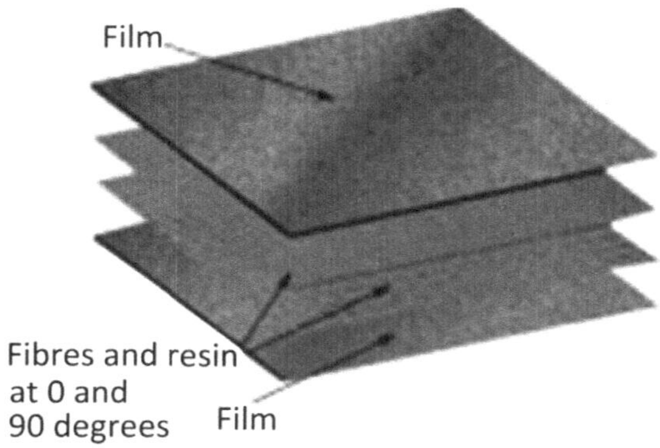

Fig. 3: Construction of Spectra Shield (Scott, 2005)

4.3 Arrangement of Fabric Materials for Multilayer Soft Armor

A non-penetrating injury (also called 'blunt trauma') occurs when the body armor retains the round, but does not effectively dissipate the energy efficiently, causing a narrow section of the armor to deform backward into the body, penetrating the skin and causing an injury. To assemble bullet proof vest, it is very important to consider the correct direction of strike face and back face materials. The arrangement of fabric materials for multilayer soft body armor is based on the three important mechanical properties, namely: work of rupture, young's modulus and bursting strength (Abdulghaffar, 2015).

5. BULLET PENETRATION MECHANICS AND FAILURE MODES

Defeating the threat is one thing and preventing the armor from failing is another. The understanding of possible failure modes in armor materials is absolutely essential when designing improved armor materials and armor systems. So, it is fundamentally important to understand each possible penetration mode, and the associated energy absorbing mechanism (Crouch, 2017). Different mechanics of bullet penetration and their failure modes are explained below as in Figures 4(a-f) respectively.

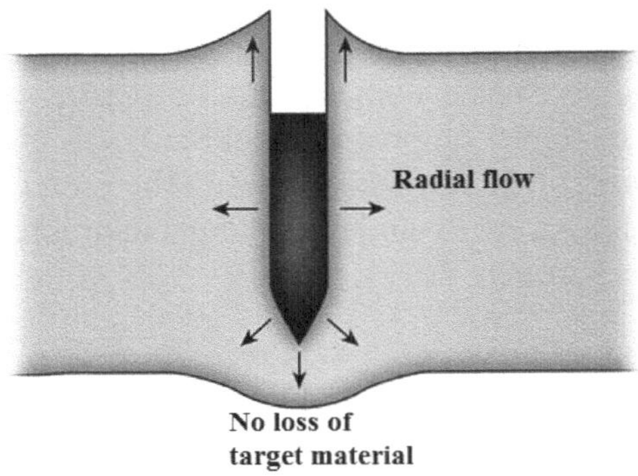

Fig. 4(a): Schematic of Ductile Hole Formation

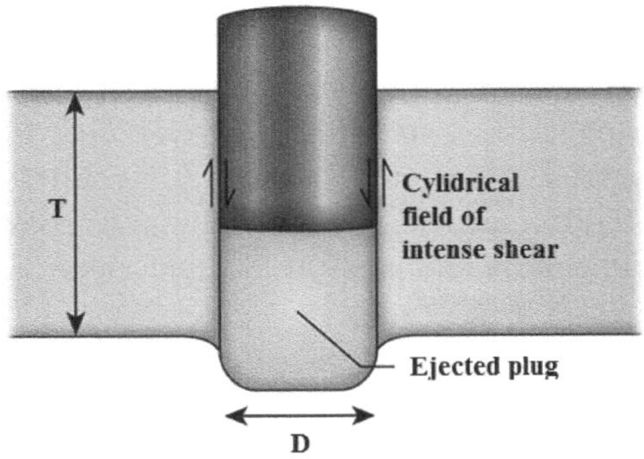

Fig. 4(b): Schematic of Plugging

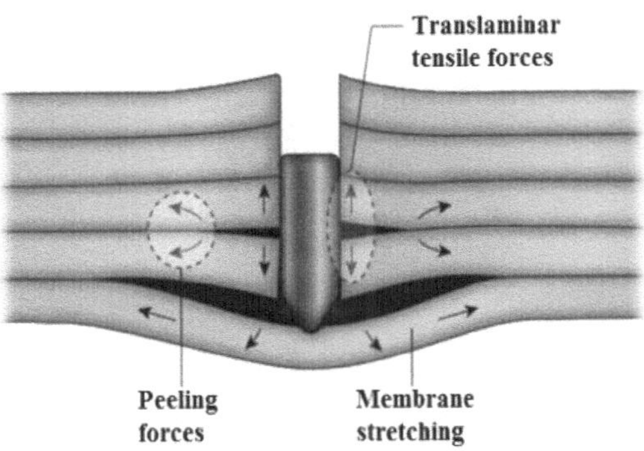

Fig. 4(c): Schematic of De-lamination

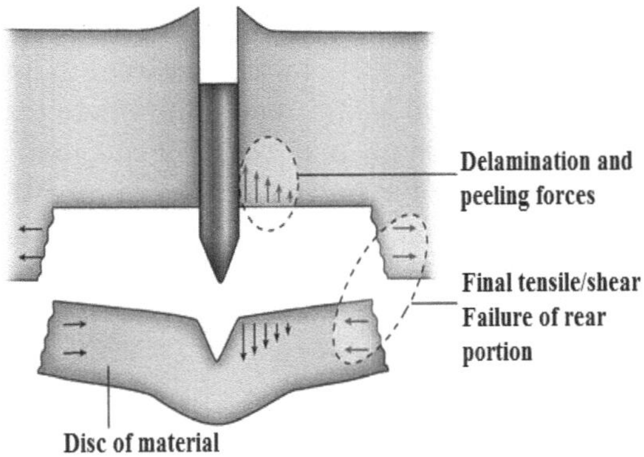

Fig. 4(d): Schematic of Discing

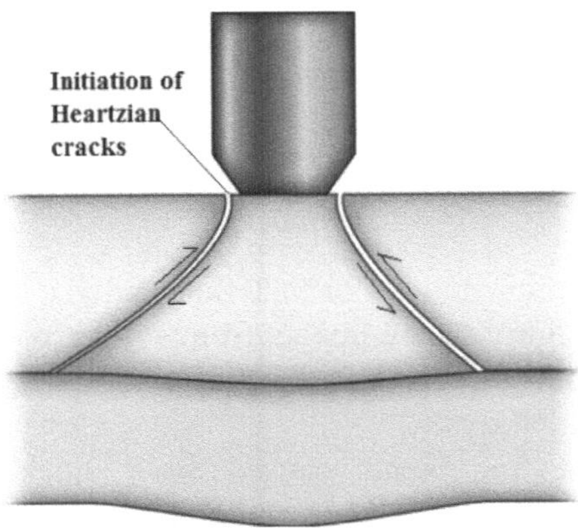

Fig. 4(e): Schematic of Conoidal Fracture

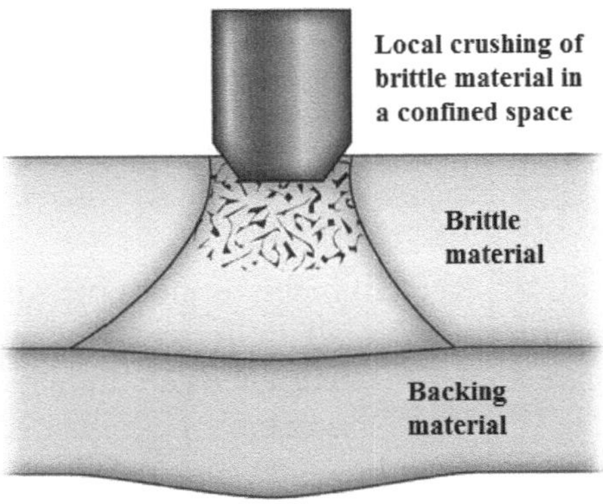

Fig. 4(f): Schematic of Comminution

5.1 Ductile Hole Formation

This is observed in monolithic structural armors or against pointed kinetic energy projectiles and considered to be efficient energy absorbing mechanism of all possible failure modes. It creates a clean hole like impression in the armor through plastic deformation, and displacement of the material (Figure 4(a)). No target material is lost. No, or little, round deformation occurs. The armor material flows outward (especially radially) from the nose of the round. The most of the resistance comes from its plastic yield characteristics and will be proportional to the in-plane compressive yield strength of the material. Also, temperature of both bullet and target material will increase. Therefore, one should consider dynamic, in-plane, compressive strength under varying conditions of temperature and pressure while designing armor (Crouch, 2017).

5.2 Plugging

This is observed in monolithic structural armors or against blunt objects and is usually associated with a low-energy absorption process. It is therefore one mechanism to prevent or suppress. When a blunt projectile strikes a target, the material ahead of the projectile accelerates, whilst the rest of the target remains relatively stationary. This leads to local compression of the material ahead of the projectile, and shear deformation in narrow shear bands under adiabatic conditions where the shear strain, strain rate and temperature may be locally very high and very little actual plastic deformation occurs (Figure 4(b)). It is controlled by the transverse shear properties of the target material. It initiated often, when ratio of the caliber of the round (core) and the thickness of the target is close to a 1:1 (Crouch, 2017).

5.3 De-lamination

This is observed in highly orthotropic solids, laminated armors, and layered structures, irrespective of projectile nose shape. Two main energy-absorbing processes accompany de-lamination: membrane stretching, which involves out-of-plane plastic deformation of the rear elements, and inter-laminar fracture, which involves either trans-laminar tensile failure, as shown in Figure 4(c), and/or inter-laminar shear failure of the inter-laminar bond (Crouch, 2017).

5.4 Discing

This is observed in highly orthotropic solids and laminated armor materials and is a special case of rear-face spalling in targets subjected to high-velocity impact from solid projectiles (Figure 4(d)). Its occurrence can dramatically reduce the ballistic limit of an armor system and is very common in ballistic failure of the light alloys.

It is caused by a mechanical fracture event, and is not caused by a shock-wave interaction. This phenomenon can be described based upon the penetration behavior of high strength event: (i) creation of intense bending moments and initiation of inter-laminar shear cracks; (ii) rapid propagation of the de-lamination crack, along the major spall plane and

(iii) final formation and separation of the disc via an in-plane, tensile-tearing fracture event (Crouch, 2017).

5.5 Conoidal Fracture

This type of through-thickness failure is observed in glasses, ceramics, ultra-high hardness steels as well as hard polymers such as the polycarbonates when impacted by blunt projectile. When projectile strikes, under the influence of a localized tensile stress, develops a cone crack that initially runs around the contact surface and then, driven by the forward momentum of the projectile, rapidly spreads downward through the material, shearing out a complete cone of separated material (Figure 4(e)). The cone angle is affected by Poisson's ratio and impact velocity, higher value of both of these results in larger diameter cone and therefore has higher ballistic efficiencies (Crouch, 2017).

5.6 Comminution

It means the reduction of solid materials, by crushing, grinding, cutting, vibrating or other processes (Figure 4(f)), it occurs at high velocities with small arm ammunition (Crouch, 2017).

5.7 Radial Cracking

This is observed in very brittle materials like glass, in tougher materials like ultrahigh-strength steels and hard plastics. Figure 5 a schematically illustrate the steps of formation of radial cracks: (i) initiation of multiple cracks near impact zone, (ii) development of a field of tensile hoop stress as the ceramic plate momentarily dishes elastically and finally (iii) relief of that hoop stress through the propagation of a finite number of radial cracks.

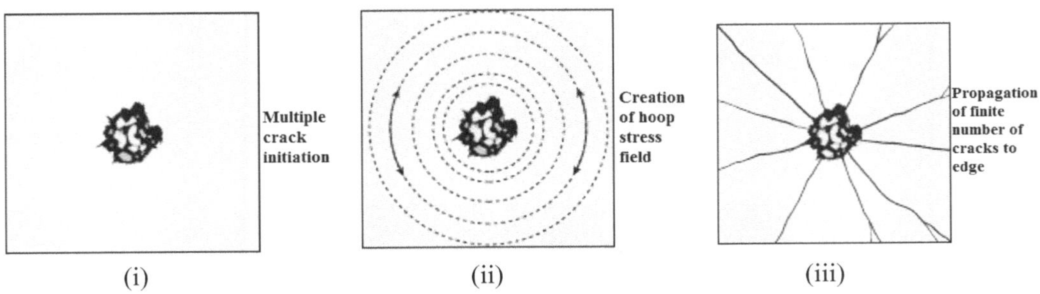

Fig. 5: Schematic of Stepwise Formation of Radial Cracks (Crouch, 2017)

5.8 Circumferential Cracking

Circumferential crack always follows the development of radial. They form as families of concentric circles around the impact site. This occurs due to thelocal dishing of the target and the backing material in particular. So, when projectile is arrested by the backing material, the support material immediately in front of the projectile gets pushed backwards

(Figure 6). This immediately creates a bending moment along the radial axis of each of the segments. Circumferential cracking can be limited by providing greater support to the brittle material in a multilayered armour system (Crouch, 2017).

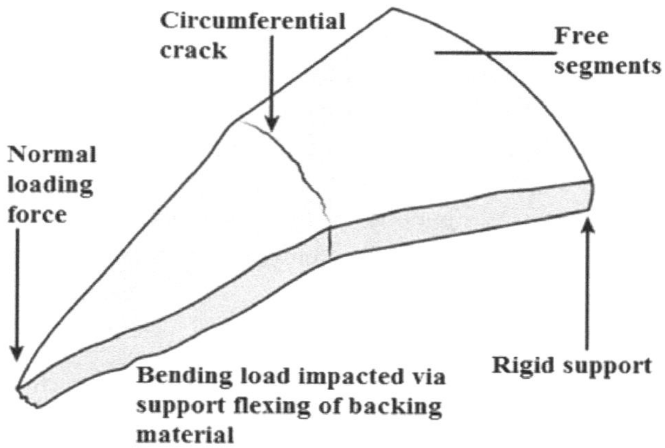

Fig. 6: Schematic of Circumferential Cracks (Crouch, 2017)

5.9 Spallation

This is a non-invasive, highly destructive failure of thick, monolithic, structural armors when attacked by high-explosive squash head rounds that detonate in contact with the armor. On impact a scab of armor material has been ejected from the back of the plate as a result of strong compressive stress waves Figure 7. A high, dynamic, through thickness, tensile strength, commonly termed spall strength, is therefore required for a material not to spall (Crouch, 2017).

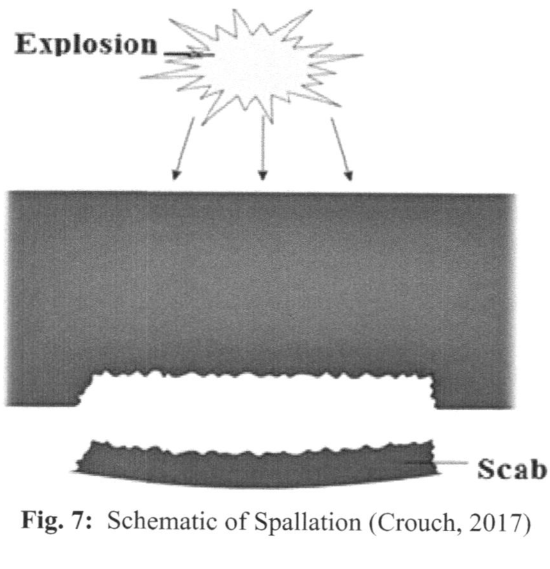

Fig. 7: Schematic of Spallation (Crouch, 2017)

6. FUTURE DEVELOPMENT

In case of soft armor, the development in fibre should be driven towards M5 fibre, which has enough capability to be used in ballistic application. Also, researchers expecting the tensile strength of M5 fibre to reach nearly 9500 Mpa and modulus over 450 Gpa. With this development the weight reduction of 50–60% of an armor system made of M5 fibre could be achieved compared to Kevlar KM2 at the same level of protection (Cunniff & Auerbach).

In case of hard armor plate inserts, the major source of failure is cracks. The improvement in the hardness of existing grade of monolithic ceramics which is very rare as the ceramics grades has been limited through its grain size (Crouch, 2017). We can use backing material such as sheets of graphene. It can absorb more impact of bullet as compare to any other textile fibre which is prime requirement of armor system. Graphene is a very thin and nearly transparent sheet of pure carbon (Kumar, 2016). It is lightweight but 100 times stronger than steel and also it has tremendous value of modulus nearly 1.1 Tpa which is very high than any ceramic grade. It also has extremely high electric conductivity, carrier mobility and ampacity which can be utilize for the application such as to keep soldiers' body warm in low temperature area.

7. CONCLUSION

The modern weapons are becoming more sophisticated, so the designing of protective clothing is really challenging for researchers. The major aim of modern body armor should keep the wearer alive and comfortable in addition to providing the desired level of protection with less fatigue. The important factors that directly influence the ballistic limit of body armor system are density, elastic modulus, strength, toughness, thickness, type and velocity of projectile, and the material configuration. The most important factor that affects the performance is how we arrange the layers of different material with respect to the mechanism of body armor. We should keep in mind every failure modes when designing an armor system.

In addition, other parameters such as environmental and weather conditions (rain, wind, dust, sunlight, cold and snow) should be taken into consideration when selecting material and designing protective clothing. The future armor should be made to provide protection against multiple strikes of modern projectiles, by utilizing materials with abundant suitable properties for the ballistic application such as M5® fibre and graphene sheets.

REFERENCES

[1] Abdulghaffar, H. (2015). Study on the Arrangement of Fabric Materials for Multi-Layer Soft Body Armor Based on their Mechanical Properties, *Journal of Fashion Technology and Textile Engineering.*

[2] Crouch, I.G. (2017). The Science of Armour Materials, *Cambridge (UK): Woodhead Publishing.*

[3] Cunniff, P.M. and Auerbach, M.A. *High Performance "M5" Fibre for Ballistic/Structural Composites.*

[4] Fernando, E., Niles, S., Morrison, A., Pranavan, P., Godakanda, I. and Mubarak, M. (2015). Design of A Bullet-Proof Vest Using Shear Thickening Fluid, *Internation Journal of Advanced Scientific and Technical Research.*

[5] Indi, Y. and Patel, A.A. (2015). A Review Paper on Bullet Proof Jacket, *Research Gate.*

[6] Kumar, N. (2016). Ballistic Vest and Its Improvement—A Review, *International Journal of Scientific Development and Research.*

[8] Nikky and Rao, P.S. (2018). A Review on Material & Ballistic Energy Absorption of Body Armour, *International Journal of Reasearch and Scientific Innovation.*

[9] Scott, A.R. (2005). Textiles for Protection, *Cambridge (UK): Woodhead Publishing.*

[10] *Twaron-A Versatile High Performance Fibre, Teijin Aramid.*

[11] Wang, L., Kanesalingam, S., Nayak, R. and Padhye, R. (2014). Recent Trends in Ballistic Protection, *Textiles and Light Industrial Science and Technology.*

Development of Orthogonal and Angle Interlock 3D Woven Fabric for Composites

S.K. Parmar[1] and P.A. Khatwani[2]

[1]Textile Technology Department, L.D. College of Engineering
(Affiliated to Gujarat Technological University), Ahmedabad, Gujarat, India
[2]Textile Technology Department, Sarvajanik College of Engineering & Technology
(Affiliated to Gujarat Technological University), Surat, Gujarat, India
E-mail: [1]skp25185@gmail.com; [2]pakhatwani@yahoo.com

ABSTRACT: *This article presents brief review of manufacturing principle and structure of orthogonal and angle interlock—3 Dimensional woven fabrics. Both these structures are not only capable to increase thickness dimension but also produce complex structure. Lucrative features are provided by 3D woven structures to be used as composite reinforcements in numerous applications like construction, automotive, ballistics, marine etc. Compared to 2D woven laminates composites, 3D woven fabrics imply higher de-lamination resistance, impact damage tolerance and lighter weight. In this paper, the variety of 3D woven structures of orthogonal and angle interlock fabrics which are developed by varying number of binder and/or stuffer yarn in conventional weaving machine is discussed. The mechanical performance of any 3D woven composite depends upon layer sequence arrange in structure and the amount and types of yarn material such as carbon, glass, Kevlar and ceramic used as binder, stuffer and weft yarn in 3D woven fabric development.*

Keywords: Angle Interlock, Composites, Industrial Applications, Orthogonal, 3D Woven.

1. INTRODUCTION

Two sets of yarns i.e. warp and weft used in conventional weaving process, intersect and interlace at right angles to form a fabric that can be seen in Figure 1(a). On contrary, 'true' 3D weaving process involves interlacement of three orthogonal set of yarns. The additional set of yarn is the binder warp or weft yarn to provide thickness to the fabric that can be seen in Figure 1(b).

Very good stiffness and strength in warp and weft directions are observed in simple 2D fabrics but still the poor result is obtained in thickness direction that should be focused. 2D fabrics also possess some inherent properties like crimp or its waviness which is most undesirable to develop composites. Thus, crimp becomes the limitation of 2D to restrict its application in the areas of space engineering, automotive engineering and Sport textiles.

3D woven fabrics are produced widely by using the multiple warps. An extension of ancient technology for forming double and triple cloth, three-dimensional weaving allows the production of fabrics up to 10 cm in thickness. The binder yarn takes path through the thickness and stuffer yarn along the length. One of the attractive features of 3D weaving is the ability to taper the shape in all three directions during the weaving process (Khokar, 2008).

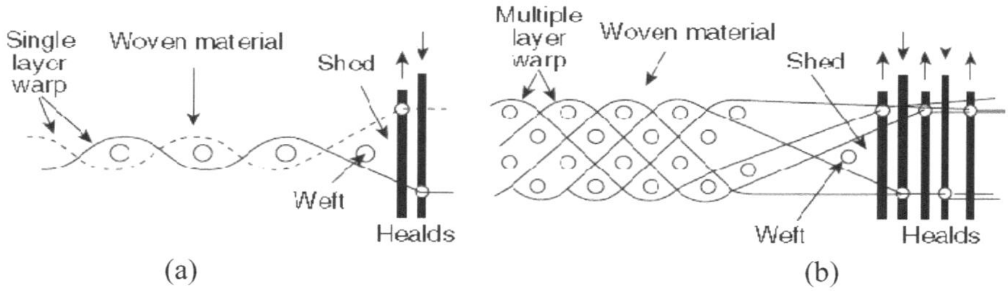

Fig. 1: (a) 2D Weaving Process (b) 3D Weaving Process

A detailed analysis of the different structures produced by these principles is needed to select the appropriate structure for particular application. Only specialist industrial fabrics like performs to be used in construction, automotive, ballistic, marine applications like high performance powerboats developed from carbon fibres, sports goods and various other applications can be targeted using this technology. Several important advantages in composite fabrication are provided by 3D woven performs. Due to reduced labour time in achieving the required thickness, where the multiple layers of 2D fabric plies are replaced with one or few layers of 3D fabric plies in manufacturing of composite structure signifies the important advantage in manufacturing thick composites.

The implementation of composite materials for use in the aerospace, automotive, marine and leisure sectors has increased steadily over the past three decades. The benefits of advanced composites such as increased design flexibility, structural gains, in particular high specific strength and stiffness to weight ratio, and increased fatigue strength have been well documented. As the limitations of conventional 2D laminates from a manufacturing and a mechanical performance perspective described and address the lack of through-the thickness mechanical performance in the composite laminate, 3D woven performs can be preferred (McIlhagger *et al.*, 2003; Stobbe & Mohamed, 2003).

3D woven fabrics are woven fabrics with solid cross-sections either in a broad panel or in a net shaped preform (Chen, 2008). Different ways are available to form such 3D solid architectures; each having its own features structurally and mechanically. In this paper development of 3D orthogonal and angle interlock fabrics and its structure- property relationship is discussed.

2. DEVELOPMENT OF 3D WOVEN FABRIC USING CONVENTIONAL WEAVING MACHINE

Three Dimensional woven fabrics can be fabricated by modifying the conventional weaving mechanisms. When the manufacturing of 3D woven fabric on conventional weaving system is concerned, it can be produced by following principles namely: Orthogonal Principle and Angle Interlock Principle.

2.1 Orthogonal Principle

In this principle, multi-warp yarn system is used to develop 3D fabric. The fundamental used here is to bind straight warp yarns and weft yarns (along 1- and 2- direction) together using binder yarns (along 3-direction). The arrangement used to develop 3D orthogonal structures is shown in Figure 2. Two warp beams were used. The upper beam is used for the warp beam, which was arranged in longitudinal direction in different layers. The lower beam is used for the binder warp, which traveled vertically and is used to bind the warp and weft yarns. High in-plane stiffness and strength are provided by the warp and weft yarns while the stability to woven structure is provided by the running binder yarns. Inter-laminar fracture toughness and impact resistance are observed on higher side in 3D orthogonal woven composites compared to laminated composites (Chen, 2008).

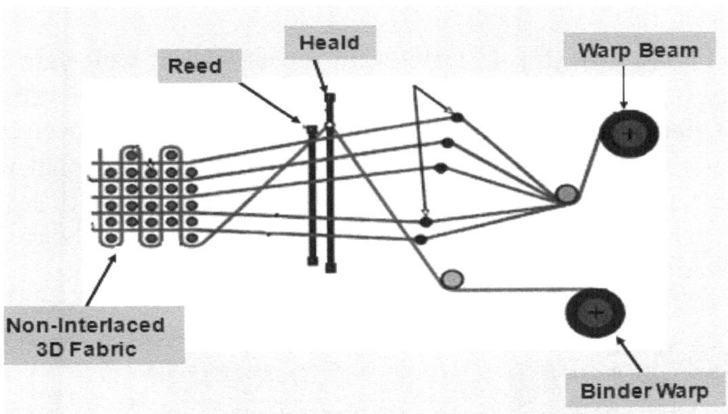

Fig. 2: Arrangement to Produce 3D Orthogonal Woven Fabric

As shown in the Figure 2, warps beam are divided into different groups and each group of thread is passed through a separate heald frame. The number of heald frames depends on number of binder yarns and warp yarn in single weave repeat. The reed profile is made in such a way that group of warp threads are passed through one dent depend on yarn count and reed count and in the next dent a single binder warp is made to pass. Lifting plan of the heald frames is decided in such a way that five-ground warp layers are arranged one above the other and the picks are inserted in between them. There is no interlacement in between the warp and weft layers. Binder threads are in vertical position in the fabric and are used for the firm gripping of the warp and weft yarns in the structure. Without binder thread, stability of the structure cannot be achieved. To weave 3D fabric, some modifications in the conventional sample weaving machine like a separate negative let-off mechanism behind the loom (to hold binder beam) are required as shown in Figure 3.

The actual weave design of orthogonal structure with 4 stuffer and 2 binder yarn are shown in Figure 4, where S1, S2, S3 and S4 indicates warp/stuffer yarn, vertical 1 to 10 indicates weft yarn and B1 and B2 indicates binder yarn. The 3D view and lifting plan of orthogonal woven structure are shown in Figure 4(a) and (c) respectively. The 3 Dimensional views of 3D woven structure are drawn using TexGen software.

Fig. 3: Arrangement of Two Extra Beams in the Sample Weaving Machine (Dash & Behera, 2018)

As shown in Figure 4(b), for 4 stuffer yarn repeat size of weft yarn is 10 pick and 5 pick are arranged one above another. Similarly if number of stuffer yarn increase to 5, 6 and 7, the repeat size of weft yarn becomes 12, 14 and 16 picks respectively in orthogonal structure. Thus the thickness and fibre volume fraction of orthogonal fabric depend on number of stuffer yarn which is incorporated in design. Selection of linear density of warp, weft and binder yarn in structure also play a very important role. Normally the count of binder yarn is always kept fine as compared to stuffer and weft yarn.

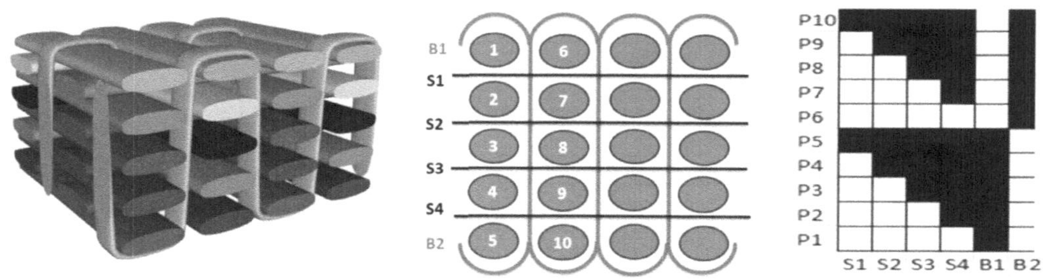

(a) 3 Dimensional View (b) Cross Sectional View (c) Lifting Plan

Fig. 4: Orthogonal Structure with 4 Stuffer Yarn

Some principle features of 3D orthogonal fabrics are as follows:

- Fabric can take the load directly and most effectively as all the three yarns i.e. warp, weft and binder yarns in all three directions are laid straight.
- Using the binding weave specifications, required amount of vertical yarns can be arranged.
- Variable perform thickness can be achieved due to easily alterable interlinking depth within the same perform.
- Structure can be either isotropic or anisotropic.
- Fibre volume fraction is 45–55%.

2.2 Angle-Interlock Principle

The basic arrangement to develop 3D angle interlock structure on conventional loom is same as used in orthogonal fabric where two warping beams are required as shown in Figure 2. Two series of warp yarns are used to produce angle interlock structure. The lower ground beam is used for the Stuffer warp, which was arranged in longitudinal direction in different layers, can be used to increase fibre volume fraction and in-plane strength. The upper beam is used for the binder warp, which travels diagonally and is used to bind several layers of weft (or warp) yarns together.

The actual weave design of angle interlock structure with 4 stuffer and five binder yarn are shown in Figure 5, where S1, S2, S3 and S4 indicate warp/stuffer yarn, B1 to B6 indicates binder yarn and vertical 1 to 30 indicates weft yarn. The 3D view and cross sectional view of angle interlock structures are shown in Figure 5(a) & (b) respectively. It is noted that for 4 stuffer yarn, repeat size of weft yarn of 30 pick and 5 pick are arranged one above another in offset way. Similarly if number of stuffer yarn increase to 6, 7 and 8, the repeat size of weft yarn becomes 42, 56 and 72 picks respectively.

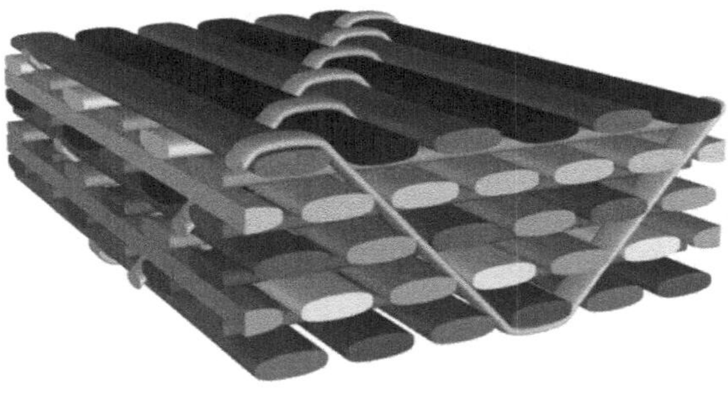

(a) 3 Dimensional View

(b) Cross Sectional View

(c) Lifting Plan

Fig. 5: Angle Interlock Structure with 4 Stuffer Yarn

Another kind of 3D angle interlock structure without stuffer yarn is shown in Figure 6, where B1 to B6 binder yarns are laid in sinusoidal wave form under repeat size of 30 pick and five layers of pick is form. Figure 6(a), (b) and (c) represents 3D view, cross sectional view and lifting plan of angle interlock structure without stuffer yarn respectively.

(a) 3 Dimensional View

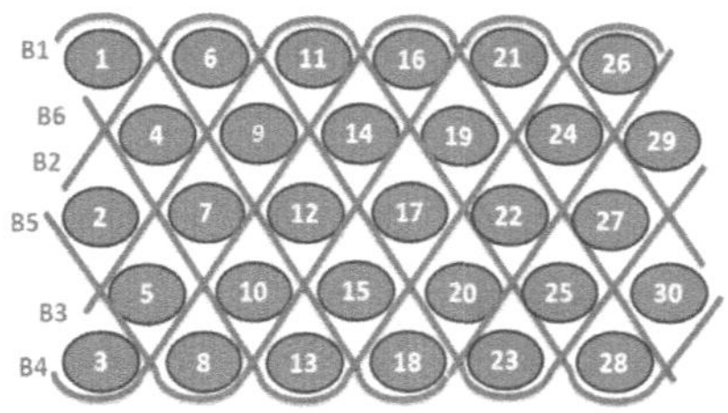

(b) Cross Sectional View

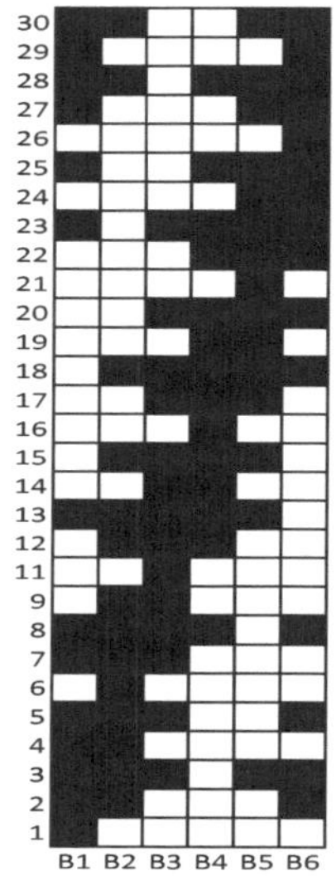

(c) Lifting Plan

Fig. 6: Angle Interlock Structure Without Stuffer Yarn

Some of the features of angle interlock 3D fabrics are:

- It is a multilayer fabric used for flat panel reinforcement.
- Warp yarns are used to bind several layers of weft yarns together or vice versa.
- Stuffer yarns can be used to increase fibre volume fraction and in plane strength (Tong *et al.*, 2002).

3. STRUCTURE PROPERTY RELATIONSHIP OF 3D WOVEN FABRIC

3.1 Role of Stuffer Yarn (Dash & Behera, 2018)

As shown in Figure 7, Orthogonal fabrics were prepared comprising of 2, 4, and 6 stuffer layers and coded as OS2, OS4, and OS6, respectively and interlock fabrics were prepared comprising of 2, 4, and 6 stuffer layers and coded as IS2, IS4, and IS6, respectively.

(a)

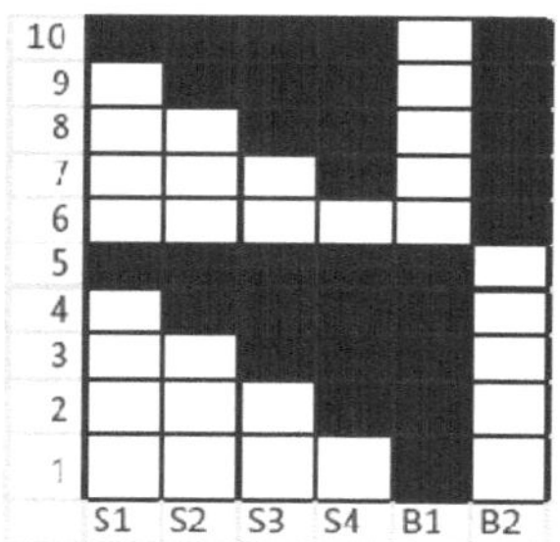

(b)

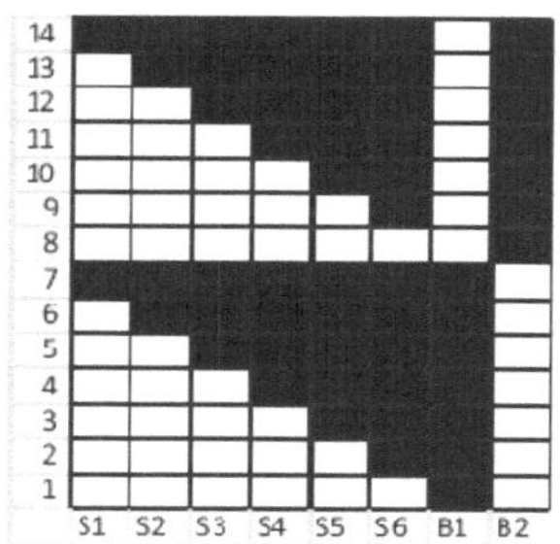

(c)

Fig. 7(a)-(c): Lifting Plan of Orthogonal Structures with 2, 4 and 6 Stuffer Yarn Respectively (Dash & Behera, 2018)

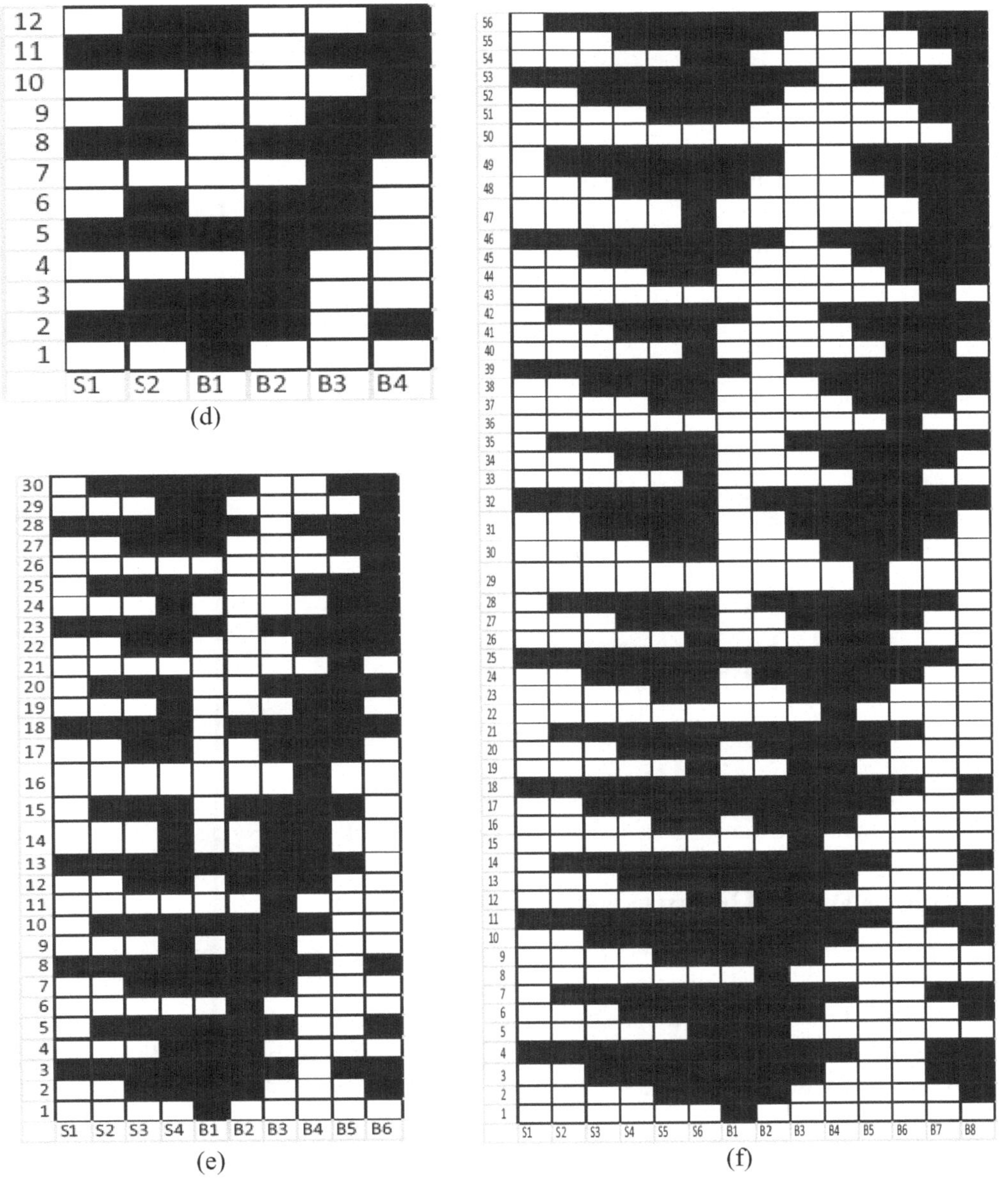

Fig. 7(d)-(f): Lifting Plan of Angle Interlock Structures with 2, 4 and 6 Stuffer Yarn Respectively (Dash & Behera, 2018)

The dominance of more number of stuffer layers for all the samples is clearly observed towards load bearing ability at all level of extensions. The ultimate tensile strength of orthogonal fabrics are noted to be higher when compared with interlock fabrics of same stuffer layers. Figures 8&9 shows the load elongation characteristic curve and normalized ultimate tensile stress of different stuffer layers fabrics.

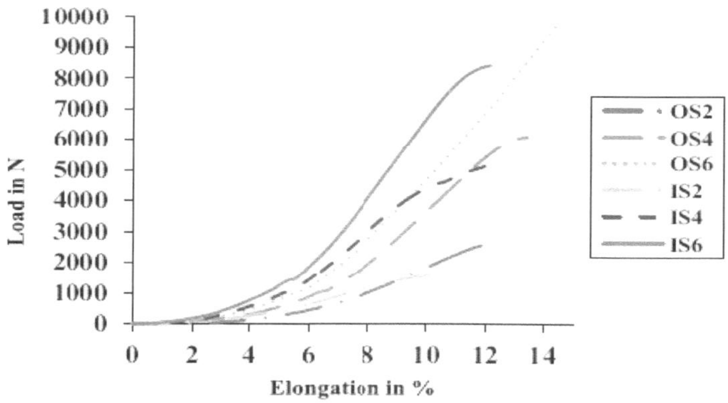

Fig. 8: Typical Load Elongation Curve of Different Stuffer Layers (Dash & Behera, 2018)

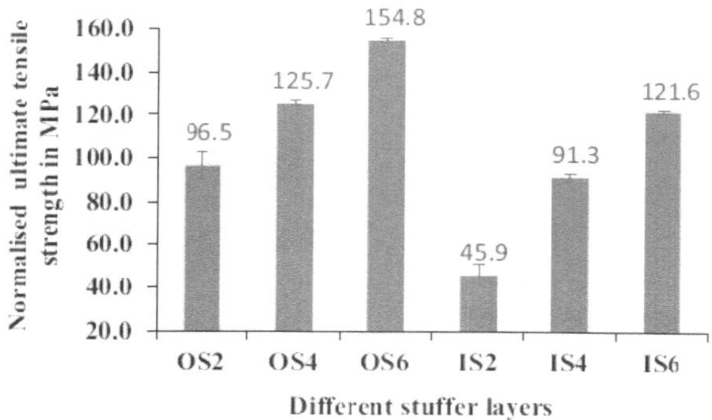

Fig. 9: Normalized Ultimate Tensile Stress of Different Stuffer Layers (Dash & Behera, 2018)

3.2 Role of Weave Design (Dash & Behera, 2017)

By varying the binder interlacement patterns and keeping the stuffer binder ratio constant, three different weave designs of 3D orthogonal and 3D interlock structures can be developed as shown in Figure 10. Orthogonal fabrics were developed with 1 × 1 plain, 4 × 4 matt and 4 × 4 twill designs naming *W*1, *W*2 and *W*3, respectively. Interlock fabrics were developed with 1 × 1 plain, 1 × 1 plain without stuffer and 4 × 4 twill designs naming *W*4, *W*5 and *W*6 respectively.

The load elongation characteristic curve of all six weave designs is shown in Figure 11. The load bearing capacity of W2 increased at higher extension percentage followed by W3 and W6 respectively. This may be attributed to higher z-yarn crimp in the structure. W1 and W5 could take lesser load compared to W4 right from the beginning to the break. Least load value is observed in W5 due to absence of straight load or less crimped warp tows. The order for the normalized ultimate strength is W6 > W3 > W2 > W4 > W1 > W5 as shown in Figure 12. Compared to the other structures, W6, W3 and W2 showed higher strength. For tensile load bearing applications, by increasing the 3D fabric thickness, the effect of binder yarn extension can be minimized in order to completely utilize stuffer tows which crimp lesser. The linear density and warp tows crimp % dominate the tensile properties.

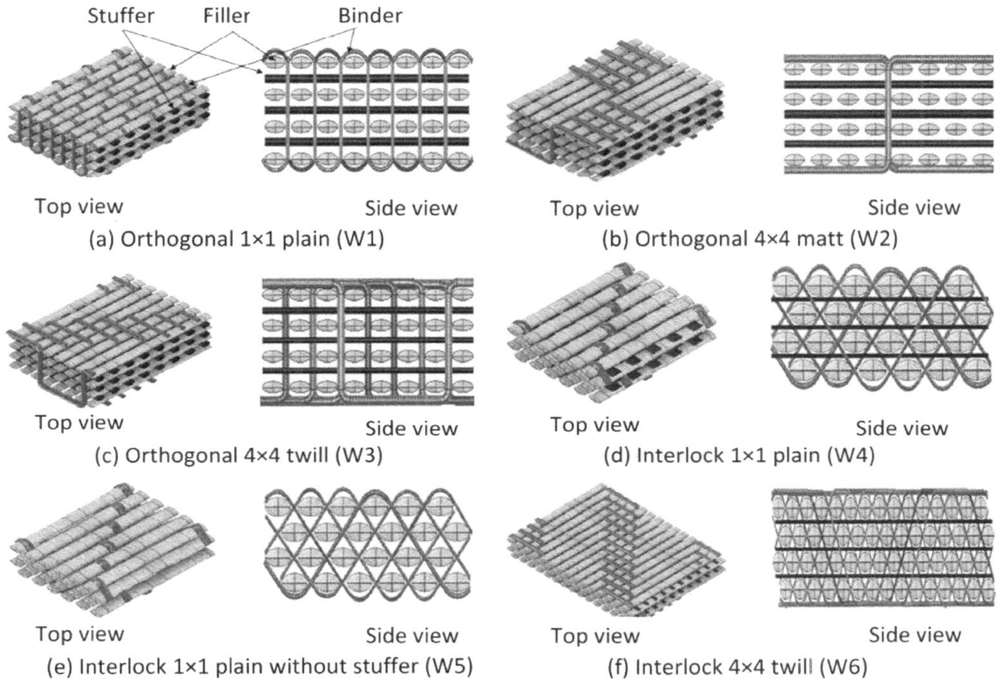

Fig. 10(a)-(f): Actual Weave Design Structure of Orthogonal and Angle Interlock Fabric (Dash & Behera, 2017)

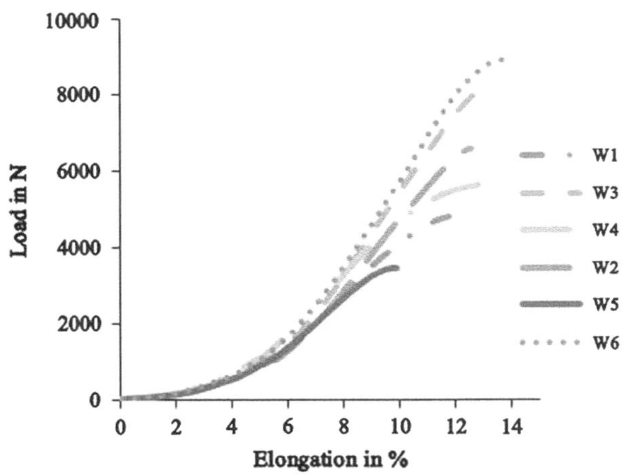

Fig. 11: Load-Elongation Curve of All Six Types of Orthogonal and Angle Interlock Fabrics (Dash & Behera, 2017)

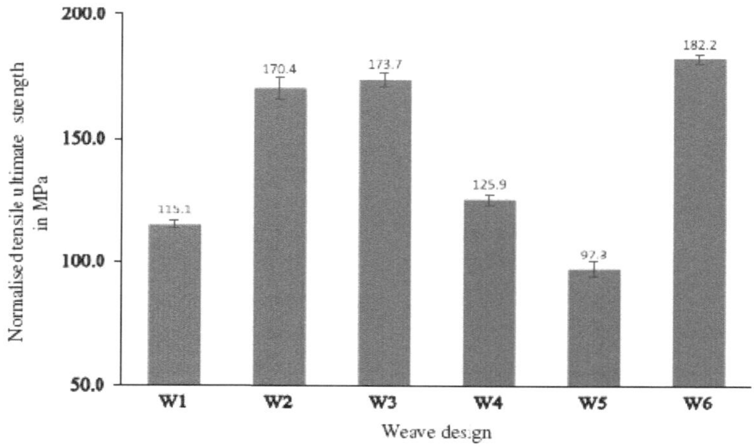

Fig. 12: Normalized Ultimate Tensile Stress of Different Structures (Dash & Behera, 2017)

3.3 Role of Stuffer Binder Ratio (Midani *et al.*, 2018)

As shown in Figure 13, the ratio of the binder yarns (Z) to stuffer/warp yarns (Y)/layers are varying to study the contribution of the Z-yarns in 3D orthogonal fabric. Four samples of

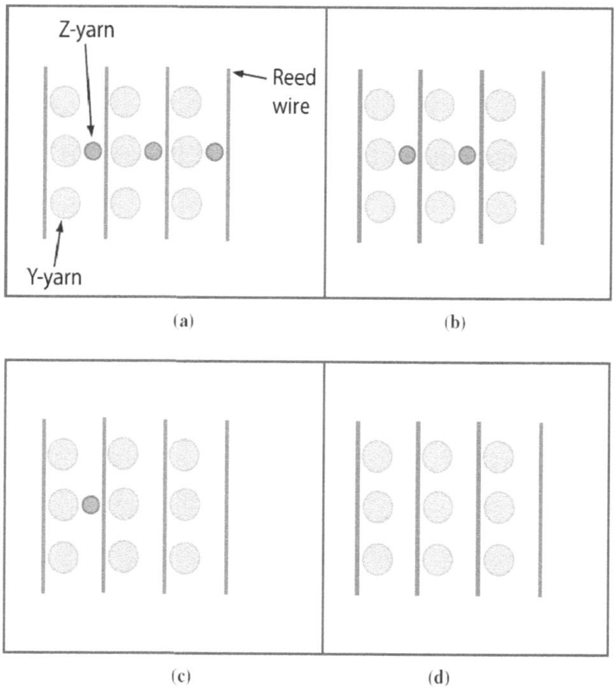

Fig. 13: Denting Plan for a Three-Layer Preform at Various Z- to Y-yarns/layer Ratios (a) 1:1, (b) 2:3, (c) 1:3, (d) 0:1. (Midani *et al.*, 2018)

3D orthogonal plain fabric is prepared by changing Z- to Y-yarns/layer ratio at 4 level,1:1, 2:3, 1:3, and 0:1. It is impossible to weave last sample (0:1), where absence of binder yarn. Practically to develop this sample, only four binder yarn are used for support and then its removed from the sample as shown in Figure 14.

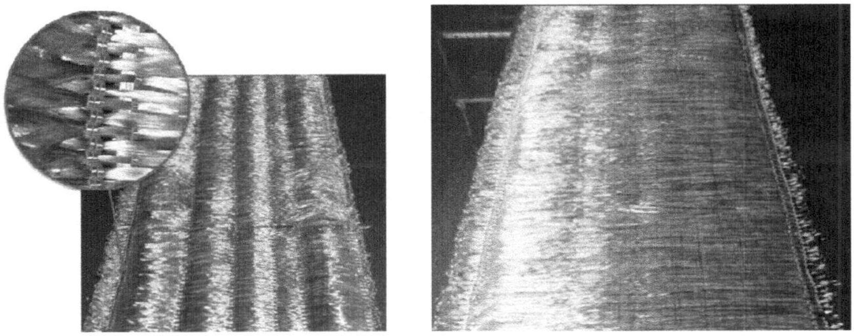

(a) (b)

Fig. 14: Non-Interlaced Orthogonal Fabric (a) With Supporting Z-yarns, and (b) Supporting Z-yarns Removed (Midani *et al.*, 2018)

It is confirmed that some of the common notions about the 3D structure, such as the fact that the presence of binder yarn component increases the out-of-plane properties and de-lamination resistance. Moreover, the study found that using binder yarn with relatively small linear density would minimize its negative effect on the in-plane tensile properties. Moreover, it had been observed that even a small amount of z-yarn in the structure such as z to y ratio 1:3 had a positive effect on the out-of-plane properties leading to 29% increase in the impact energy compared to the sample with no Z-yarn interlacements. Moreover, the Z-yarn played a significant role in maintaining the orientation, alignment and spacing of the X- and Y-yarns. Thus, it is recommended to adopt orthogonal structures when designing composites for enhanced out-of-plane properties, de-lamination resistance, and damage tolerance.

4. APPLICATION

Recently 3D woven fabrics have been finding increased usage in textile composites for commercial structural applications. 3D woven fabrics are being used in a number of applications as composite reinforcements for construction, automotive, ballistic and various industrial uses: for marine applications like carbon fibre performs for high performance power boats; lightweight constructions like reinforced section in automotive engineering and aeronautics; pipeline construction; in sports like shin guard for soccer, protective headgear for sky diving, high speed water sports etc. 3D Woven composites have been provided ideal for security applications, where a high level of protection is expected from the lightest possible components. In transport application and vehicle construction composites have major advantages because lower weight of composites result in higher payload and lower fuel consumption (Behera & Mishra, 2008).

3D woven composites are ideal material for aircraft and aerospace applications where high-strength to weight ratio is required. These are used in space structures such as missiles, rockets and satellites. Space structures require low weight, high stiffness, a low coefficient of thermal expansion and dimensional stability, which are the main properties offered by the 3-D fabrics. Instead of that their use is also increasing in sporting goods like hockey sticks, snow and water skis, bows, arrows, skateboards, bats, tennis rackets, helmets etc. (Tong *et al.,* 2002).

5. CONCLUSION

Various weave structure design of orthogonal and angle interlock fabric can be easily produced for composite as per end use application by doing some modification on conventional loom. Orthogonal structure has received extensive attention specially in characterizing the tensile response. In both structure orthogonal and angle interlock, it is clear that binder yarn in thickness direction can suppress de-lamination and delay final failure. The frequency and repeat size of binder yarn in weave structure also play very important role. In conclusion, 3D orthogonal and angle interlock fabric composites have proven that not only they have enhanced out-of-plane properties, but also better damage resistance/tolerance and enhanced in-plane properties when compared to 2D woven laminates.

REFERENCES

[1] Behera, B.K. and Mishra, R. (2008). 3-Dimensional Weaving, *Indian Journal of Fibre and Textile Research,* 33, 274–287.

[2] Chen, X. (2008). CAD/CAM of 3D Woven Fabrics for Conventional Loom, *The First World Conference on 3D Fabrics and Their Application*, 10–11.

[3] Dash, A.K. and Behera, B.K. (2017). Role of Weave Design on the Mechanical Properties of 3D Woven Fabrics as Reinforcements for Structural Composites, *Journal of the Textile Institute,* 109 (7), 952–960, DOI:10.1080/00405000.2017.1393787.

[4] Dash, A.K. and Behera, B.K. (2018). Role of Stuffer Layers and Fibre Volume Fractions on the Mechanical Properties of 3D Woven Fabrics for Structural Composites Applications, *Journal of the Textile Institute*, 110 (4), 614–624, DOI:10.1080/00405000.2018.1502502.

[5] Khokar, N. (2008). Second Generation Woven Profiled 3D Fabrics from 3D Weaving, *In: Proceedings of The 1st World Conferences in 3D Fabrics & Their Applications,* Manchester.

[6] McIlhagger, A.T., Quinn, J.P. and McIlhagger, R. (2003). The Tailoring of Composite Mechanical Properties by Textile Reinforcement Design, *In: Proceedings of the She Met Ser, University of Ulster,* Jordanstown, Ireland.

[7] Midani, M., Seyam, A.F., Saleh, M.N. and Pankow, M. (2018). The Effect of the Through-Thickness Yarn Component on the In-and Out-of-Plane Properties of Composites from 3D Orthogonal Woven Preforms, *Journal of the Textile Institute*, 110 (3), 317–327, DOI:10.1080/00405000.2018.1481722.

[8] Stobbe, D. and Mohamed, M. (2003). 3D Woven Composites: Cost and Performance Viability in Commercial Applications, *In: International SAMPE Symposium and Exhibition,* California US: Long Beach, 1372–1381.

[9] Tong, L., Mouritz, A.P. and Bannister, M.K. (2002). 3D Fibre Reinforced Composite Materials, *Elsevier Applied Science*, London.

Designing of Functional Clothes for Arthritic Men and its Acceptability

Medha S. Umrikar

Department of Textiles and Apparel Designing, College of Community Science, VNMKV, Parbhani, Maharashtra, India

E-mail: mbukalkar@yahoo.co.in

ABSTRACT: *Arthritis is an age-related, bone-degenerative ailment where in the protective tissue on top of the joints gets damaged. Osteo Arthritis is 9.6% among men and 18% among women. It accounts for the decrease in activities of daily living (ADL) in elderly dependent population in the community. The key factor in the most serious form of arthritis is inflammation, evidenced by heat, redness and retardation of growth, swelling and pain in the fingers, wrists, elbows, knees, hips and feet. As the disease progresses the joints may stiffen and in some cases, so distorted that movement becomes difficult or even impossible. Functional garments are those which can be put on and taken off (donned and doffed) with little or no help of other individual, permit enough freedom of movement and are comfortable when worn on body. The garments must be simple in style, attractive in design and enhance the appearance of the wearer (Tatte & Glisson, 1961). This study was conducted to design clothes for men with arthritis with the objective to determine the problems faced by arthritic men in present clothing practices, to alter design and reconstruct the garments according to their requirement, to assess the suitability of constructed garments. For this study 30 arthritic men between 55–65 years of age were selected. An interview schedule was prepared and the respondents were interviewed personally. In accordance to the requirements and problems expressed by the respondent, the shirt and pant were selected and functional designs were sketched. Additional features such as knitted and stitched patches on shoulders, elbow in shirts and on knee portion in pants pads were incorporated. These were given to 5 respondents for trial. An acceptability index was calculated on three-point rating scale with the score of 3, 2, 1 for acceptable, least acceptable and not acceptable out of total score 90. It was concluded from this study that, immobility of joints while donning and doffing of the garments was most frequent with highest percentage i.e., 83.33. Retention of heat was found in pads stuffed with rice grain for more time i.e., 1 hour 30 minutes. This design was suitable for arthritic patients with therapeutic treatment. Design DP2 scored highest points i.e., 81 with acceptability index 90 followed by DS2 scored 76 with acceptability index 84.44.*

Keywords: Clothes for Arthritic Men, Donning and Doffing, Functional Garments, Garment Designing, Inflammation.

1. INTRODUCTION

Arthritis is an age-related, bone-degenerative ailment where in the protective tissue on top of the joints gets damaged. Osteo Arthritis is 9.6% among men and 18% among women. It accounts for the decrease in activities of daily living (ADL) in elderly dependent population in the community. It is referred to as 'wear and tear' arthritis and is usually associated with aging. Rheumatoid arthritis, an inflammatory disease affects the joints as well as other parts of body. The key factor in the most serious form of arthritis is inflammation, evidenced by heat, redness and retardation of growth, swelling and pain in the fingers, wrists, elbows, knees, hips and feet. As the disease progresses the joints may stiffen and, in some cases, so distorted that movement becomes difficult or even impossible. Symptoms of both types include stiffness joint pain, joint swelling bony enlargement and subcutaneous nodules often leading to the inability to carry out daily living activities that were once routine (Ford & Tolmar, 1983).

The garments must be simple in style, attractive in design and enhance the appearance of the wearer. (Tatte & Glisson, 1961) The dressing skills and clothing needs and preferences of an 85 years old arthritic woman who used a wheel chair were assessed by Park & Janice (1989) in a case study and it was concluded that garments could be designed to be comfortable when the body was in to variety of positions. The features of the garments designed for the study were one and one-half length waist band, raglan sleeves, knit fabrics, pressure tape closures, front openings and action pleat. Functional garments are those which can be put on and taken off (donned & doffed) with little or no help of other individual, permit enough freedom of movement and are comfortable when worn on body with this goal.

2. MATERIALS AND METHOD

For this study 30 arthritic men between 55–65 years of age were selected. An interview schedule was prepared and the respondents were interviewed personally. In accordance to the requirements and problems expressed by the respondent, the shirt and pant were selected and functional designs were sketched. Additional features such as knitted and stitched patches on shoulders, elbow in shirts and in pants, on knee portion, pads were incorporated. Based on this, two types of shirts DS1 (shirt with stitched cloth pads on elbow and shoulder for stuffing and DS2 shirt with stretchable knitted pad on elbow portion). In pants, DP1 pant with stitched patch for heating pads on knee portion and DP2 pant with knitted patch on knee portion were designed. For stuffing rice grains, fine sand and fleece were used. The duration for retention of heat in pads with this material were studied. These were given for trial to five respondents. Six parameters were asked. Total score for each design was calculated on three-point rating scale with the score of 3, 2, 1 for acceptable, least acceptable and not acceptable out of total score 90. An acceptability index was calculated by calculating percentage value of total score.

3. RESULTS AND DISCUSSION

Table 1 shows type of stuff use and period for heat retention. Three types of stuffs were used for pads-sand, fleece, and rice grains. These were heated to temperature of 100°. Cooling time was noted. It is clear that retention of heat with rice grains was for more time i.e., 1 hour 30 minutes.

Table 1: Type of Stuff Used in Pads and Time of Retention of Heat

Sr. No.	*Type of Stuff Used for Heating Pads*	*Time of Retention of Heat H/M*
1.	Sand	1/15
2.	Rice grains	1/30
3.	Fleece	1/20

Table 2 shows frequency and percentage of distribution of respondents by type of clothing problem. It is clear from the table that bunching of shirt sleeve at elbows was 15 i.e., 50% which was least than other problems such as bunching of pants at knees for which the frequency was noted to be 16 and percentage was 53.33. Lack of support on joints was the problem most frequently occurred and it was 20 i.e., 66.66%.

Table 2: Frequency and Percentage Distribution of the Respondents by Type of Clothing Problem N-30

Sr. No.	*Problem*	*F*	*%*
1.	Bunching of shirt sleeve at elbow	15	50.00
2.	Bunching of pants at knee	16	53.33
3.	Lack of support on joints	20	66.66
4.	Immobility of joints	25	83.33

Table 3 indicates acceptability index of functional clothes. This was calculated on three-point rating scale with the score of 3, 2 and 1 for acceptable, least acceptable and not acceptable. Scores obtained by design DS2 (Shirt with knitted pads on elbow and shoulder) scored highest i.e., 91 than design DS1 (Shirt with stitched cloth patch) for which the score was 70. In case of pant design DP2 (pant with knitted pad at knee portion scored highest) scored highest i.e., 96 than design DP2 (pant with stitched cloth patch).

Table 3: Acceptability Index and Total Score Obtained by Functional Clothes

Sr. No.	*Parameters of the Functional Garments*	*Upper Garment*		*Lower Garment*	
		DS1	*DS2*	*DP1*	*DP2*
1.	Comfortable	8	12	8	15
2.	Supportive to joints	7	15	8	15
3.	Ease in movements	10	10	10	12
4.	Relief in swelling	15	12	15	12
5.	Ease in donning	10	12	10	12
6.	Aesthetic appeal	10	15	8	15
Total Score		**60**	**76**	**59**	**81**
Acceptability Index		66.66	84.44	65.55	90

Functional Garments—Shirt and Pant Designed for Arthritic Men

Shirt Design DS1:
Stitched Cloth Patch on Shoulder and Elbow

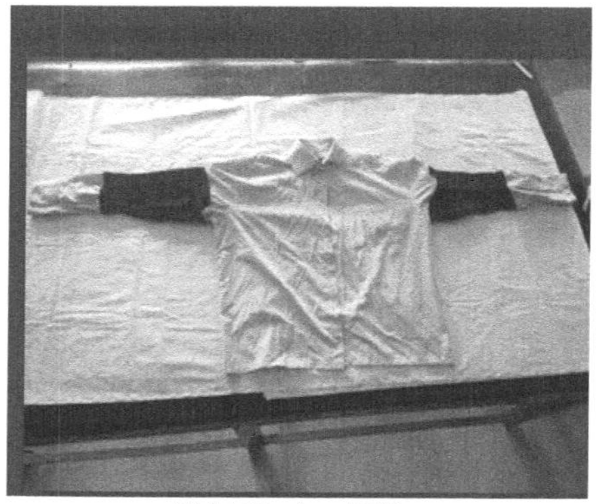

Shirt Design DS2:
Stitched Cloth Patch on Elbow

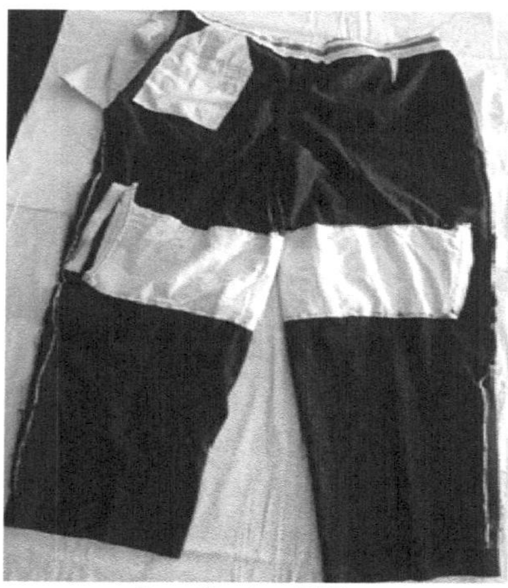

Pant Design DP1:
Stitched Cloth Patch on Knee

Pant Design DP2:
Stitched Knitted Pad on Knee

4. CONCLUSION

It was concluded from this study that immobility of joints while donning and doffing of the garments was most frequent with highest percentage i.e., 83.33.

Retention of heat was found in pads stuffed with rice grain for more time i.e., 1 hour 30 minutes. This design was suitable for arthritic patients with therapeutic treatment. Design DP2 i.e., Pant with knitted patch on knee portion scored highest points i.e., 81 with acceptability index 90 followed by DS2 i.e., shirt with knitted patch at elbow and shoulder with 72 score and acceptability index 84.44.

REFERENCES

[1] Ford, A.B. and Tolmar, S. J. (1988*). Journal of Am. Geriatrics Soc.* (36), 187–197.

[2] Koestor and Laker. (1984*). Family and Consumer Sci. Res.* J. (13), 153–158.

[3] Medha, B.U. (2008). Designing of Functional Sari Blouse for Arthritic Women, *Journal of Dairying, Foods & H.S.,* 27 (3/4), 234–237.

[4] Park, C. and Janice, M. (1989). *Journal of Home Eco.,* 81(3), 6–10.

[5] Reich and Otten, (1991). *Textile Research Journal,* 9(4), 34–36.

[6] Tate, M.J. and Glisson, O. (1961). *Family Clothing, John Wiley and Sons, Inc. New York,* 166–167.

Section-V

Environment and Sustainability

Microfibre Pollution and Sustainability

S. Raja Balasaraswathi[1] **and R. Rathinamoorthy**[2]
Department of Fashion Technology, P.S.G. College of Technology,
Coimbatore, Tamil Nadu, India
E-mail: [1]s.rajabalasaraswathi@gmail.com; [2]r.rathinamoorthy@gmail.com

ABSTRACT: *Microplastic pollution in the marine environment has got greater attention in recent years. Deep sea, coastal areas, sea ice, beach sands were found contaminated by microplastics (plastics less than 5 mm in size). The analysis of various sources of microplastics found in the samples have confirmed that synthetic textiles are the major source of microplastics that are contributing in the form of microfibres that shed during the wearing and washing of synthetic textiles. It has been reported that synthetic textiles used in the apparels account for 35% of microplastics in the marine environment. It has been estimated that by the year 2050, around 22 million tons of microfibres will be entering the ocean annually due to the increasing trend of synthetic textile production and consumption. The microfibre pollution is not less when comes to creating negative impacts on the environment. These microfibres, being very small, are mistaken as food by many small living organisms in the sea and they are reaching up to humans through food chain. Though alterations of washing parameters and laundry aids can effectively reduce microfibre pollution, they are short term, and the modification of textile materials can be the long-term solution where the researches are very less. This paper clearly elaborates the impact of microfibre pollution, need for the awareness among the people and the research gap in the textile industry in controlling microfibre pollution.*

Keywords: Microfibre Pollution, Microplastic, Sustainability, Synthetic Textiles, Textile Modification.

1. INTRODUCTION

Sustainability has become the important aspect which is grabbing more attention in every field as the concern towards the environment and people is given more attention. Sustainability, as defined by World Commission on Environment and Development, is the development that is made to meet the current generation's needs without affecting those of future generations (Joy & Pena, 2017). Being one of the largest polluting industries, the Textile and Fashion industry is paying more attention to sustainable development. The fashion and textile industry are not sustainable at present. The large uptake of natural resources without having alternate ideas to preserve is exploiting the people of future generation impacting the profitability and business opportunities. Sustainable fashion will be a key factor to protect the earth from destruction to a greater extent as the fashion industry is one of the major threats to the Earth's resources. In recent years, the most fascinating topic which is getting greater acceleration in the global fashion market is 'fast fashion'. Actually, the term 'fast fashion' refers to the clothing that quickly moves from catwalk to stores with the purpose of mimicking the current luxury trends. It is an approach to make the fashion trends available for the normal consumers quickly and cheaply (Jang *et al.*, 2012). The 'fast fashion' is one of the major reasons for the higher consumption of synthetic textiles. Fast fashion focuses on faster production at affordable prices. And synthetic textiles will be a good choice in achieving the low cost and quick production concept of fast fashion. Though their production processes are much simpler than the processing of natural fibres in terms of cultivation land and water requirement, the risks with these materials are relatively very high (*https://superegoworld.com/blogs/the-world/the-dangers-of-synthetic-fibres-and-fabrics-on-the-environment*, 2020). The use of synthetic materials is a huge threat to the sustainability as the impacts with these materials are high. As the production of synthetic fibres depends on the fossil fuel, it greatly results in the depletion of non-renewable resources. For the production of polyester fibres alone, it consumes around of 70 million barrels of oil every year (*https://www.sustainyourstyle.org/en/fibre-ecoreview*, 2020). In addition to this, the impacts of production process are also significant. The production process is energy intensive (almost 125 MJ of energy for the production of 1 kg of polyester) (*https://www.commonobjective.co/article/fibre-briefing-polyester*, 2020) and there are chances for the emission of undesirable by-products during the production (Slater, 2008). Apart from these, the most unnoticed impact of synthetic textiles is microfibre emission. These micro-sized fibres that shed from the synthetic textiles can add up to microplastic sources that contaminate different levels of environment. The non-biodegradability nature of the synthetic textiles makes is more hazardous. This microfibre pollution has become a major concern for sustainability in textile and apparel business (Louisa, 2017). It has been estimated that nearly 1.02 kg of fibres can be produced everyday by the population of 100,000 people. Though it seems smaller than microplastic wastes, the major concern is that these micro plastics are difficult to trace and recover (Resnick, 2019). This paper aims to elaborate the sustainability issues with the microplastics, sources of microplastics and the predominance of textiles in the microplastics pollution.

2. MICROPLASTICS—AN EMERGING MARINE POLLUTANT

Plastics that are less than 5 mm will come under the category of microplastics which need to be addressed because of its ubiquitous nature (Browne *et al.*, 2007). The increased usage of plastics in the past few decades accelerates the emission of microplastics into the marine environment. The global market size of plastics in the year 2019 is 568.9 billion USD (*https://www.grandviewresearch.com/industry-analysis/global-plastics-market*, 2020). The source of microplastics that are found in the marine environment can include abrasion of tyres, synthetic textiles, marine coating, road marking, personal care products like cosmetics and plastic pellets (Boucher & Friot, 2017). The microplastic has become an emerging problem in marine environment. It has been reported that marine debris concentration varies from 0–101000 pieces per square kilometer, out of which 70% are made of plastics (Linn, 2016). Microplastics are found in deep sea as well as sea surface. It has been estimated that almost 93–268 kilotons of microplastics are floating in the oceans which only includes the plastics that are lighter than the water and floating on the surface. If the denser plastics were also taken into account, then there will be significant amount of microplastics contaminating the marine environment. Microplastics were also found in coastal regions (Browne *et al.*, 2011) beach sands (Noren *et al.*, 2014) and in deep sea (Woodall *et al.*, 2014) A recent research has estimated 8–14 million tons of microplastics on the sea floor (Britta *et al.*, 2020).

3. MICROPLASTICS IN FOOD CHAIN

Microplastics, which are being very small in size, are often mistaken as food by small organisms in the water bodies. Almost all levels of marine food chain are found with microplastic contamination. Microplastics were found in different sea-based foods which can be a huge threat to human beings. A researcher has examined the edible portion of various sea food organisms including oyster, prawns, squid, crabs, and sardines. All the samples were contaminated by poly vinyl chloride particles and polyethylene is found in higher concentration. Among different organisms analyzed sardines are contaminated more by microplastics with 0.3 mg/g of tissue concentration (Francisca *et al.*, 2020). In another study, two different fish species from lakes were examined and reported 100% of samples were contaminated with microplastics. They have also reported 1–49 microplastic particles/fish (Raven *et al.*, 2020). The table salt, which is a common seasoning in our food, is found to be contaminated with microplastics. The salt brands across world are examined and out of 39 brands analyzed only 3 were found to be free from microplastics while all others are contaminated. Asian brands are found to have more microplastic particles. It has been estimated that an average adult is consuming approximately 2000 microplastics every year through salt (Parker, 2018). Moreover, honey, beer, some vegetables and fruits are found with microplastics. The researchers studied the prevalence of microplastics in vegetables and fruits have reported 223,000 and 97,800 high median level of microplastics in fruits and vegetables respectively with apple and carrot as the most contaminated fruit and vegetable (Gea *et al.*, 2020).

As a direct threat to human life, microplastics have been found in drinking water. The tap water from 5 different continents were examined for the presence of microplastics and the report says that 83% of the samples are contaminated with the microplastic particles. Among the samples examined, the water sample from India is ranked 3 for highly polluted water (*http://www.beachapedia.org/Microfibres*, 2020). Moreover, to make the situation worse, the bottled water was also found with microplastic particles. The drinking water samples of bottled water from 9 different countries were analyzed. The research has reported approximately 325 particles/litre of bottled water. Out of 259 bottles, they have found microplastic particles in 242 bottles (Priti & Mukherjee).

The smaller size of microplastics leads to large surface to volume ratio which can accommodate various pollutants on its surface particularly persistent organic pollutants. The ingestion of microplastics cause increased toxicity due to the pollutants on its surface as well as the various chemical additives in the plastics (Linn, 2016). The ingestion of microplastics can cause potential effects on human beings yet the effects are not clearly defined or proven (Raven *et al.*, 2020).

4. MICROPLASTICS IN ATMOSPHERE

Many researchers have investigated the presence of microplastics in the air. A study that has been made to analyze the microfibres in the atmosphere by sampling outdoors reported that the quantity of microfibres in the atmosphere varies with various factors including consumption habits, socio-economic status, traffic, urbanization (Ahmet *et al.*, 2018) Another study on the analysis of microfibres in the air sampled in indoor and outdoor sites have confirmed the presence of microfibres in the atmosphere and a considerable number of microfibres ranging from 1.0–60.0 fibres per cubic meter are found. Among the fibres found, 33% were synthetics (plastics) (Rachid *et al.*, 2016). The researchers have found inhaled microplastic fibres in the human lungs while assessing cancerous lung specimens. With this, they have concluded that these plastic fibres could possibly contribute to the risk of lung cancer (Darmion, 2019).

5. SYNTHETIC TEXTILES—MAJOR SOURCE OF MICROPLASTICS

Synthetic Textiles are the dominant source of microplastic emission. They can contribute to microplastics in the form of short fibres. The use of synthetic textiles has become inevitable because of its ultimate properties. As discussed before, the fast fashion system fuels the demand of synthetic textiles and the problem get serious with low quality synthetic textiles. Synthetic textiles correspond to 25–30% of microplastics in the marine environment (Jianli *et al.*, 2019). It has been estimated that around 0.19 million tons of microfibres are entering the marine environment annually (Leonas, 2018). It is estimated to raise to 22 million tons by the year 2050 (*http://www.ellenmacarthur foundation.org/ publications,* 2020). Synthetic microfibres are found in different levels of environment including coastal region (Browne, 2011), surface water (Rachael *et al.*, 2017), sea ice (Obbard *et al.*, 2014), and even in atmosphere (Francesca, 2020). The analysis of samples collected from different coastal regions across six continents has proven that the contamination of coastal region with synthetic fibres which resembles those used in the apparels (Browne, 2011). The other researcher found microplastics in sea ice in arctic region where rayon (54%) was predominant followed by polyester (21%) and nylon (16%) (Obbard *et al.*, 2014) Around 300 million individual microfibres were found to be discharged along the surface of Hudson River, USA (Rachael *et al.*, 2017). Similarly, in Saigon River, a huge proportion of synthetic fibres are found to contribute to microfibres present. And polyester accounts for 70% of total synthetic fibres found (Lisa *et al.*, 2018). The analysis of sewage effluent has reported that flake hold a huge proportion of 67.3% which is followed by fibrous particles (18.5%). Among various fibrous materials, polyester and polyamide were dominant by contributing 28% and 20% respectively (Kerstin *et al.*, 2016). Figure 1 shows the dominance of microfibres in the source of microplastics in different levels of environment.

As various research on the microplastics have reported that the fibrous materials found resemble synthetic fibres that are being used in the textiles and apparels, various researchers have come forward in analyzing the microfibre emission from the synthetic textiles. Napper and Thompson (2016) have reported domestic washing of synthetic

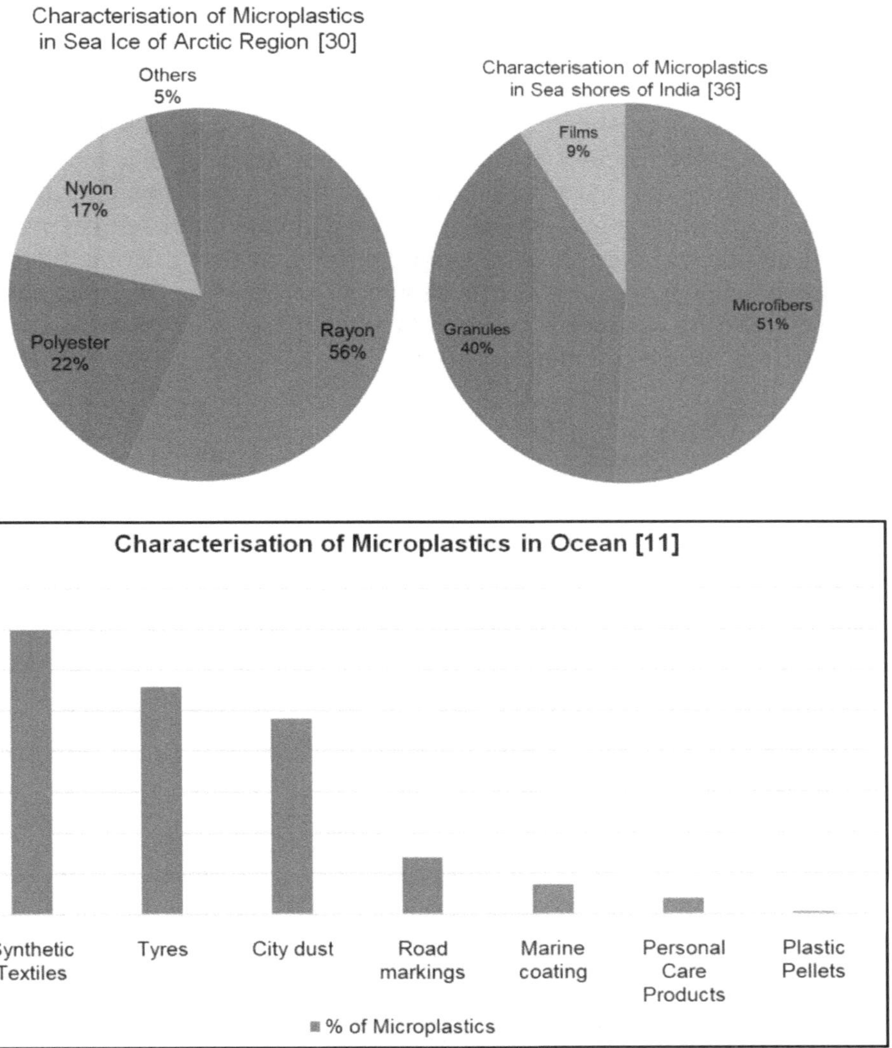

Fig. 1: Dominance of Microfibres as the Source of Microplastic

textiles are the major cause for microfibre emission. They have estimated that around 728,789 fibres per 6 kg of wash load (Napper & Thompson, 2016). In other report, the researcher estimated the microfibre release by assuming that 30–50% of clothing that has been used as synthetic textiles and the per capita amount of laundry as 200–300 kg. Their estimation results showed that 9.85 million populations can release 8–945 tons of microfibres every year by domestic laundering. They have also estimated that washing of 1 kg of synthetic fabric can shed 12–640 mg of microfibre (Tiwari *et al.*, 2019). Moreover, various research has been made on analyzing the shedding behavior of textile materials under different washing conditions.

6. CONCLUSION

Microplastic pollution is emerging as a major concern for sustainability. The ubiquitous nature of microplastics makes the issue more alarming. Microplastic particles were found in different levels of environment. These microplastics are being ingested by various living organisms and being transferred to human beings through food chain. Textile and Fashion industry is contributing more to the microplastic pollution in the form of synthetic fibres. The synthetic fibres that are shedding from the textile materials during different activities particularly washing is noted as the major source of microplastic pollution. The research on the synthetic textiles is needed to reduce the microfibre emission that can contribute to control the microplastic pollution.

REFERENCES

[1] Ahmet, T.K., Meral, Y. and Senemçiftiçi, B. (2018). Ubiquitous Exposure to Microfibre Pollution in the Air, *Eur. Phys. Journal of Plus,* 133:488, DOI: 10.1140/epjp/i2018-12372.

[2] Boucher, J. and Friot, D. (2017). Primary Micro Plastics in the Oceans: A Global Evaluation of Sources, IUCN, Gland, Switzerland, *https://doi.org/10.2305/IUCN.CH.2017.01,* 43.

[3] Britta, D.H., Chris, W. and Justine, B. (2020). There are Up to 14 Million Tons of Micro Plastics on the Seafloor-Worse than We Thought Plastics Have Been Found even in the Mariana Trench, *https://theprint.in/environment/there-are-up-to-14-million-tonnes-of-micro plastics-on-the-seafloor-worse-than-we-thought/521213/* (Accessed on 14th December 2020).

[4] Browne, M.A., Galloway, T. and Thompson, R. (2007), Microplastic—An Emerging Contaminant of Potential Concern?, *Integrated Environmental Assessment and Management,* 3(4), 559–561.

[5] Browne, M.A., Phillip C., Stewart J.N., Teuten, E., Tonkin, A., Galloway, T. and Thompson, R. (2011). Accumulation of Micro Plastic on Shorelines Worldwide: *Sources and Sinks, Environ. Sci. Technol.,* 45, 9175–9179. *https://dx.doi.org/10.1021/es201811s.*

[6] Damion, C. (2019). Micro Plastics 'Significantly Contaminating the Air', Scientists Warn, *https://www.theguardian.com/environment/2019/aug/14/microplastics-found-at-profuse-levels-in-snow-from-arctic-to-alps-contamination* (Accessed on 14th December 2020).

[7] Ellen MacArthur Foundation (2017). A New Textiles Economy: Redesigning Fashion's Future, *http://www.ellenmacarthurfoundation.org/publications (*Accessed on 15th April 2020).

[8] Fibre Briefing: Polyester, *https://www.commonobjective.co/article/fibre-briefing-polyester* (Accessed on 13th December 2020).

[9] Francesca, D.F., Maria, C.C., Maurizio, A. and Thompson, R.C. (2020). Micro Fibre Release to Water, Via Laundering, and to Air, Via Everyday Use: A Comparison Between Polyester Clothing with Differing Textile Parameters, *Environ. Sci. Technology,* DOI: 10.1021/ acs.est.9b06892.

[10] Francisca, R., Elvis, D.O., O'Brien, J.W., Fraissinet-Tachet, S., O'Brien, S., Gallen, M., Samanipour, S., Kaserzon, S., Mueller, J.F., Galloway, T. and Thomas K.V. (2020).

Quantitative Analysis of Selected Plastics in High-Commercial-Value Australian Seafood by Pyrolysis Gas Chromatography Mass Spectrometry, *Environ. Sci. Technol.* 54, 9408−9417.

[11] Gea, O.C., Ferrante, M., Banni, M. and Favara, C. (2020). Micro and Nano-Plastics in Edible Fruit and Vegetables, The First Diet Risks Assessment for the General Population, *Environmental Research,* DOI: 10.1016/j.envres.2020.109677.

[12] Jang, J., Ko, E., Chun, E. and Lee, E. (2012). A Study of a Social Content Model for Sustainable Development in the Fast Fashion Industry, *Journal of Global Fashion Marketing,* 3–2, 61–70. DOI: 10.1080/20932685.2012.10593108.

[13] Jianli, L., Yunfei, Y., Jiannan, D., Bo, Z. and Weidong, G. (2019). Micro fibres: A Preliminary Discussion on their Definition and Sources, *Environmental Science and Pollution Research*, 26, 29497–29501, DOI: 10.1007/s11356-019-06265-w.

[14] Joy, A. and Pena, C. (2017), Sustainability and the Fashion Industry, *Conceptualizing Nature and Traceability,* DOI: 10.1007/978-3-319-51253-2_3.

[15] Kerstin, M., Karin, E, Anna, F., Kalle, H., Johan, H., Mikael, O., Johanna, S. and Anais, V. (2016). Swedish Sources and Pathways for Micro Plastics to the Marine Environment A Review of Existing Data, IVL, *Swedish Environmental Institute Report* no. 2016, C 183, Stockholm, Sweden. *https://www.ivl.se/webdav/files/Rapporter/C183.pdf.*

[16] Leonas, K.K. (2018). Textile and Apparel Industry Addresses Emerging Issue of Microfibre Pollution, *Journal of Textile and Apparel Technology and Management,* 10(4).

[17] Linn, A. (2016), Shedding of Synthetic Microfibres from Textiles, Gothenburg University. *https://bioenv.gu.se/digitalAssets/1568/1568686_linn---str--m.pdf* (Accessed on 15th April 2020).

[18] Lisa, L., Emilie, S., Thuy-Chung, K.L., Rachid, D., Boukerma, K., Emmanuel, R., Johnny, G. and Bruno, T. (2018), Macro Plastic and Micro Plastic Contamination Assessment of a Tropical River (Saigon River, Vietnam) Transversed by a Developing Megacity, *Environmental Pollution,* 236, 661–671.

[19] Louisa, S. (2017), Rivers and Oceans Being Polluted by Microfibres-Microfibre Pollution: The Latest Topic in Sustainability for Brands and Textile Suppliers, *https://www.ispo.com/en/trends/microfibre-pollution-latest-topic-sustainability* (Accessed on 13th December 2020).

[20] Microfibres (2019). Beachapedia, *http://www.beachapedia.org/Microfibres.*

[21] Murphy, F., Ciaran, E., Frederic, C. and Brian, Q. (2016). Wastewater Treatment Works (WWTW) as a Source of Micro Plastics in the Aquatic Environment, *Environmental Science & Technology,* DOI: 10.1021/acs.est.5b05416.

[22] Napper, I.E. and Thompson, R.C. (2016). Release of Synthetic Micro-Plastic Plasticsfibres from Domestic Washing Machines, Effects of Fabric Type and Washing Conditions, *Marine Pollution Bulletin,* Vol. 112, No. 1–2, 39–45. *https://doi.org/10.1016/j.marpolbul.2016.09.025*

[23] Noren, F., Noren, K. and Magnusson, K. (2014). Marine Microscopic Debris, Survey along Swedish West Coast, IVL, Swedish Environmental Institute Report No. 2014: 52, County Administrative Board, West Gotaland. *https://sverigesradio.se/diverse/appdata/isidor/files/406/14638.pdf* (Accessed on 27th August 2020).

[24] Obbard, R.W., Sadri, S., Wong, Y.Q., Khitun, A.A., Baker, I. and Thompson, R.C. (2014). Global Warming Releases Micro Plastic Legacy Frozen in Arctic Sea Ice, *Earth's Future*, 2, 315–320, DOI:10.1002/2014EF000240.

[25] Our Fibre Eco Review, *https://www.sustainyourstyle.org/en/fibre-ecoreview* (Accessed on 13th December 2020).

[26] Parker, L. (2018). Micro Plastics Found in 90 Percent of Table Salt, *https://www.nationalgeographic.com/environment/2018/10/microplastics-found-90-percent-table-salt-sea-salt* (Accessed on 14th December 2020).

[27] Plastic Market Size, Share & Trends Analysis Report By Product (PE, PP, PU, PVC, PET, Polystyrene, ABS, PBT, PPO, Epoxy Polymers, LCP, PC, Polyamide), by Application, by Region, and Segment Forecasts, 2020–2027, *https://www.grandviewresearch.com/industry-analysis/global-plastics-market* (Accessed on 28th August 2020).

[28] Priti, M. and Mukherjee, M., Micro Plastics the Most Widespread and Persistent Hazards of Plastic, *Toxics Link Factsheet Number 56.*

[29] Rachael, Z.M., Andrew, J.R.W., Brooke, O.W., Galloway, T.S. and Abigail, P.W.B. (2017). Mountains to the Sea: River Study of Plastic and Non-Plastic Microfibre Pollution in the Northeast USA, *Marine Pollution Bulletin, http://dx.doi.org/10.1016/j.marpolbul.2017.07.028.*

[30] Rachid, D., Johnny, G., Cècile, M., Corinne, M., Mohamed G.C., Valèrie, L. and Bruno, T. (2016). A First Overview of Textile Fibres, Including Micro Plastics, in Indoor and Outdoor Environments, *Environmental Pollution*, 1–6, *http://dx.doi.org/10.1016/j.envpol.2016.12.013.*

[31] Raven, H., Catherine, M.O. and William, L.P. (2020). Micro Plastic Prevalence in Two Fish Species in Two U.S. Reservoirs, *Limnology and Oceanography Letters,* 5, 147–153 DOI:10.1002/lol2.10140.

[32] Resnick, B. (2019). More Than Ever Our Clothes Are Made of Plastic, Just Washing Them Can Pollute the Oceans, *https://www.vox.com/the-goods/2018/9/19/17800654/clothes-plastic-pollution-polyester-washing-machine* (Accessed on 13th December 2020).

[33] Slater, K. (2008). Environmental Impact Polyester and Polyamide Textiles, *Polyesters and Polyamides*, 171– 199, *https://doi.org/10.1533/9781845694609.1.171.*

[34] The Dangers of Synthetic Fibres and Fabrics on the Environment (2020), *https://superego world.com/blogs/the-world/the-dangers-of-synthetic-fibres-and-fabrics-on-the-environment* (Accessed on 13th December 2020).

[35] Tiwari, M., Rathod, T.D., Ajmal, P.Y., Bhangare, R.C. and Sahu, S.K. (2019). *Distribution and Characterization of Micro Plastics in Beach Sand from Three Different Indian Coastal Environments.*

[36] Woodall, L.C., Sanchez-Vidal, A., Canals, M., Paterson, G.L.J., Coppock, R., Sleight, V., Calafat, A., Rogers, A.D., Narayanaswamy, B.E. and Thompson, R.C. (2014), The Deep Sea is a Major Sink for Micro Plastic Debris, *Royal Society Open Sciences,* 1(4), 140317. DOI: 10.1098/rsos.140317.

Water Conservative and Environmental Friendly Dyeing Systems

S.R. Shah[1] **and A.J. Vaidya**[2]
Department of Tex. Chemistry, Faculty of Tech. & Eng., M.S. University, Baroda, Gujarat, India
E-mail: [1]shailesh.r.shah-tchem@msubaroda.ac.in; [2]ankur.vaidya-tchem@msubaroda.ac.in

ABSTRACT: *Today all over the globe, people are talking about the crisis of water. Textile industry is the second largest industry after agriculture, consumes the purest water for its various processing. Many literatures pertaining to reduction in water quantity in textile processing and particularly in dyeing are available and all are mainly in scattered manner and most of them are patented. Waterless dyeing or zero water dyeing is the new concept for colouration of textiles. Many concepts such as, modification of machineries, dyes and chemical developments, process cycle variations, super critical* CO_2 *dyeing, air dyeing, application using atmospheric plasma, digital printing, recycle of effluent water, etc. have been developed in textile industries for rationalization of water quantity. The waterless dyeing has many benefits, namely, zero effluent, natural water resources savings, reduction in processing cycle and so on. In the present study, various technological developments of water conservative dyeing systems have been elaborated and compared. The future scope of developments in the technology of water conservation has also been reported in the study.*

Keywords: Air Dyeing, Atmospheric Plasma treatment, Digital Printing, Eco-friendly, Supercritical CO_2 Dyeing.

1. INTRODUCTION

All the chemical industries, consuming pure form of water, are facing problems and significantly concerned for the rationalization of use or alternative resources in place of the water. Next to agriculture, the textile and allied industries are second largest consuming the volume of fresh water. Water is the prime natural resource found on earth, much of which (> 97%) is saline and rest of them (approximately one third) available in frozen state in the form of ice (Gleick, 1993). Intake of fresh water, in the form of liquid or ice for the protection of humans and all other living species is a prime prerequisite for scientific practices and regulation of bureaucrats. However, a lot of water in its purest form needed for the textile industries which creates scarcity for the humans and other living species (Table 1 and Figure 1). Several techniques have been applied for streamlining the use of water in chemical wet treatment of textile treatment. The numerous approaches developed and marketed by technocrats and researchers for rationalization of water such as, modification of equipments/machineries, new invention and developments in dyes and chemicals, changes or alteration of the process cycle, supercritical CO_2

dyeing and air dyeing. In addition, the recycling of effluent from textiles wet processing after proper treatment and reusing in general and particularly for washing process has also been attempted by many scientists (Elmaaty & Eman, 2017). Various literature pertaining to water rationalization in textile wet processing are available, however, all are in scattered manner and/or patented. In the study, different techniques used for water consumption have been systematically elaborated with their scientific theories.

Table 1: Water Quantity Used in Textile Wet Processing

Processing Cycle	*Water Used (lit/kg)*
Preparatory Processes	
Desizing	2.5 to 20
Scouring	19 to 42
Continuous Bleaching	2.5 to 125
Mercerization	1
Dyeing Processes	
Beam	165
Back	235
Jet	200
Paddle	290
Jigger	100
Skein	250
Stock	165
Pad-batch	17
Package	180
Continuous Dyeing	165
Indigo Dyeing	8 to 50
Printing Processes	
Printing	25
After Treatments	110
Finishing Processes	
Chemical Finishing	5

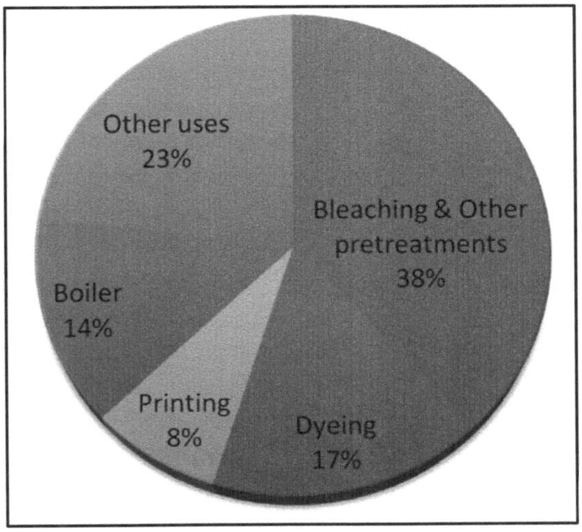

Fig. 1: Water Consumption Percentage in Textile Wet Processing Industries

2. DEVELOPMENTS (MODIFICATION) OF MACHINERIES

The most commonly used conventional dyeing devices in textile wet processing for various forms are jigger, winch, jet, padding mangle and package dyeing. The dyeing of various textile fibres are carried out at 1:15 material to liquor ratio with about 20 average steps (including pre-treatment) required about 300 liters of water for 1 kg of fibre. The machineries are modified or developed in such a way that the liquor ratio and subsequently the total quantity of water can be reduced for dyeing process. The closed winch and jigger prevents evaporation of water, trough less padding mangle reduces the liquor waste and double beam jet dyeing consumes the water quantity per kilogram of fibre. The various means used in dyeing with respect to developments (modifications) in machineries are summarized in Table 2.

Table 2: Machineries Development (modification) for Water Consumption in Dyeing Process

Dyeing Machine	*Used for*	*Modifications*	*Remarks*
Jigger	Batch wise dyeing of open width fabric	• Closed jigger • Double beam jigger	• Prevent evaporation • Reduce M:L ratio
Winch	Batch wise dyeing of rope form of fabric	• Closed Winch • Multiple Rope Winch	• Prevent evaporation • Reduce M:L ratio
Jet	Batch wise dyeing of synthetic fabric	• Multi nozzle soft flow jet dyeing machine • Double rope jet dyeing machine	• Reduce M:L ratio • Reduce M:L ratio
Padding Mangle	Continuous dyeing of open width fabric	• Trough-less Padding • Kiss-roll Padding	• Reduce M:L ratio • Reduce M:L ratio
Package dyeing machine	Batch wise dyeing of yarn and loose fibre	• SLR Package dyeing machine	• Reduce M:L ratio

3. DYES AND CHEMICALS DEVELOPMENTS

Dyes are used to improve textiles' aestheticity and marketability, and chemicals are used to maintain suitable media for dyeing. Many developments are carried out by dyestuff industries to increase exhaustion percentage of dye, so that less water quantity are required in subsequent steps. Chemicals and auxiliaries are also developed in such a way that the maximum amounts of dyes are transferred on fibre with low quantity of water and reduced water quantity in washing. Important development in dyestuff and chemical industries with respect to water consumption is summarized in Table 3.

Table 3: Dyes and Chemical Development for Consumption of Water

Name of Dyes and Chemicals	*Applicable on*	*Development*
Disperse dye	Polyester, Acetate Rayon, Acrylic fibre dyeing and printing	RD Dyes High Sublime Dyes
Reactive dye	Cellulosic fibre dyeing and printing	HE Dyes
Solvent dye	Polyester, Nylon, Acetate Rayon, Acrylic fibre dyeing and printing, Inkjet Printing Ink	Basic Solvent dyes Metal complex Solvent dyes Disperse solvent dyes
Surfactants	Synthetic fabrics dyeing and printing	Multi function surfactants

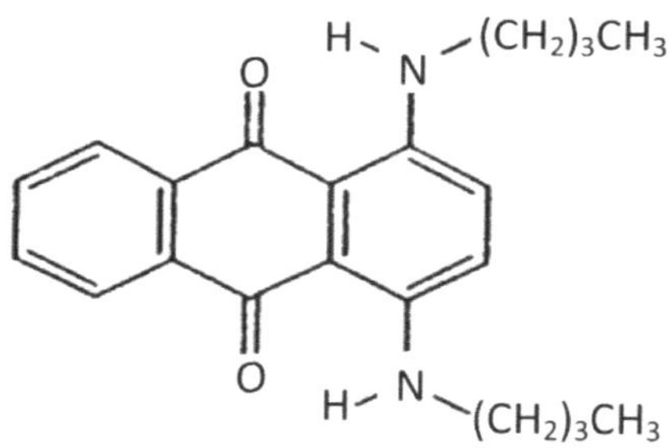

Fig. 2: Solvent Blue (C.I. Solvent Blue 35)

4. DYEING CYCLE ALTERATION

Normally, batch wise dyeing consumes high amount of water, compared to continuous dyeing (Figures 3 and 4). The dyeing of fabric or yarn carried out in continuous manner can reduce subsequent amount of water. Further, in reactive colour instead of doing exhaust method of dyeing, cold-pad batch using silicate exhaust more amount of dye. Such dyeing required less amount of water in subsequent washing and effluent processes.

Fig. 3: Indigo Sheet Dyeing Machine

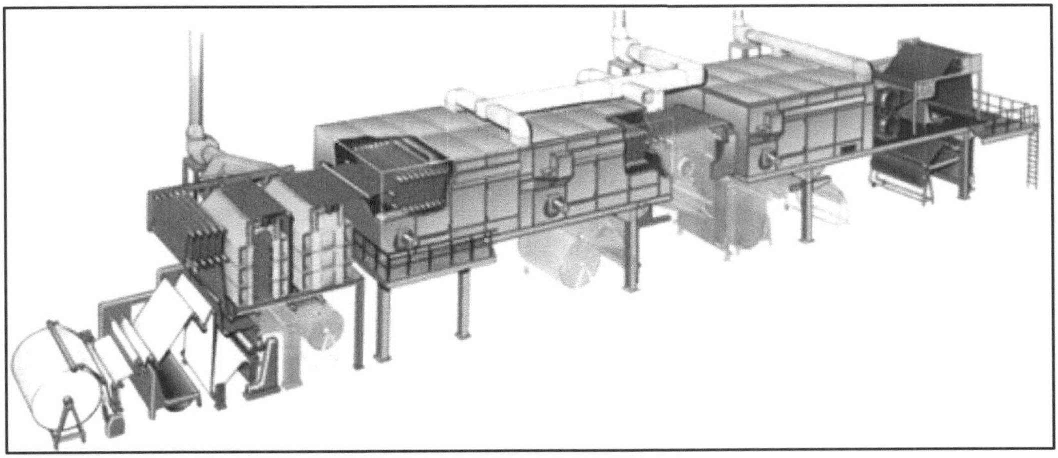

Fig. 4: Continuous Fabric Dyeing

5. SUPER CRITICAL CO_2 DYEING

In the late 1980s, super critical fluid (SCF) dyeing was first developed by German professor Schollmeyr and his research groups. Because of its non-flammability, non-toxicity and simple usability, CO_2 is used as a fluid, according to studies. In comparison, the critical temperature (Tc = 31.1°C) and pressure (Pc = 74 bar) are lesser than many other fluids, making CO_2 more appealing (Banchero, 2020). At above 31.1°C and 74 bar, the super critical CO_2 behaves as both liquid and steam. In the supercritical step, CO_2 can be used to dissolve hydrophobic dyes because of its solving abilities against non-polar or mildly polar organic molecules, and gas like low viscosities and diffusion properties can result in shorter dyeing time relative to water.

Fig. 5: DyeCoo Supercritical CO_2 Dyeing Machine

In 2008, Dyecoo developed a method of waterless dyeing where supercritical CO_2 is used as a solvent for different groups of dyes (DyeCoo, 2019). The benefits of this method are shorter time, 50 percent lower operating costs relative to the traditional form of water dyeing. In addition, it is possible to reuse CO_2 and dyes, making the process commercially viable and environmentally sustainable. However, a strong hindrance to polar hydrophilic fibre is the low polarity of supercritical CO_2; thus, these dyeing systems are only applicable to synthetic materials (Shen-Kung & Pi-Shiun, 2012). Modern research and development tends to work on addressing a variety of problems such as this. For illustration, there is typically a vertical configuration of the most supercritical CO_2 machinery, resulting in non-homogeneous fluid distribution, dye aggregation and thus irregular gravity-induced dyeing. Latest attempts have been made at a comparatively low temperature for horizontal equipment for zipper tap dyeing in supercritical CO_2 and suggest that commercialization of waterless dyeing will have significant practical and strategic significance.

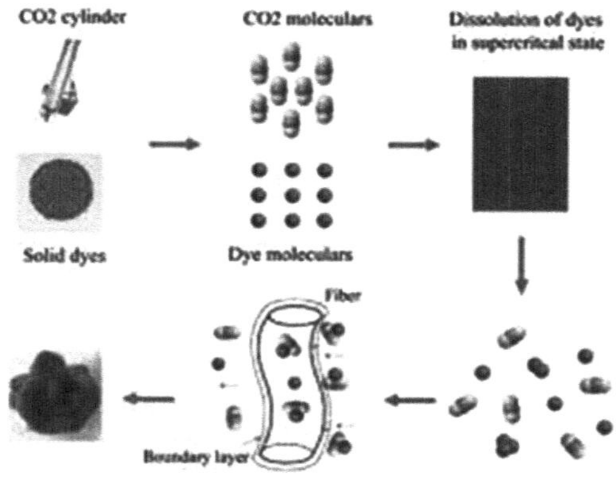

Fig. 6: Supercritical CO_2 Dyeing Mechanism

Recently, special dye has been developed and its use in supercritical CO_2 ecological silk dyeing produces excellent adsorption, absorption and fixation behavior with the highest colour fastness. Using supercritical CO_2, Elmaaty synthesized five unique disperse dyes and its application for coloration that demonstrates strong coloration behaviors, characteristics and colour fastness. Supercritical CO_2 dyeing attracts significant interest overall and several academic papers have been published in recent years. It is also clear that supercritical textile dyeing assisted by CO_2 has a promising future for freshwater restoration and waste reduction (Penthala *et al.*, 2020).

6. AIR DYEING

The airflow dyeing principle of Machine GmbH was invented by Wilhelm Christ and later patented in 1978. It was developed and copyrighted by Colorep, a renewable technology corporation headquartered in California. After undertaking and independent review, the firm reported that the fabric's air dyeing cuts water use by 95 percent and energy usage by 86 percent based on the method of dyeing and fabric used. This lowers the share of global warming in the sector by 84 percent.

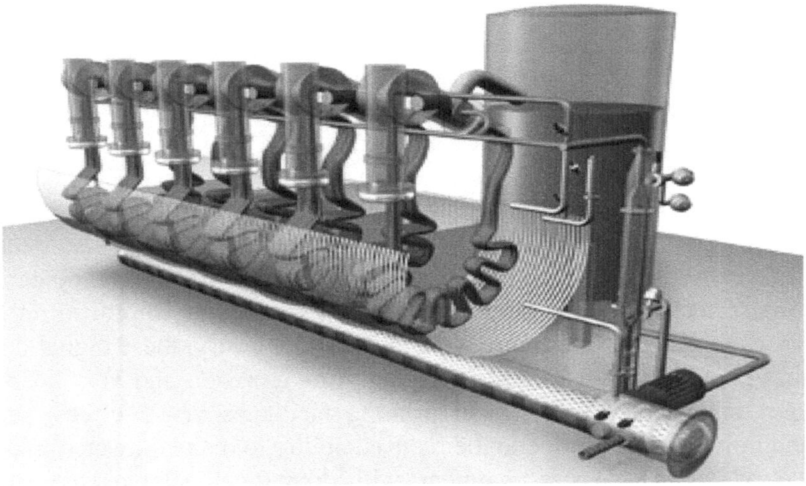

Fig. 7: Airflow Dyeing Machine

Dyeing liquor is first atomized in airflow dyeing equipment, then combined with high pressure airflow and eventually sprayed onto the cloth to be dyed. Since water acts as a dyeing liquor solvent and dye chemicals touch the cloth directly, relatively little water is ingested. Although the air dye procedure does not contaminate water and uses air to transport dye instead of water, no toxic waste is produced and no waste of water.

In the last few years, couples of companies have invented their Air Dye and Colorzen waterless air dyeing technologies. In the Air Dye technology, instead of dipping them in the conventional bath filled with water and dye, fabrics are put into a printing device. Specially formulated color from paper on to polyester cloth is used in this process. Air Dye's essential benefits include less releases of radioactive waste, reducing water use, and many factors are described in many scientific studies. However, the limited use of this technology is that it can be applied to synthetic materials such as polyester.

A group of apparel experts invented colorzen technology in 2012 to color cotton fabrics by changing the cotton fibre structure using the pre-treatment process. Cleantech solution International Inc., recently founded in China, develops and manufactures airflow dyeing machines that use air instead of water. The water is atomized by this dyeing system under high pressure. Therefore, up to two thirds of the water used by conventional models will be preserved (Vignesh & Sukanya, 2015).

The process of air dyeing is relatively recent and the expense of installing machines is costly. The potential interest of researchers in textile dyeing may therefore be the production of low-cost, quick air dye colouring approaches for natural as well as synthetic fabrics with less usage and virtually zero pollutants.

7. APPLICATION USING ATMOSPHERIC PLASMA

In the case of dyeing, plasma processing may be used to enhance the diffusion of dye molecules through the fabrics, improving the strength of colour and wash ability of several natural and synthetic textile products. Plasma application increases dye saturation, dyeing uniformity, lowers the volume of dyestuff and water added, and enables effluent reuse to lead to a substantial cost and environmental effect mitigation (Elabid *et al.*, 2016).

Atmospheric pressure plasmas have proven to be an efficient alternative to low-pressure plasma for dyeing in the past 10 years. By dyeing polyamides, polyester and wool fabrics using various dyestuffs, excellent results were obtained. The dyeing properties of plasma treated fibres are associated with the chemical structure of the surface and modifications of the surface. The increase in dye affinity, however, also depends on the exposure time, composition of the gas mixture and applied energy (Helmy *et al.*, 2016).

Over the past decade, inkjet printing technology has shown improved properties over conventional methods of textile printing. In the textile industry, these digital developments are becoming popular with excellent efficiency, low emission and very adaptable to the rapid changes in fashion. While inkjet printing facilitates visual effects such as tonal gradients and infinite patterns, due to the reduced ability to maintain water, ink, finish, etc., the absence of a suitable pre-treatment will dramatically lower the print quality. Atmospheric plasma is a very powerful pre-treatment process to increase the adoption of inkjet printing. Overall, it is clear that atmospheric plasma developments have greatly improved the dyeing and printing of textiles successfully over the past decade.

8. CONCLUSION

Dyeing is an important step in terms of creating aesthetic and marketing values of textiles. However, the said step requires a lot of water (approx. 300 lit/kg) for process. The rationalization is the prime requirement due to the annihilation of water resources from earth. Various approaches, such as developments of machineries, dyes and chemical developments, dyeing cycle alteration, super critical carbon dioxide dyeing, air dyeing, etc. have been developed by different technocrats. All the methods are useful in consumption of water with few constraints. Zero water dyeing process is essential in future to survive, and more researches should be emphasized on the said techniques.

REFERENCES

[1] De Giorgi, M.R., Cadoni, E., Maricca, D. and Piras (2000). Dyeing Polyester Fibres with Disperse Dyes in Supercritical CO_2, *Dyes and Pigments*, 45, 75–79.

[2] DyeCoo (2019). Waterless dyeing *(http://www.dyecoo.com).*

[3] Elabid, A.E.A., Zhang, J., Shi, J., Guo, Y., Ding, K. *et al.* (2016). Improving the Low Temperature Dyeability of Polyethylene Terephthalate Fabric with Dispersive Dyes by Atmospheric Pressure Plasma Discharge. *Appl. Surf. Sci.* 375, 26.

[4] Elmaaty, T. and Eman, A. (2017). Supercritical Carbon Dioxide as a Green Media in Textile Dyeing: A Review. *Textile Research Journal,* 1184–1212.

[5] Gleick (1993). Water and Conflict: Fresh Water Resources and International Security, *International Security*, 79–112.

[6] Helmy, H.M., Hauser, P. and El-Shafei (2016). A. Influence of Atmospheric Plasma-Induced Graft Polymerization of DADMAC into Cotton on Dyeing with Acid Dyes, *Journal of Textile Institute* 108, 1871.

[7] Hendrix, W.A. (2001). Progress in Supercritical CO_2 Dyeing, *Journal of Industrial Textiles*, 31(1), 43–55.

[8] Mauro, B. (2020). Recent Advances in Supercritical Fluid Dyeing, *Coloration Technology*, 115–120.

[9] Montero, G.A., Smith, C.B., Hendrix, W.A. and Butcher, D.L. (2000). Supercritical Fluid Technology in Textile Processing: An Overview, *Indian Engineering Chemical Research*, ol. 39(12), 4806–4812.

[10] Nandhakumar, R., Kaviyarasu, R. and Kalidass, M. (2012). Dyeing of Fabrics without Water: A Review, *Indian Textile Journal*, 224–226.

[11] Odabasoglu, Avinc and Yavas (2013). Waterless Textile Dyeing, *Journal of Textiles and Engineer,* 62–79.

[12] Ozcan, A.S., Clifford, A.A., Bartle, K.D. and Lewis, D.M. (1998). Dyeing of Cotton Fibres with Disperse Dyes in Supercritical Carbon Dioxide, *Dyes and Pigments*, 36(2), 103–110.

[13] Perkins (1991). A Review of Textile Dyeing Processes, *American Association of Textile Chemists and Colorists*, 23–27.

[14] Raju, P., GisuHeo, H.K., In-Yeol, Lee *et al.* (2020). Synthesis of Azo and Anthraquinone Dyes and Dyeing of Nylon-6, 6 In Supercritical Carbon Dioxide, *Journal of CO_2 Utilization*: 49–58.

[15] Shen-Kung, L. and Pi-Shiun, C. (2012). Literatures on Dyeing Technique of Supercritical Fluid Carbon Dioxide, *American Journal of Analytical Chemistry*, Vol. 3, No. 12A Article ID: 26157, 8.

[16] Textile Dyes Industry Trends (2019). *The Advent of Environment Friendly Textile Dye.*

[17] Van-der, K., Fernandez, M., Cid, M.V., Woerlee, G.F., Veugelers, W.J.T. and Witkamp, G.J. (2007). Dyeing of Natural and Synthetic Textiles in Supercritical Carbon Dioxide with Disperse Reactive Dyes, *The Journal of Supercritical Fluids,* Vol. 40, Issue 3, 470–476.

[18] Vignesh, D. and Sukanya, L. (2015). Air-Dyeing Technology—A Review, *Textile Today,* 142–145.

Khadi: A Successful Journey of Sustainable Textile Materials from Local to Global

V.D. Gotmare* and S.S. Kole
Textile Manufactures Department, VJTI, Mumbai, Maharashtra, India
E-mail: *vdgotmare@gmail.com

ABSTRACT: *Khadi, an eco-friendly textile material and known signature fabric of India, has gained immense importance worldwide. A hand spun and hand-woven sustainable textile material sometimes working on solar energy also reached to fashion shows and became a magnet in young Indians. It is popularized as Khadi for nation and Khadi for fashion. KVIC, statutory body of government of India has continuously adopted various tools to promote and enhance its quality product and adopted e-marketing concept for its growth. Khadi has once again come out of its customary veil, marketing its presence in the exclusive HS code bracket to categorize its products in export. CKMC, an eco-mark developed by KVIC which controls genuineness of Khadi fabric production. Khadi has successfully traveled from local to global with its innovative face and is in great demand in International Market. This paper discusses the journey of Khadi fabric from local to global besides its present status, future growth, sustainability characteristics and its potential to boost economic growth in India.*

Keywords: Eco-Khadi, HS Code, KVIC, Solar Energy, Sustainability.

1. INTRODUCTION

Textile industry is the one which uses large amount of energy, water and natural resources. This industry is now focusing on the sustainable practices from fibre to finished product to reduce impact on environment. Today, there is significant increase in the level of the greenhouse emission and it contributes towards increase in global warming and carbon footprint, due to high level of textile emissions from the industry. Implementing sustainability in the production and processing of textiles not only helps to reduce the hazardous impacts to the environment including the workers working on the shop floor to earn fair wages and ensure proper working conditions. The finished garment manufactures now taking one step ahead to promote their brands using sustainability tool and which is a need of the hours. The concept of eco-fashion sometimes gets neglected in the midst of fashion trends by the designers even though the products are available at cheap and stylish clothing but the scenario is gradually changing because of eco branding in fashion industry. In the first decade of the 21st century many young industrialists, considered sustainable fashion as an important aspect of their business. To promote the sustainable clothing many brands are being showcased in international fashion shows. Today, the consumer is also eager to wear the cloths with eco-brands and this becomes a driving factor to increase the consumption of green clothing in retail garment. The sustainability concept is nowadays very important part for many national and international players to sustain their business to fulfil the new demands by the customers all over the world. The Khadi organization is a very old brand which uses the sustainable practice in the production of textiles right from the Fibre to the fabric and has created a benchmark for the production of green garments.

2. KHADI STATUS IN INDIA

The Khadi commonly known the "fabric of Indian independence," what's more, has both an image of India's possible monetary independence and a vehicle for imparting to the British the poise of destitution and the equity of Indian human advancement. The symbol of nation promoted by Mahatma Gandhi across India and the Globe with essential messages and people accepted the product in post independent era. India has created self-supporting sustainable business of clothing. By fashioning himself, Gandhi created a symmetry and transparency in promoting khadi as a symbol and an example for others to follow the norms. Gandhi became a very credible leader for successful implementation of the khadi production in the nation. Gandhi viewed visual experience as an unbiased and straight forward sort of correspondence that was available to everybody. Gonsalves assumes on comparative lines when he advances the contention that Gandhi's trials with his dress shows his slow comprehension of apparel as a ground-breaking visual talk in itself, which could be utilized to impact individuals and convince them into the opportunity battle.

3. PRODUCTION AND PROPERTIES OF KHADI FABRIC

The unique property of Khadi which keeps wearer warm in winter and cool in summer has made it a very versatile fabric which covers all class of applications in fashion industry. The fabric is coarse in nature with natural texture which facilities ease of crumpling hence to keep fabric intact small amount of starch is added in the production process. This fabric value gets enhanced after every washing cycle so every wash creates different effect on the fabric and losts for 4 to 5 years even after vigorous washing. Hand woven design enables Khadi to look very attractive and designers add new design in development of value-added products to promote the sale of Khadi. Natural hand-made weaving of handspun yarn into the fabric provides passage of air circulation in the fabric.

Fig. 1: Khadi Drive in India

Fig. 2: Process of Making Khadi Fabric

Fig. 3: The Procedure from Cotton to Khadi

3.1 Central Khadi Mark Committee at Khadi Centers

To improve the customer awareness about the originality of Khadi products the Government of India has established CKMC committee which will certify various organization of genuine Khadi production under the KVIC rules and regulations. The official

Khadi INDIA

Fig. 4: CKMC Mark on Khadi Fabric and Dress Materials

communication regarding the CKMC is published in the Gazette of India dated 22nd July 2013. To identify the originality of Khadi, the Khadi Mark brand was placed in the khadi products and regularized from 30th September 2013 boosting the sale of khadi across the globe. It is mandatory to use the logo of Khadi mark on every Khadi product before delivering the products in the market.

3.2 Khadi Denim Apparel by Levis India

Levi's is one of the pioneer and global leader of production of blue denim apparel. Continuous development in the denim production with different attractive and comfortable effects is now essential to meet the customer challenges. The Levis India launched the *Khadi by Levi's* creation to promote the Khadi products exclusively on the occasion of 15th August 2014. The new class of five pocket denim was designed in Arvind Mills with Khadi effect and the product is successfully accepted by the customers (Akankshaya, 2013).

3.3 Khadi Fashion Fabric by Raymond's Ltd.

The Raymond launched the Khadi fashion fabric across the globe in association with The Khadi & Industries Commission (KVIC). The new business has created many opportunities with respect to employment of around 2.10 lakhs man-hours for spinners and weavers. The developed Khadi brand is now available at KVIC outlets and Raymond shops across the country (Sharma, 2013).

3.4 Khadi 'Ready-to-Use Mission'

Major Khadi institution is now working as lead institution for production of garments including covering, designing, finishing and packaging of Khadi products under the regulation of, "Ready-to-Use" Mission during 2005–2006. As per the rules, other institutions working in nearby areas are required to associate and to conduct the various activities of the lead institutions. The joint activity of institutes helps to exchange the new innovative ideas which will help to upgrade product quality (Gupta, 2012). The National Institute of Fashion Technology (NIFT), Kolkata and Chennai have been engaged as Service Providers for conceptualization and development of project, identification of location, installation of machinery, training, commissioning of project and post installation support.

4. GLOBAL KHADI PROMOTION BY GOVERNMENT OF INDIA

To promote "KHADI: Local to Global" mission under the Union Commerce and Industry Ministry in association with FICCI, IIFT and Khadi India the Indian government is planning to promote economic growth of Khadi by implementing the new designs and styles in Khadi. The mission has started to implement the strategy of making Khadi promotion at least six districts in five states as a pilot project to promote Khadi as global fabric. The mission helps to deploy more than 80 lakh people and needs to create demand in the Indian fashion industry and textile sector. The Indian railways now demanding more than four lakh bedrolls to promote the Khadi as symbol of national pride and become a regular partner with KVIC in developing a strategy to provide disposable Khadi bedrolls and towels at economical rates on daily basis.

Fig. 5: Khadi in Fashion Show

5. KHADI'S SHARE IN TEXTILE PRODUCTION, INTERNATIONAL STATUS AND ITS COUTURE STATUS

The Khadi and Village Industries Commission expressed that "The portion of Khadi texture in the absolute material factory creation has multiplied to over 8.49% in 5 years to 2019–20." The ascent in Khadi texture creation from 4.23% to 8.49% could be conceivable because of Prime Minister's clarion call to embrace Khadi. Indian Khadi textures are sold in both home grown and fare market. None the less, the portion of "khadi trade" in all out-material area is extremely low, under 0.22% of absolute material fares. This might be because of its moderate assembling measure through hands bringing about less creation, more requested all around the world and likewise the greater cost. Subsequently center around fares of this area gets significant. Khadi promotes and helps to increase the awareness amongst the people across the globe about the sustainability in the area in latest fashion trends. With the happening to 21st century Khadi got a more elevated level of consideration from the originators; they began investigating with this flexible texture and ensured that the local texture stays stylish.

6. KHADI FOR ECONOMIC DEVELOPMENTS OF INDIA

Textile industry is known for providing both direct and indirect employment to more than 35 and 50 million Indians and largest employment sector after agriculture. This is added advantage to promote the Khadi for export promotion. Today, the changing supply change management of Khadi products for tracing the new trends in the market for promotional strategies to enter newer markets with guidelines of on trade policy, market access at WTO front, compliances, design interventions and possibilities of CSR interventions from corporates and PSUs. The fare possibilities of Khadi are having confidence with world moving towards characteristic and eco- friendly items. An ever-increasing number of individuals are getting aware of obligation towards the climate and environment and more prominent interest is being created for bio-degradable and eco-accommodating items.

7. PRESENT STATUS OF KHADI MARKET IN INDIA

The requirement and sale of Khadi products is increased by 28% to about Rs. 3,215 crore in 2018–19, this is considered as most rapid rise in recent years. The demand of ready-made garments and solar vastra is gaining very good response in the market. In India, the textile industry is one of the major polluting and energy consuming industry, normally textile process consumes about 34% in the spinning, 23% in weaving, 38% in the chemical processing and about 5% for miscellaneous purposes. It is found that the electrical energy in the textile sector is consumed in different processer from Fibre to finished garment and other utility operations like machinery, cooling and temperature control systems, lighting, equipment etc. (Jain, 2014).

Fig. 6: Khadi in Menswear

7. INNOVATIVE ACTIVITIES IN KHADI TEXTILES

1. Khadi and Village Industries Commission (KVIC) launch India's first-ever Khadi fabric footwear 3,000 spinners now weaving stories of Khadi in UK.
2. Khadi may soon open its maiden foreign venture beginning with Bhutan.
3. Khadi gets a makeover, Eyes ₹ 5,000 crores in business. Khadi saw rapid double digit growth over the last five years. From driving employment in the country, to getting young designers to come out with chic products, Khadi is seeing a transformation in its fortunes.

4. Key to promote Khadi as a global fabric which in turn will promote economic growth.
5. Khadi taps Amazon for garment e-sales.
6. The government plans to create a "super premium" segment for Khadi in a bid to increase sales of products made by using the hand-woven fabric by tapping the luxury customer base.

8. ROLE OF KVIC IN CURRENT PANDEMIC SITUATION AND ITS GLOBAL MARKETING

8.1 Khadi Mask for Corona Virus in India

Currently various developments are taking place in the designing of face masks to protect from the COVID-19. The KVIC officials have invented low-cost triple layer masks for protecting the people from harmful corona virus.

8.2 Features of Khadi Masks

1. The mask is produced from handcrafted Khadi fabric with three pleats and is piped on two sides.
2. The khadi silk mask is triple layered in which two inner layers of cotton and uppermost layer is of silk.
3. Dimensions of the mask: 8.25 inches by 6.5 inches.
4. The mask has been equipped with adjustable ear loops for the convenience of the users.
5. It has been designed using a variety of silk designs and hence it is available in multiple colors.
6. The mask is biodegradable, skin friendly, washable, reusable and breathable.

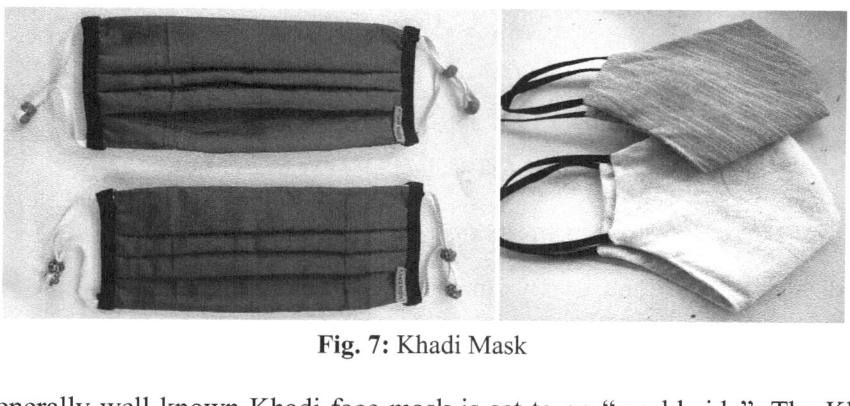

Fig. 7: Khadi Mask

The generally well-known Khadi face mask is set to go "worldwide". The Khadi and Village Industries Commission (KVIC) will currently investigate the chance of trading the Khadi cotton and silk face masks to far-off nations after the Ministry of Commerce and Industries lifted the restriction on fare of non-clinical/non-careful masks, all things considered. Under the Prime Minister "Aatmanirbhar Bharat Abhiyan". Increasing in huge demand of face masks during the global Covid-19 pandemic, the KVIC has developed double-layered and triple-layered Cotton as well as Silk face masks respectively. The KVIC plans to export the newly developed Khadi face masks across various countries like Dubai, USA, Mauritius and several European and Middle East countries where Khadi's popularity has significantly grown over the last few years. KVIC plans to sell Khadi masks in these countries through the Indian embassies.

9. SUSTAINABLE APPROACHES IN KHADI MAKING

9.1 Low Carbon Footprint of Khadi Textiles

The greatest advantage of Khadi textile is that it carries the lowest carbon footprint. Khadi consumes only 3 liters of waste in the process against 55 liters in conventional textile process. Use of natural colors in the khadi process helps to reduce the carbon emissions up to 28% when compared to conventional textile process.

9.2 Energy Conservation in Khadi Making

Out of one kilo of cotton, thinking about 10% wastage, one hand of 1000 yarns can be produces on the Ambar Charkha with two shafts with human energy, which would take 50 min. Then again, to weave one meter of Khadi, six to seven hands are required that would require 1.33 human hours. Altogether, it requires 2.25 human hours to deliver one meter of Khadi. If we apply human work output in agriculture, that is equal to 0.1 HP or 0.074 kWh-to Khadi production, we would get 0.225 HP or 0.17 kWh energy equivalents for producing one meter of Khadi. Khadi is approximately 3.24 times energy efficient than mill cloth (Parel, 1997).

9.3 Renewable Energy for Khadi Production

Solar energy is considered as an alternative which can replace current fossil fuels like coal and gas in the heating of water and many other chemicals in the textiles. Solar energy utilization in running the khadi charkha not only helps to reduce green-house emissions but also increase in the process efficiency for getting desired quality output as compared to manual charkha. The World Wide Fund for Nature, also known as the World Wildlife Fund (WWF), notes that electricity generation from fossil fuels causes pollution of air leading to acid rain, damaged forest areas, and affected agricultural production leading to loss of billions of dollars worldwide. Nuclear power pollutes water and land and has caused environmental catastrophes. Use of solar energy will eliminate these unsafe, unclean consequences from using conventional fossil fuels (Kripalani, 1982; Venkatesh, 2016).

10. CONCLUSION

India is widely regarded as a country associated with the world's richest and most diverse textile legacies. The today's positioning of Khadi apparel as a truly indigenous, nature-culture product in a niche segment is an innovative way of thinking about fashion, environmental impact and sustainable practices. The government initiative for promoting the growth of Khadi such as registration of new khadi institutions, signing MoU with private players and Khadi "ready to use mission" is now playing a crucial role for expanding the growth of Khadi sector but, it is still expected to have a continuous research in association with institutions for development of innovative products which enables the growth of Khadi sector. The government should support new emerging sustainable manufacturing and material science technologies in the Khadi sector for redefining the Khadi both as identity, encompassing ecology and lifecycle of clothing as part of everyday material and to make eco-Khadi production and consumption more sustainable.

REFERENCES

[1] Akankshya (2013). Khadi the Miraculous Indian Fabric.

[2] Anthony Parel (ed.) Gandhi (1997). Hind Swaraj and other Writings Cambridge University Press.

[3] Gupta, C. Fashioning Swadeshi, Clothing Women in Colonial North India, Economic and political weekly, 47, (2012), 76–84.

[4] Jain, E. (2014). Khadi and Contemporary Fashion: The National Movement and Post modern Context (M. Phil thesis).

[5] Krishna, K. (1982). Gandhi: a life (National Book Trust, 1982), 121.

[6] Sharma, S. Energy Management in Textile Industry, Indian Journal of Power systems and Operation Energy, 2, 45–49.

[7] Venkatesh, S. (2016). KVIC inks partnership with Raymonds ltd to promote khadi. http://www.forbesindia.com/article/special/kvic-inks-partnership-with-raymonds-ltd-to-promote-khadi/45049/1.

Section-VI

Management in Textiles

"Vocal for Local"—A Swadeshi Movement for Next Generation Manufacturing

N.D. Mhatre
ITAMMA, Mumbai, Maharashtra, India
E-mail: ndmhatre@gmail.com

ABSTRACT: *The quote of Mr. Thomas L. Friedman, Editor of New York Times some years back that the 'World is flat' has proved true today. Today COVID-19 has demanded a change in the existing geopolitical system, calling for restructuring of all the arrangements of trade and barter agreements, MoUs, rules and regulations, policies, planning, legal laws, etc. In brief the scenario of Indian Textile Engineering Industry states that India's textile machinery trade with the world was US$ 4857.91 million (12.31%) in 2018, whereby major contribution came from imports of about 83% i.e. US$ 4032 million, mainly from China, Japan, Germany, Singapore and Italy; on the other hand Exports were only 17.29% i.e. US$ 839.78 million mainly to the countries Bangladesh, Germany, Turkey, Vietnam and the Netherlands. India is producing "Textile Machines, Spares and Accessories worth ₹ 6,865 Crores; however it is fulfilling only 46.61% of Home-consumption to a tune of ₹ 3,200 Crores against the demand of ₹ 13,004 Crores which is 52.71%. Further the involvement of Entrepreneurship states that about 40–45% are manufacturers and the rest are Traders/gents/Dealers. Based on the above scenario an attempt is made in following topics to explain the possible opportunities to Indian TEI as created through new normal status developed by COVID-19. Preparing Indian TEI to be ATMA NIRBHAR for Domestic requirement (Vocal for Local through Make-in-India) and Export requirement (through Made for World).*

Keywords: COVID-19, Export Markets, Geopolitical System, Textile Machines, TUF Scheme.

1. INTRODUCTION

1.1 Preparing Indian Textile Engineering Industry (TEI) to be ATMA NIRBHAR for domestic requirement ('Vocal for Local' through "Make-in-India") and export requirement (through "Made for World")

European technologies are preferred globally due to their sustainability and Artificial Intelligence (AI) systems, supported by user-friendly and environment friendly in-built features. The new machines being expensive, many Indian Entrepreneurs had preferred pre-owned machines under the TUF scheme of the Government of India. China offered workshop facilities for many European machinery manufacturers to assemble their machines in China due to availability of cheap labour. Taking this opportunity, China developed very cost effective machines with European technology standards under reverse engineering techniques and targeted India as their major export market. However, due to lack of good quality metallurgical aspects in order to deliver a low-cost Product, the Chinese machines could not give sustainable performance.

2. ROLE OF INDUSTRY ASSOCIATIONS GIVING IMPORTANCE TO "ADDING TO SKILLS"

Adding to skills" in past Industry Associations played a pivotal role as a facilitator in connecting the Industry, Government and Academicians. However today, the Digitalization has brought out a excellent connectivity among the above identities and so the role of Industry Associations has now changed as a proactive body interacting with its members in helping them to grow in business and technology.

Today 80% of Textile Engineering units are MSMEs, whereby 75% are still having a conventional set-up. Industry Associations need to play a proactive role of helping these units to develop in technological and infrastructural set-up in order to compete in the Global Market. Programmers like Technology scouting missions by collaborating with research institutes/bodies, trainings on 5s, Kaizen, Lean, Six Sigma-Yellow & Green belt, waste management, personal productivity tool, energy conservation, etc. needs to be organized to make them competent in delivering their products at world class standards. Also skilling the members for internal audits on ISO, ZED, TUV, etc. and implementation of Lean Competitive schemes will help them in upgrading their factory set-up at state-of-the-art level. Organize Business Scouting Missions, technology/skill display meetings/trials, at National and International levels during Delegations and Exhibitions, and give them opportunity to learn and observe the HR and Skills of Sales and Marketing Teams of other Entrepreneurs and also encourage them to display their own skills on the same platform, thus reducing the skill gaps/short falls of newly developed technology/skills.

While taking an advantage of TUF subsidy scheme of the Government of India, many Entrepreneurs have purchased pre-owned looms with updated technologies. Unfortunately, these MSME Units were not able to get Preventive maintenance catalogues including Lubrication schedules from the owners of the old machines. Lubrication activity becomes of vital importance in maintaining the condition and performance of these pre-owned imported machines and calls for the manufacturing of Lubricants and its application gadgets (centralized lubricating systems, oil reservoirs/baths) by the Indigenous lubricant manufacturers, providing an appropriate substitute for imported products ('ATMA NIRBHAR BHARAT' through 'Make in India' initiative).

3. ROLE OF GOVERNMENT IN IMPLEMENTATION OF 'ATMA NIRBHAR BHARAT'

3.1 Smart Data Clinic

Today due to Digitalization, the DATA in regard with every industry's stability/growth can be available to the Students and accordingly they prepare themselves to take appropriate decision of their future carrier. Also, it is observed that the students prefer white caller job with User-friendly, healthy and environment-friendly work atmosphere having a set-up of state-of-the-art technology. The concept of Smart data clinic can improve and encourage the employment status of this field. Whereby the DATA needs to be presented and linked with the opportunities in new developing areas of technology and techno-commercial, scope and availability of turnkey projects, scope of employment through Government schemes/policies, etc. at national and international levels. Thus serving the purpose of a Clinic, giving a platform of solution to each one right from student, employee, employer, Industry expert, etc. and ultimately attracting the appropriate man-power in this field.

3.2 TUF Scheme for TEI

Under TUF scheme for Textile Manufacturer Industry, huge number of pre-owned imported looms equipped with computerized gadgets has been installed in India during last decade. However it is noticed that many entrepreneurs of decentralized sector could not afford to avail the after sales services as well as employ a special technical man power to attend this Hi-tech imported pre-owned Shuttleless looms. Such activities are always being offered by the new machine manufacturers and so it was the need of the hour to involve these activities as the part and parcel of this subsidy scheme. It is suggested to have the same type of TUF scheme for Indian TEI, which will create a scope to Students, technical manpower (Industry Expert/Consultant) and machine manufacturers to work jointly to deliver these services and help in reducing the skilling gaps further sustaining the performance of these looms.

1. Creating an Eco-system for innovations and technology development in Textile industry by Connecting the TEI and the Textile Manufacturing Industry under a common platform and further helping in appropriate distribution of research and development funds as presently both these Industries are governed by different Ministries.
2. Creating Textile Accessories and Spares Depot in the Textile Parks leading to availability of the same at doorstep and further helping the Entrepreneurs in ease of doing business.
3. Organizing Technology Scouting missions with TRAs like NRDC, CSIR, ISRO, etc. and helping the Entrepreneurs for Product/Design development through in-house Projects.
4. Formation of Textile Centre's at decentralized textile clusters and providing them with after Sales and Technical Services, thus creating in-house textile clinics.
5. Formation of Common Facilitation Centre's (CFCs) to facilitate, conceptualize and implement new turnkey projects, training, workshops, seminars, etc.

In order to take the slogan 'Vocal for Local' to the next stage "Focus on Local for the World" following suggestions may be considered:

1. Progress to a stage of 'plug and play' concept for new projects especially Textile Parks.

2. Develop Export Cells of Textile Engineers to promote Indian Textile Machines, Spares and Accessories in the Textile Producing Countries.
3. Develop Special Pavilions of Indian Textile Machines, Spares and Accessories in National and International Exhibitions.
4. Develop Business Scouting Missions during International Events through Virtual Events, Display Centre's in other countries and Barter Schemes.

4. ROLE OF EDUCATIONAL INSTITUTE FOR THE DEVELOPMENT OF SMART STUDENTS

Today's era of Digitalization and Smart manufacturing, have resulted in the fast Technological development of Textile machines, whereby frequent changes are seen in metallurgy, design, process, IT, Computerization aspects of the machines. However in the past these developments were projected during ITMA International Exhibitions every 4 years. Also such frequently changing technological developments can not be involved/adopted in the regular Degree/Diploma syllabus, but can be further updated to the students through various initiatives as given below at appropriate intervals, helping them to become SMART Students:

1. Visits to International exhibitions,
2. Visit of Industry Experts to Institutes as Guest lecturers,
3. Visits to state-of-the-art Factories with latest technology machines,
4. Portable Technical Clinics involving machine manufacturers and Industry Experts,
5. Technology Display Centre's in the Institutes through tie-ups with machine manufacturers,
6. What's app technology group of students roping in machine manufacturers, Industry Experts and research and development Institutes/Associations,
7. Student's Corner inviting presentations of students on new technology/process on weekly/fortnightly basis,
8. On-line webinars on new technology/process through tie-ups with machine manufacturers,
9. Installation of cut-models of important assemblies of newly developed machine for training purpose,
10. Installation of latest technology machines through tie-ups with machine manufacturers under research and development/Training/Sample development purpose. These activities will help the students in getting exposure to new skills by overcoming the lacunae of conventional technologies as per the regular syllabus, reducing the technology/skill gaps especially in the Indian Textile Engineering sector.

5. HR AND SKILLS PROMOTING INDIAN TEI

Following activities needs to be focused for promoting Innovative HR and Skills in Indian TEI:

1. Fashion show specifically highlighting the features of machines, whereby today most of the fashion shows highlight the patterns of textiles worn by the models on the ramp.

2. Special tie-ups with the Manufacturers of Fibre, yarn, fabric and garment, needs to be done by the Machine manufacturers for branding the machine features like safety, eco-friendly, environment-friendly, user-friendly in their marketing policy.
3. Machine manufacturers to be proactively involved in the pilot plant projects and thesis of the students at the Institute level.
4. Awareness of the important features of the machines needed to be displayed at the factories/mills through attractive display boards/posters resulting in the mouth to mouth publicity among the man-power.

6. CONCLUSION

As stated in the earlier topic that since both the Textile Industries are governed by different Ministries the distribution of the R&D funds is improper and not resulting in the delivery of the state-of-the-art machines/spares to cater in-house production and import substitution and capture the global market. However the study by UNESCO of Gross expenditure in R&D in various sectors spent by the Government, Business Enterprises and Educational Institutes, state that Indian Government is 2nd highest in the contribution when compared with other 9 countries, whereby Japanese Government is spending the least on R&D among all and still in Technology aspects Japan is considered at top among all. Thus an appropriate Eco-system with a thorough evaluation and allocation of Government funds in Indian Textile Manufacturing and Engineering Industry with a result orientated strategy for R&D and Skilling projects of Applied Research, JVs, Technology transfers, Skill-cum development, etc. will lead in making the Indian Textile machines/Spares a brand for Global market. This initiative will ultimately encourage new skills/start-ups in Indian TEI, and will also make it ready for HR and Skills considering the requirement of 35 million skilled workers, by 2024–25 as per the sector growth rates. In this Era of Digitalization, new technologies like going green, smart manufacturing, industry 4.0, etc. will place a challenge for 'Vocal for Local' as it demands the suppliers and consumers of the Swadeshi Products to be acquainted with the smart technologies. However, realizing this problem, Government of India has thoughtfully introduced well in advance 'Make-in-India,' 'Skill-in-India' and 'Start-up India' simultaneously, whereby the slogan 'Vocal for Local' will go hand in hand with all three, helping India to be "ATMA NIRBHAR BHARAT".

REFERENCES

[1] ITAMMA's Platinum Jubilee Annual Report, 2017–2018.

[2] Ministry of Science & Technology India's R&D Expenditure & Scientific Publications, posted on 1st May, 2020.

[3] NCTC Study on Improving Competitiveness of the Textiles and Clothing Industry in India-By E. & Y., 18th November, 2020.

[4] Textile Excellence, 15th May, 2019.

[5] Textile Value Chain, Oct. 2020.

Impact of COVID-19 on Indian Textile Industry and Growth Opportunities Post COVID-19

Abhimanyu Kaushik* and Ishita Agarwal
Veermata Jijabai Technological Institute, H.R. Mahajani Road, Matunga, Mumbai, Maharashtra, India
E-mail: *askaushik_b18@tx.vjti.ac.in

ABSTRACT: *The onset of COVID-19 has had a severe impact on global as well as the Indian Textile Industry. It has disrupted the Textile supply chain which has resulted in a steep decline in consumer demand, imports and exports. The growth projection of the textile and apparel industry in India, which was once projected to reach USD 220 billion by 2025–26, is now shrinking by 28–30% in 2020. The industry is facing a workforce crunch, broken supply chain, lack of funds and loss of confidence. So as the pandemic comes to an end, we should have our focus on utilizing the opportunities and mobilizing the workforce. The prime opportunity is fulfilling the export orders that have been diverted from Chinese Textile Industry to India and to take full advantage of the ongoing trade tensions between the US and China. Indian Industry can be even more profitable by unifying the scattered supply chain. By 2025–26, significant technological changes are expected in the industry, which will create a need for upskilling of employees. The government implementations and policies will help the Indian Textile and apparel industry, such as the allowance of 100% FDI. This will attract INR 10000 Billion as foreign investments in the coming years under various schemes like Integrated Textile Parks. The Medical Textile sector has had a huge demand in the domestic and global market. Sensing this opportunity many domestic manufacturers have taken the lead in developments of PPE kits and antiviral fabrics. But still, we need to keep a check and train our engineers accordingly to sustain the future of our Indian Textile Industry.*

Keywords: Apparel Industry, COVID-19, Medical Textile, Textile Supply Chain, Upskilling.

1. INTRODUCTION

COVID-19 a pandemic that shook the foundations of even the most well settled industries all over the world. The Textile Industry has not been an exception, it has taken a backseat enduring severe damages incurred over its functioning, productivity, and employee relations. This has resulted in a relentless distortion of the supply chain and a steep decline in consumer demand, imports and exports.

The labour intensive industry had to suffer a lot due to unpredicted mass migration of skilled and unskilled labours. This situation brought the production to a standstill, which was followed by instant cancellations and lack of orders in the market. With heavy liabilities and no income, the small scale industries suffered the most and some of them had to be shut down permanently. The effect has been such that the growth projection of the textile and apparel industry in India which was once projected to reach USD 220 billion by 2025–26 is now shrinking by 28–30%.

1.1 Statistics

Impact of COVID-19 on the textile industry can be understood by having a look at these following statistics, which throw some light on the position that our Indian industry holds in the global arena. Also they depict the numerical impact on the import and export values of some essential Textile products. Our Textile Industry accounts for about 2.3% contribution to India's GDP. Apart from that Textile Industry in India has a share of 11.4% in total export earnings with 7% of industry output in value terms. It holds a global share of 6%. This industry has a compounded annual growth rate of 7% since 2004–2005. From the total share of exports from Indian Textile industry, 60% is exported to the United States and European Union. Being the second largest employer with over 2000 spinning mills, the Textile and the Apparel Industry provides direct employment to over 45 million people and 60 million in the allied industry.

Table 1: Monthly Export Data for Textile and Clothing (Value in USD million)

Export Category	*April 2019*	*April 2020*	*% Change*
Cotton yarn/Fabs/made ups/handloom products	844.22	148.11	–82.46
Man-made yarn	388.61	61.76	–84.11
Jute mfg.	22.15	2.08	–90.61
Carpet	107.26	8.94	–91.67
Handicrafts, handmade carpet	141.3	11.53	–91.84
Apparel	1,409.53	126.31	–91.04
Textile & Clothing	2,913.07	358.73	–87.69
% of textile & Clothing in total Exports	11.17%	3.46%	

Source: Directorate general of commercial intelligence and statistics, ministry of commerce.

This economic backbone and sustaining industry took a toll during the pandemic, causing it great adversity. From the archives of the Ministry of Statistics Planning and Implementation and The Directorate General of Commercial Intelligence and Statistics, Ministry of Commerce it is found that there has been a drastic decline in the export volumes of the Textile and Clothing sector. The Handicraft and Handmade carpet sector is worst hit with a fall of about 91.67% in April 2020 compared to April 2019. The sudden lack of demand struck the Indian export numbers of Cotton and Jute so hard, that the leading producer saw a sharp decline of about 80–90% in the year 2020.

Table 2: Textile & Clothing Index of Industrial Production Growth Rate (%, Y-o-Y)

Sector	*March 2019 (%)*	*March 2020 (%)*
Textiles	1.10	–13.10
Wearing Apparel	14.40	–20.50

Source: Ministry of statistics planning & implementation.

All these factors have collectively resulted in an uncomprehending and a negative growth rate with respect to the Textile and Clothing Index of Industrial Production.

The wearing apparel reflected a 20.50% degrowth in March 2020.

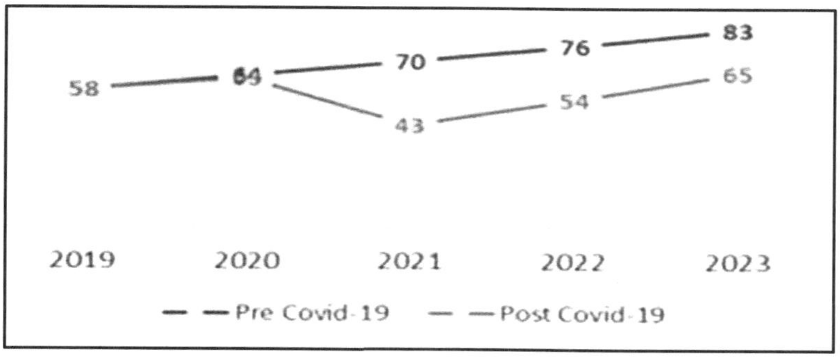

Fig. 1: Apparel Domestic Market (USD billion)

Source: Technopark analysis.

According to the data released by Technopak Analysis, the growth projection of the apparel domestic market, which was once projected to reach USD 83 billion by 2023, will now shrink to USD 65 billion.

2. OPPORTUNITIES

Every cloud has a silver lining, similarly this event definitely has something for the industry to buckle up and be on the right track towards growth.

There are opportunities that Indian Textile Industry can make full use of, some are at hand and some that the industry needs to create.

2.1 Opportunities at Hand

2.1.1 *China and Trade Tension*

There has been a diversion of orders from China as it is the epicenter of the pandemic. Various brands are increasingly looking for a China alternative, South India is facing some fresh first time orders. India should strive hard to become the preferred sourcing destination for International brands. An excellent example of this is German Leisure wear brand Marc O'Polo placing a huge order for jerseys to Indian vendor Warsaw International.

On September 14, the US imposed restrictions on the import of certain products originating from the Xinjiang Autonomous Region in China, citing concerns on illegal and inhumane forced labour in the region, rating agency Icra said in a report. President Donald Trump made it harder to produce fabric, yarn and apparel in the US by imposing tariffs on Chinese fabric, yarn and apparel. If apparel makers used US cotton and sent it to China, they are liable to pay a 25% punitive tariff and if they made clothes there and transported it back, then it will be 15%, this means the apparel brands get hit twice.

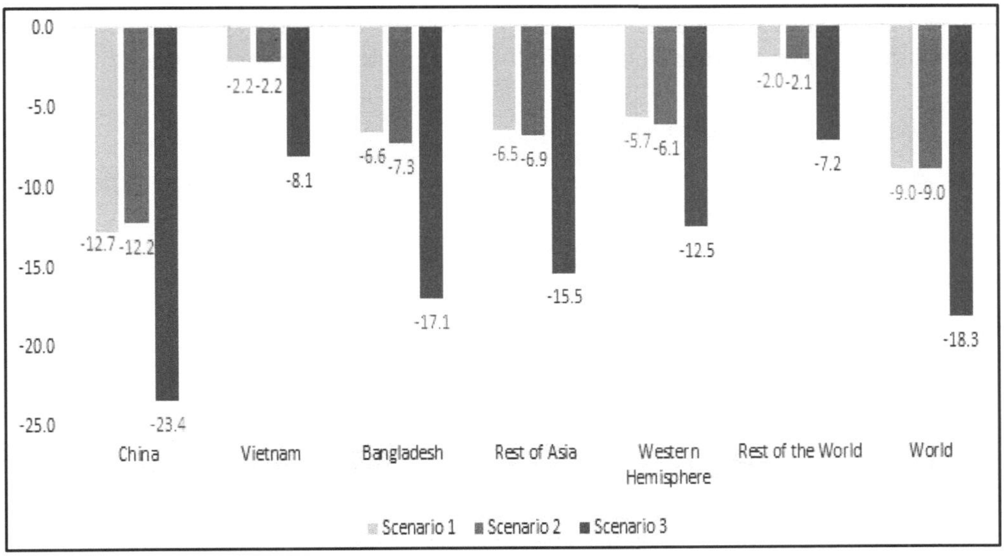

Fig. 2: Impact of Covid-19: Percentage Change of US Apparel Imports in 2020

Source: shenglufashion.com

Full advantage of the Trade tension between the US and China can be taken by attracting maximum orders towards India. Different trade policies drafted by the US have increased the cost of trade between the two nations making it the least favorable for the industry and brands. This is the best time for the Indian players to present themselves at their best ability and quality in front of global trade.

2.1.2 *Government Policies*

The government of India has initiated various policies to support the textile and apparel sector's growth for the long-term horizon. The government has announced INR 3,000 billion collateral free automatic loan for businesses, including the Medium Small and Micro Enterprises (MSME) which will help a lot of small Textile and apparel units to thrive in such difficult situations. With the extension granted by the Government and RBI, pertaining to 6 months moratorium for interests and for 4 years repayment period including one-year postponement for MSME segments for Covid-19 loans, would provide the much-needed upliftment to the textile industry.

The allowance of 100% Foreign Direct Investment in the Textiles Sector is expected to attract INR 10,485 billion (USD 140 billion) foreign investments in the coming years and will also carry out high investments under various schemes like Scheme for Integrated Textile Parks (SITP) and Technology Upgradation Fund Scheme (TUFS) to encourage the flow of more private equity and to train workforce.

To further accelerate the growth in the textile industry, the Textile Ministry assigned INR 6,900 million (USD 106.58 million) for the set-up of 21 readymade garment manufacturing units in seven states for the development and modernization of Indian Textile Sector. Opportunities to create to sustain and keep up with the rising competition in the industry at a global level the Indian Textile Industry has to create some opportunities.

2.1.3 *Fixing the Supply Chain*

Supply chain is the backbone of the Textile Industry, and a healthy and consolidated supply chain is what earns profits. While India has the advantage of every value chain in the industry but currently it is quite uncompetitive. India has many small scale industries and thus a scattered supply chain.

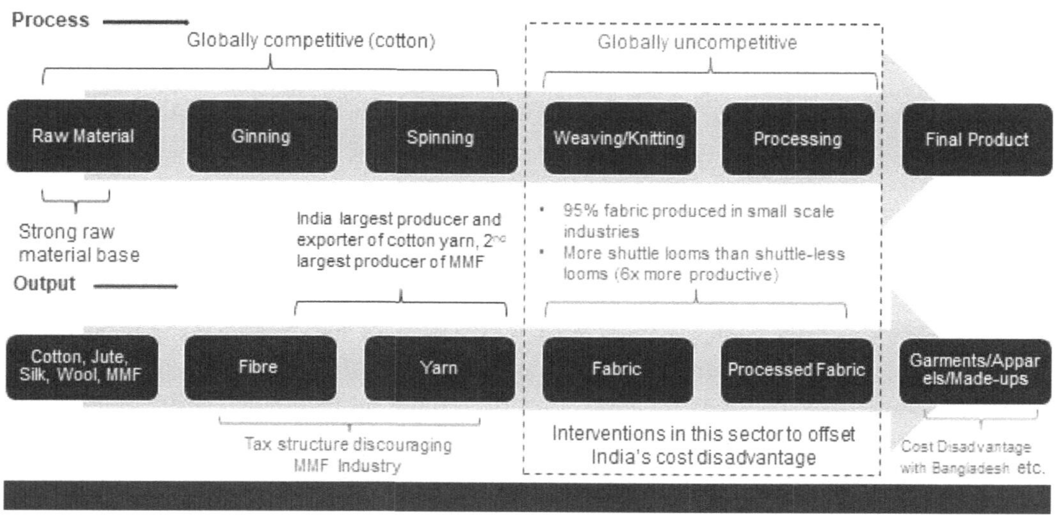

Fig. 3: Uncompetitive Global Value Chain

As quoted by Dr. G.V. Aras, "Cost is where India takes a beating. The scattered supply chain increases the logistics cost, but this can be overcome by setting up large capacities in the states where the manufacturing and logistics cost is low." India has a strong raw material base, both in case of natural and man-made fibres. Along with the production rates of fibre and yarn, high contamination level and poor quality of fibre, are major concerns that need focused attention. Focusing on the quality along with the quantity of our produce will help us in competing in the global market. When coming to weaving, knitting and processing, the value chain is globally uncompetitive. This is what the industry needs to fix up. More than 95% of the weaving sector is unorganized and we use more shuttle looms than shuttle less. Shuttle less looms are 6 times more productive than the conventional shuttle looms. India is among the leading exporters of raw materials to countries like Bangladesh and China. Further, these countries process and then export value-added products that earns them greater revenues.

Indian Textile Industry has to focus on these value-added products to come at par with the competition at the global level.

2.1.4 *Upskilling*

The COVID-19 pandemic induced lockdowns caused all the major economies to fall into recession and lead to widespread unemployment. It has changed the way the industry works and its future demands. With the advancement of technology, a need for upskilling will be created which will be addressed by the companies and employees to ensure fruitful employment and a stable labour market. According to the World Economic Forum Survey, with the advancement of technology, 43% of businesses will reduce their workforce and 34% will require a workforce with updated skills. The adoption of cloud computing, big data and e-commerce remain high priorities for business leaders. Companies are expected to witness change in their locations, value chains and size of the workforce in the next five years catering to the advancements.

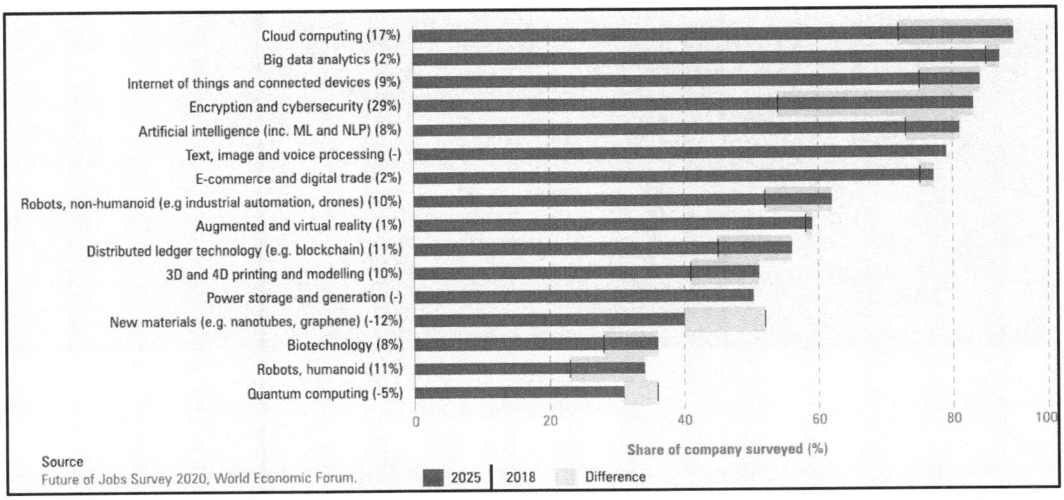

Fig. 4: Technologies Likely to be Adopted by 2025 (by share of companies surveyed)

Based on their study, WEF has estimated that by 2025, 85 million jobs will be displaced by a shift in the division of labour in humans and machines while 97 million new roles may be created according to the new division of labour between machines, humans and algorithms. As demand will change in the next five years, the skill gap will continue to be high by 2025. Skill groups such as critical thinking and analysis, problem solving and self-management skills will be expected by the employers. On average, company's estimate 40% of workers will require reskilling of six months or less. Companies believe that reskilling their existing employees will be more cost-effective and will benefit the society at large. Here the government will play a bigger role in upskilling. Currently only 21% businesses are able to utilize their public funds for upskilling. The public sector needs to provide strong support by creating incentives and investments for the jobs of tomorrow.

2.1.5 *Medical Textiles*

Considering the growing need of PPE kits, India has become the 2nd largest manufacturer of PPE in the world with around 4 lakh PPEs being produced every day. Poor quality of China kits has been in the news and also there's a trust deficit. Indian businessmen can take the right advantage of this situation. An astonishing 600 companies in India are certified to produce PPEs today. Well established large cap companies like Vardhman, Arvind Ltd., Welspun, & D'Decor have launched their range of anti viral fabrics. This has opened the road to heavy investments in the Medical Textile sector.

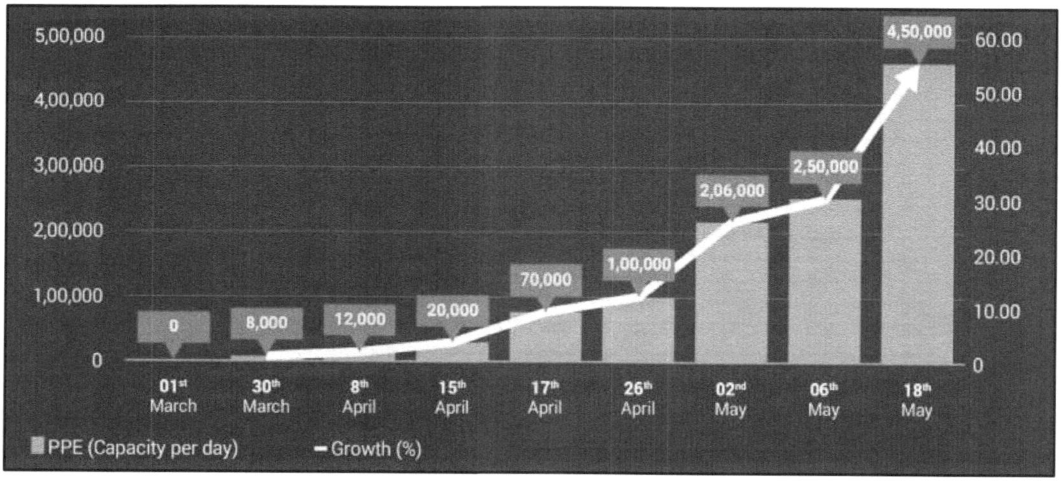

Fig. 5: PPE Production in India Per Day

Source: Investindia.com

According to a study, there will be domestic demand of PPE worth INR 10,000 crore for the next one year and internationally there will be a $60 billion business in 2025, whereas India has done only $260 million so far last year. The guidelines listed by the Ministry of Health and Family welfare emphasize the need to select the right type of equipment. Along with PPEs, disinfectant wipes are set to witness a two-fold increase in demand. Overall, the disinfectant wipes market is likely to create an absolute value opportunity of more than USD 685 million during the forecast period. At this stage, the Indian Medical Textile Industry should chip in and book profits by establishing themselves in the global arena.

3. CONCLUSION

It is still estimated that India has the potential to reach USD 70 Billion (INR 5,242 billion) in exports and achieve a much higher share of the global market by 2024. India is blessed with an exceptional and hardworking workforce. If all the above mentioned points are taken care of, Indian Textile Industry will soon witness its golden days.

REFERENCES

[1] A. (2020). International Buyers Shifting Apparel Sourcing to India, *Textile Excellence,* 18(5), 9.

[2] COVID-19 Lockdown: Impact on Textile Industry (2020, April). Retrieved from, *https://www.fibre2fashion.com/industry-article/8635/covid-19-lockdown-impact-on-textile-industry.*

[3] Dis-infecting Wipes to Witness 2X Increase in 2020 (9th April, 2020). Retrieved from-*https://www.factmr.com/media-release/1448/global-disinfecting-wipes-market.*

[4] Mittal, R.A. (2020). Companies, Employees Need to Up-skill for Jobs of the Future. *Textile Excellence,* 18(9), 1–2.

[5] Overview of Textiles Industry in India and Impact of Covid-19. (2020, September 2). Retrieved from *https://www.televisory.com/blogs/-/blogs/overview-of-textiles-industry-in-india-and-impact-of-covid-19.*

[6] PPE in India: An INR 7000 Cr. Industry in Making (2020, May 25th). Retrieved from-*https://www.investindia.gov.in/siru/personal-protective-equipment-india-INR-7000-cr-industry-in-the-making.*

[7] Weaving the Way for Indian Textile Industry (2020, January 29). Retrieved from https://niti.gov.in/weaving-way-indian-*indiantextileindustry#:~:txt=India%20is%20also%20the%20second,registering%20a%20growth%20of%203%25.*

Epilogue

This part of the book is to draw the attention of the readers to take a ways from the chapters of the book. The topic of the papers has been selected according to the need of the present days to help the reader.

1. In this study a uniform, porous and bulky mechanically entangled fibrous sheet from eco-friendly and natural, low-grade/process waste jute fibre has been designed for strawberry cultivation. The agricultural performance of this engineered eco-friendly textile material has been compared with the commonly used plastic sheet and traditionally used sawdust, straw and banana leaf in the same strawberry field under same conditions using random complete block design with three replicas. It was observed that process waste jute nonwoven shows better performance than other materials under experimentation.
2. This paper reported about the study of the synergetic effects on decorticated hemp fibre and describes the method of ecofriendly treatment with enzymes. The report says that the individual enzymes, act only on the specific component of non-cellulosic impurities keeping others intact. However, mix enzymes removed all the impurities in considerable amount. The removal of non-cellulosic components was supported by improved wettability, whiteness and dyeability.
3. In this study okra Fibre was extracted from okra plant harvest waste and subjected to the different retting conditions. Okra Fibre was extracted at different time intervals (7–35 day) in two different water conditions i.e. stagnant and running water. The conducted study revealed the fact that the whiteness index is not only dependent upon the retting time but also upon the retting condition.
4. The composites were prepared using compression moulding technique with various fibres loading. The effects of fibre loading on mechanical properties such as tensile, flexural and impact and water absorption were investigated. The experiment shows that the composite with 50% of fibre loading gives excellent mechanical properties in terms of tensile strength, impact strength and flexural strength.
5. In this study, natural bio-mordants (alum + harda) were used for subsequent dyeing of silk with Punica and same pre-mordanted cotton fabrics for dyeing with Tesu leaves. These two varieties of natural dyed fabrics were subjected to post dyeing after-treatments with chitosan and CTAB to improve their wash fastness and also post-treated with aqueous and MeOH extract of eucalyptus leaves and few Commercial UV absorbers with or without citric acid/lemon juice as acid catalyst to enhance light fastness and UV protection factor (UPF) respectively.
6. The study emphasises on modifying the surface morphology of polyester by introducing some protein micro particles on the surface to enhance its dyeability with protein dyes. Change in the properties (absorbency, strength, dyeability, etc.) of the polyester fabric with the applications of the solubilized keratin extracted from waste wool fibre has also been studied. The major advantage of this modification is that polyester fabric can also be dyed with dyes applicable on protein fibres.

7. This study explains the effect of mordanting on the dyeing of wool fabric by metallic salts (Mordant's). Meta mordanting technique has been employed for the application of mordant. The remarkable effect of mordant has been visualized on the colour strength and fastness properties of the dyed sample.
8. This research paper is directed towards suitable finishing formulation of Natural Biocide and Chitosan on cotton fabrics by exhaust application method to evaluate its use as a durable antimicrobial finish to meet functional and environmental demands to extend their high impart applicable properties.
9. This study is to explore the impact of COVID-19 on purchase pattern for fashion apparel and also to understand current purchase behaviour for fashion & apparel. This understanding will be useful for Fashion Apparel manufacturers, brands & retailers for making their market planning.
10. This paper presents an overview of changing paradigms of apparel manufacturing in the context of Industry 4.0. It also discusses, some key advancements such as real-time data monitoring, Intelligent Quality Management Systems (IQMS), and Smart Machine Maintenance Systems (SMMS) that are impactful and bringing a new wave of significant improvements in apparel manufacturing.
11. In this paper, study of the vapour moisture transmission behaviour through the seamed and un-seamed fabric are carried out and observed that seamed fabrics when tested in presence of sweat and water showed greater water vapour transmission rate as compared to un-seamed fabric.
12. This paper focuses on a comparative study on different types of masks available for prevention of COVID-19. It concluded that 3- Layered cotton mask is a good mask although it is not much effective for corona virus but it is comfortable, washable, reusable and available at reasonable price and can be made at home also.
13. This investigation gives a broad overview of impulse buying behavior and other aspects of purchase behavior related to apparel. This review paper focuses on numerous characteristics which influence impulse buying like consumer, retail store, E-commerce website, situational and product related characteristics.
14. The paper is focused on the production of low-cost bacterial cellulose by using different media composition and waste material. This study establishes a new protocol to synthesize bacterial Cellulose from Acetobacter xylinus 2526 NCIM to achieve maximum production yield.
15. In this work needle-punched non-woven fabrics from textile waste for thermal insulation application are developed and analyzed the influence of density and porosity of the nonwovens on the thermal insulation performance. The test results showed that all the developed nonwovens have an excellent thermal insulation performance and are comparable with that of conventional insulation materials.
16. In this study, jute fibre has been used as reinforcement in the form of fibre and fabric. The fibre fraction was used in three different proportions viz. 0%, 1% and 2%. Layers of the jute fabric were varied as 1, 2 and 3. The composites were tested for thermal insulation, load bearing capacity and moisture absorption. It was found

that jute fibre, in both the forms, helps in improving thermal insulation of the composites.

17. This paper reviewed the science of body armor material with emphasis upon current understanding of relevant energy absorbing mechanism in fibres, fabrics, polymer laminates and ceramics. The trends in ongoing developments in ballistic fibres are then reviewed, analyzed and future projection offered. It has also reviewed possible failure modes in armor materials which is essential in designing improved armor materials and armor systems. Study on the arrangement of fabric layers for soft body armor with the drive for lighter and more protection vest has been reported.
18. This article presents a brief review of manufacturing principle and structure of orthogonal and angle interlock 3 Dimensional woven fabrics. Both the structures are not only capable to increase thickness but also produce complex structure. Lucrative features are provided by 3D woven structures may be used as composite reinforcements in numerous applications like construction, automotive, ballistics, marine etc.
19. This study was carried out to design clothes for men with arthritis with the objective to determine the problems faced by arthritic men in present clothing practices. This paper deals with the alteration of design and reconstructs the garments according to their requirement.
20. This paper clearly elaborates the impact of microfibre pollution, need for the awareness among the people and the research gap in the textile industry in controlling microfibre pollution.
21. This paper reports the various technological developments of water conservative dyeing systems have been elaborated and compared. The future scope of developments in the technology of water conservation has also been tried to report in the study.
22. This paper discusses the journey of Khadi fabric from local to global besides its present status, future growth, sustainability characteristics and its potential to boost economic growth in India.
23. This paper explains the possible opportunities to Indian TEI as created through new normal status developed by COVID-19. Preparing Indian TEI to be ATMA NIRBHAR for Domestic requirement (Vocal for Local through Make-in-India) and Export requirement (through Made for World).
24. This paper reviewed the impact of COVID-19 on Indian Textile Industry and its growth opportunities in post COVID situation. It also explains the situation of Indian textile industries with respect to global scenario.

Author Index

9 788194 290391

Printed at Repro India Ltd.